Marine Radioeco

Radioactive Risk Set

coordinated by
Jean-Claude Amiard

Volume 6

Marine Radioecology

Jean-Claude Amiard

ISTE

WILEY

First published 2022 in Great Britain and the United States by ISTE Ltd and John Wiley & Sons, Inc.

ISTE Ltd
27-37 St George's Road
London SW19 4EU
UK

www.iste.co.uk

John Wiley & Sons, Inc.
111 River Street
Hoboken, NJ 07030
USA

www.wiley.com

Library of Congress Control Number: 2022941924

British Library Cataloguing-in-Publication Data
A CIP record for this book is available from the British Library
ISBN 978-1-78630-779-8

Contents

Preface

The marine environment is home to a large number of marine species. Moreover, as the marine environment is the domain where life on planet earth appeared, the number of taxonomic classes present is considerably greater than in others. This considerable marine biodiversity, associated with numerous ecosystems and the complexity of food webs, explains the great stability of this environment. It is also a current and future source for humans, both from the point of view of food and from the point of view of biomolecules of various interests (nutritional, therapeutic, diagnostic, anti-cancer and anti-viral, antioxidants, etc.).

However, the marine environment is the final repository of a large number of forms of physical, chemical and biological pollution that negatively affect marine life. The chemical pollution of rivers and oceans has become for the public an unavoidable subject, because it is measured in thousands of tons discharged per day. It overshadows all other forms of pollution which are more modest, but which also raise questions of concern. Among the latter, radioactive pollution is not negligible.

Radioecology, a scientific discipline born in the middle of the 20th century, has two objectives, the estimation of the fate of radionuclides in the environment and the evaluation of their radioactive risk for living organisms. This discipline is relevant to all the earth's environments.

The design of this book follows the classical approach to risk estimation, which is divided into four steps: hazard identification, hazard exposure assessment, hazard characterization or effects assessment and risk characterization.

Chapter 1 is devoted to the objectives of radioecology and the approach to estimating radioactive risk. The assurances given by the ICRP (International Commission on Radiological Protection) until 2000 that "if Man is adequately protected then other living things are also likely to be sufficiently protected" were not at all well founded.

Chapter 2 identifies radioactive hazards. They are mainly due to the double origin of radionuclides in the marine environment. Their natural origin is essentially to the contribution of the lithosphere caused by the leaching of primeval rocks or the upwelling of magma. The unnatural origin of radionuclides is due to anthropogenic activities in the nuclear field, both military and industrial or medical, both during programmed activities and those resulting from accidents. The dominant source is however that of nuclear fallout.

Chapter 3 deals with the fate of radionuclides, which are never fixed, in the major marine compartments, i.e. water and sediments, with case studies of the impacts due to the nuclear facilities at La Hague and Sellafield.

Chapter 4 presents the fate of radionuclides in marine organisms. In particular, this chapter reports on the various routes of penetration, the modes of contamination by adsorption and absorption and the mechanisms of bioaccumulation of radionuclides in cells, their distribution into the bodies of organisms as well as their elimination. The factors influencing bioaccumulation and trophic transfers are also detailed.

Chapter 5 summarizes the radioactive contamination of various marine sites. These are mainly the marine sites of the American, British, French and Soviet atomic bomb tests, the sites impacted by atmospheric fallout, the sites near reprocessing plants for radioactive nuclear fuel and the marine sites affected by civilian (Chernobyl, Fukushima) and military (Palomares, Thule) nuclear accidents and the dumping of solid radioactive waste. The various environmental monitoring networks are detailed.

Chapter 6 provides an inventory of radiation doses to marine organisms from nuclear accidents and releases from civil nuclear facilities and the main factors influencing these doses. The systematic underestimation of absorbed doses resulting from the poverty of the available calculation tools is underlined, a deficit that leads to the use of simplifying assumptions and to the neglect of radiation sources.

Chapter 7 details the harmful effects of ionizing radiation (mortality, alteration of reproduction, effect of age on irradiation effects) at the various levels of biological organization of marine organisms from the molecular to the ecosystem. It underlines the lack of knowledge about doses received at the organ level and the impacts on marine biodiversity.

Chapter 8 evaluates the radioactive risk to marine organisms with the significance of uncertainties that affect both the doses received and their effects.

The conclusion summarizes all the information in the book and discusses the main gaps in our knowledge of radioecology so that a more realistic assessment is possible.

Nuclear power is a complex scientific and technical field that has also been invited into the social debate, following the serious accidents at Chernobyl and Fukushima, as well as the environmental problems posed by the various links in the so-called "fuel cycle", which ranges from uranium mining to the final storage of the various types of radioactive waste.

In order to remain within the scientific domain, and in keeping with the other volumes of the "radioactive risk" series, I have endeavored to adhere to the scientific truth as closely as possible. To do this, each statement is supported by at least one bibliographic reference to a scientific work published in an international peer-reviewed journal (i.e. where the texts are reviewed and corrected by peers) or by official national or international organizations working in the nuclear field.

August 2022

Acknowledgments

Jean-Claude Zerbib, former CEA Senior Expert and radiation protection engineer, had the difficult task of reviewing, annotating and criticizing this manuscript. He also provided me with precious documents to complete the abundant literature that was used in the writing of this work. I would like to thank him very much for this. Professor Philip Rainbow (former Keeper of Zoology, Natural History Museum, London, United Kingdom) has done the same for the English version. I warmly thank them both for their time and efforts.

Pierre Barbey, from the University of Caen, member of ACRO and expert to the *Haut Comité pour la transparence et l'information sur la sécurité nucléaire* (French Committee for Transparency and Information on Nuclear Safety, HCTSIN), was kind enough to proofread and correct the last two chapters of the French book. His contribution has been very useful. I also wish to thank Professor Philip Rainbow (former Keeper of Zoology, Natural History Museum, London, UK) who did the same for the English version. I warmly thank them both for their time and efforts.

Several colleagues answered my questions. They include Karine Beaugelin-Seiller (IRSN), Henri Métivier (former IRSN) and Scott Fowler (former IAEA researcher in Monaco).

I would also like to thank the members of the Scientific Council of the ANCCLI who helped me in the understanding of certain subjects. The same goes for all the members of the *Groupe Radioécologie Nord Cotentin* (North Cotentin Radioecology Group, GRNC), a pluralist group, for the remarkable work carried out together with every courtesy.

1

General Information on Marine Radioecology and on Risk Assessment

1.1. General information on radioecology

Radioecology has three objectives: the first is to detect the presence of radionuclides in the environment and to investigate their origins; the second is to understand their transfer and concentration processes in ecosystems and the third is to evaluate the impact of natural and anthropogenic radioactivity on the environment (radioecological impact) and on the exposure of the human population (dosimetric impact). These studies follow the same approach as ecotoxicology, which is concerned with stable chemical pollutants such as metals (cadmium, mercury, etc.), nitrates or persistent organic pollutants, whereas radioecology is concerned only with unstable isotopes of elements or radionuclides. Given the presence of radionuclides in all ecosystems and the complexity of transfers, radioecologists work on the three main environmental media: the atmosphere, the hydrosphere (marine and aquatic environments) and the lithosphere (terrestrial environment). In this book, only the marine environment will be discussed.

The term "radioecology" appeared in 1935 but did not become widespread until the 1950s. It results from the fusion of the words "radioactivity" and "ecology". The first work on radioecology was made public in 1955 in Geneva, during the first international meeting on the peaceful use of nuclear energy. The first international congress on radioecology was held in the United States (1961) and was edited by Schultz and Klement [SCH 63]. Many other radioecology congresses followed, such as the one organized by the IAEA in Vienna in 1966, the first one organized in France in 1969 (Cadarache) and the one coordinated by the National Research Council of the National Academy of Science in 1967–1970 [NAS 71].

Radioecological studies are carried out using two main approaches, field (in situ) and experimental approaches [CAQ 01]. The former are the most realistic, but they do not allow for the determination of transfer mechanisms. The objective of the experiments is to build radionuclide transfer models [THI 99] that can be used to quickly predict the impact of radioactive contamination on a territory, based on a limited number of measurements of the radioactivity of sediment samples taken in the field.

In the field, radioecologists choose physical environments that fix radionuclides, such as sediments, or species called bioaccumulators that concentrate radionuclides, such as mosses, lichens, algae or mollusks (mussels or oysters). In the laboratory, radioecologists reconstitute simplified ecosystems such as, for example, the contamination of crops by various radionuclides by varying the nature of the soil and climatic conditions. This work leads to the development of models that simulate the transfer of radionuclides in the various environmental media.

1.1.1. *History of radioecology*

Initially, radioecological studies were oriented to provide data from a health point of view. The corollary (introduced in particular by the ICRP in 1977, ICPR26) that prevailed was that Man was the most sensitive species to the effects of ionizing radiation and that by protecting Man, one protected all living organisms. Therefore, the first "environmental" studies were aimed at monitoring food sources (of wild or anthropogenic origin). In the environment, the measures concerned mainly edible species and very often only the consumed parts of wild species such as fish muscles. Since environmental contamination in the 1950s and 1960s was mainly due to atmospheric fallout from atomic weapons testing, studies were devoted mainly to the two relatively long-lived radionuclides resulting from these explosions, namely cesium-137 and strontium-90, and to a lesser extent to the associated transuranics (notably $^{239,240}Pu$).

In the 1970s, with the decrease in atmospheric fallout, interest focused on fission products that were released into the environment around spent fuel reprocessing plants, such as cobalt-60, ruthenium-106, antimony-125, silver-110m and manganese-54. Following the Chernobyl accident (1986), radioecological studies were essentially focused on cesium-137 and secondarily on some transuranics. It was also at this time that the number of studies on the effects of radiation on wildlife increased. In the 2000s, the European Union launched two important programs for radioecology. The first one being "Environmental Protection for Ionizing Contaminants in the Arctic" where the sentinel species are lichens, soil invertebrates, as well as mammals (lemmings, reindeer) and birds (marine species are in the minority). The second program is called FASSET (Framework for ASSessment of Environmental ImpacT), and it is responsible for assessing the effects of radiation in particular on mortality, morbidity, its impact on reproduction and cellular damage in various species [STO 02]. After the war in Kosovo, several studies were interested in the effects of depleted uranium used in military equipment (in particular "shaped charge" rockets).

The European project ERICA (Environmental Risk for Ionising Contaminants: Assessment and Management) of the 6th PCRD-Euratom launched in March 2004, was completed in February 2007. It allowed for the updating of the database concerning the effects of ionizing radiation on non-human organisms, the exploitation of this database to define criteria for the protection of ecosystems and the design of a method that enables the characterization of ecological risk by analyzing the exposure of fauna and flora to ionizing radiation and the effects of this exposure.

1.1.2. *The main marine radioecology laboratories*

For about 40 years, the vast majority of laboratories working in the field of marine radioecology were state laboratories. For example, in the United States, these were USAEC laboratories, and in the Soviet Union, laboratories dependent on the Academy of Sciences such as MMBI (Murmansk Marine Biological Institute of the Russian Academy of Sciences) in Murmansk or the Sevastopol Biological Station in Sevastopol in Crimea (USSR). In the United Kingdom, the main laboratory at Lowestoft (CEFAS) is dependent on the Ministry of Agriculture, Fisheries and Food (MAFF) [HUN 98]. In France, the marine radioecology laboratory was under the authority of the *Commissariat à l'énergie atomique* (CEA, Central Commission for Nuclear Energy), which itself was originally directly attached to the French Prime Minister. This laboratory was located on the site of the spent fuel reprocessing plant at La Hague. As a result, public information was highly controlled. With its independence, it moved to Cherbourg-Octeville.

In addition, in the United States, various academic institutions have worked in the field of marine radioecology such as the School of Marine Sciences at Stony Brook University, the School of Oceanography at Oregon State or the Department of Fisheries at the University of Washington (Seattle, WA).

At the international level, the IAEA created a laboratory called MEL (Marine Environment Laboratory) located in Monaco [FOW 88]. This laboratory has provided valuable data on the measurement of contaminants in the sea.

After the Chernobyl accident and especially after the Fukushima accident, a large number of academic teams became interested in marine radioecology topics, renewing and diversifying the studied themes.

1.1.3. *The triad of radioecology: exposure, bioaccumulation and adverse effect*

As with most pollutants, the impact of radionuclides on living beings depends on an interaction at three levels, exposure, bioaccumulation and adverse effect.

1.1.3.1. *Transfers from physical environments to organisms*

Radionuclides in the marine environment have multiple origins, whether they are natural or anthropogenic radionuclides (Chapter 2). Their dispersal in all marine environments leads to an exposure of living organisms to radioactive elements. However, the exposure will not systematically induce an accumulation of environmental contaminants within organisms because different barriers will limit their intake[1] (Figure 1.1). Even though the circulation of contaminants is a worldwide phenomenon, the more distant a site is from the source of contamination, the greater the dilution in the water. This contamination gradient will be the first factor to control the intake. Under the influence of environmental conditions, many radionuclides, and the molecules that accompany them in physical media, are likely to be transformed: chelation[2], hydrolysis, photodegradation, biodegradation, etc. Though the majority of radionuclides are likely to cross biological membranes, these can however play a barrier role, for example, by altering the speciation and the bioavailability of a radioactive metal by the secretion of chemical modifiers in the external medium or by precipitating it on the cell surface in the form of metal sulfide microcrystals. Many organisms (bacteria, animals, plants) secrete mucopolysaccharides of highly variable composition, and subtle changes in charges and reactive groups are likely to interfere with contaminant intake. Another way to limit intake is the existence of impervious extracellular barriers such as carapaces, integuments, tests, shells and scales that help to reduce the surface area of exposed living epithelium available for cutaneous transport [MAS 95a].

Once the radionuclide is incorporated (Figure 1.1), it will either be stored within the tissues or excreted. Storage in intra- or extracellular compartments will not automatically result in an adverse effect on the body. Thus, with respect to transition metals and their radioactive isotopes, detoxification processes are present in many organisms. This can be the synthesis of particular metalloproteins, for example,

1 The term "intake" means incorporation of the radionuclide into the body.

2 Chelation is a physico-chemical process during which a complex is formed between a ligand and a radionuclide.

metallothioneins in animals and phytochelatins/metallothioneins in plants, which sequester the metals with their thiol groups, preventing them from interfering with the enzymes whose activity they would otherwise disrupt. This is the most common defense process used in vertebrates. It also exists in most taxa [AMI 06], but in invertebrates, the preferred pathway of detoxification is the biomineralization of metals in various inclusions [MAR 02]. However, detoxification by storage is not effective for radionuclides as it is for their stable counterpart for the ionizing radiation remains within the tissues.

Unlike most other pollutants, the rule "no effect without bioaccumulation and no bioaccumulation without exposure" does not apply to radionuclides. Because of the production of various types of ionizing radiation, an organism externally exposed to a radionuclide will experience external irradiation that may cause adverse effects. Bioaccumulation leads to internal irradiation.

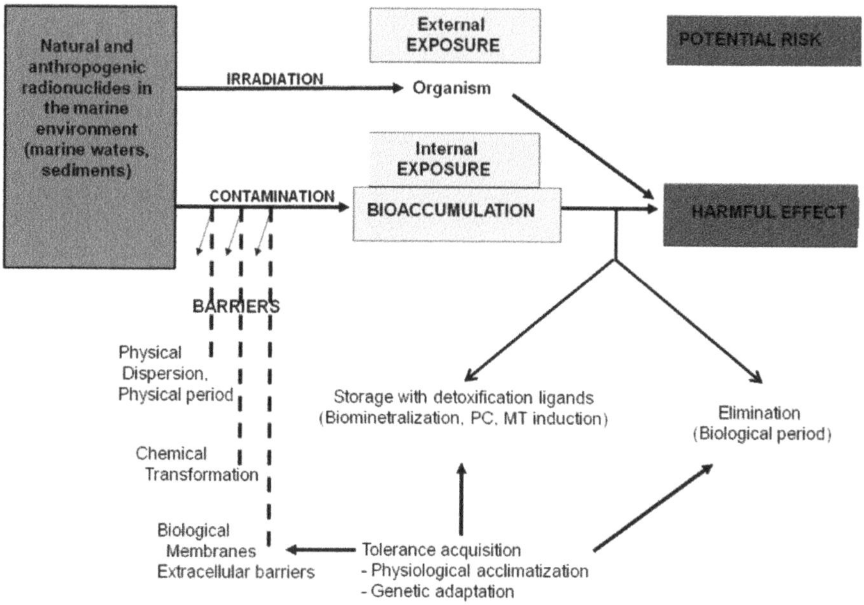

Figure 1.1. *The triad of marine radioecology. For a color version of this figure, see www.iste.co.uk/amiard/radioecology.zip*

1.1.3.2. *Contamination and irradiation, specificities of radionuclides*

Transfers of radionuclides to marine organisms can be direct from the various physical environments (marine waters and sediments) or indirect through food chains.

In living organisms, radioactive contamination can be external or internal. This contamination will remain external if the radionuclide adsorbs onto the external parts of the organism such as its skin, its gills or its lungs. It will be considered internal if the radionuclide is absorbed, crossing the biological barriers that are the skin, the gills or lungs, the digestive tract. In the same way, the ingestion of a contaminated prey item (or food), its digestion and the passage through the wall of the digestive tract of part of its contents will lead to a more or less intense internal contamination and irradiation (Figure 1.2).

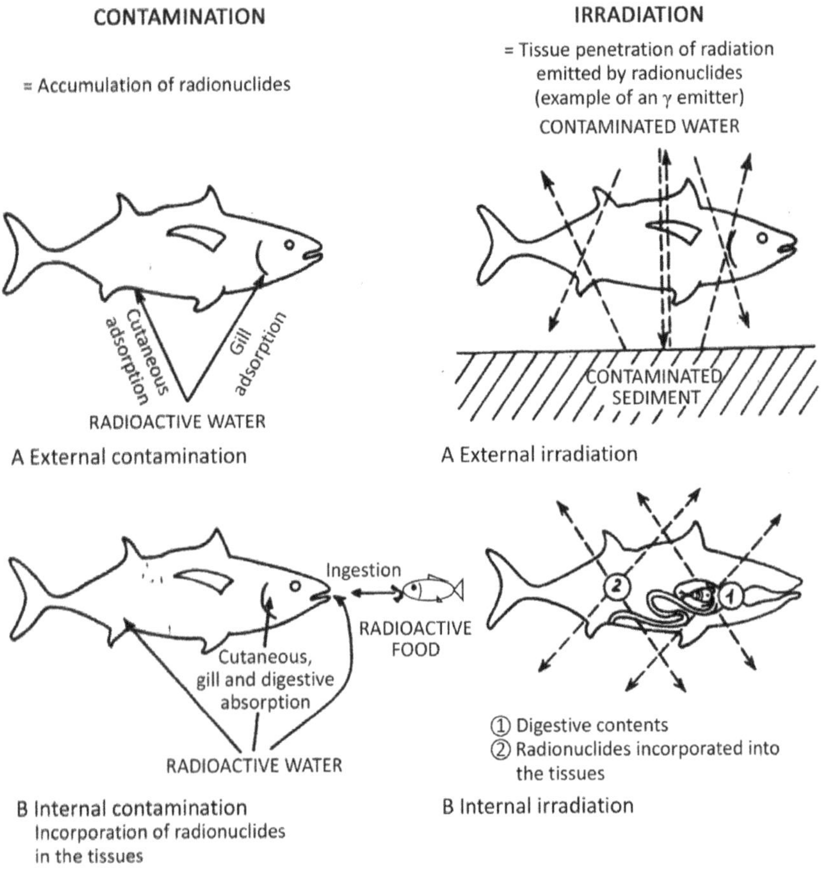

Figure 1.2. *Contamination and irradiation of organisms (from [AMI 80])*

The contamination of the physical environment (air, water, soil, sediment), external to the organism, induces an external irradiation. Conversely, internal contamination of the organism induces internal irradiation. By extension, the radioactive air inhaled during breathing, the absorbed water and contaminated food ingested during feeding will be considered as internal irradiation.

1.2. The principle of risk assessment

Living things are subject to three major classes of hazards: physical stresses, chemical stresses and biological stresses [AMI 17a]. Frequently the notions of hazard and risk are confused by the public.

A risk is the probability that a person will suffer harm or adverse health effects from exposure to a hazard. The risk is therefore the probability of the hazard occurring. This risk depends on two probability distributions, that of the hazard and that of the exposure. In the case of ionizing radiation, the hazard is a function of the dose (dose–response function). On the other hand, the probability of exposure is an inverse function of the dose. Indeed, the probability of suffering a high dose is lower than that of suffering a low dose. As a result, the risk zone is located at the overlap p(D) x p(E/D). The risk is therefore the sum of the hazard and the exposure. This reasoning can be applied in a similar way when considering an organism exposed to a potentially toxic compound in its environment.

Risk analysis involves three steps: hazard assessment, risk management and risk communication. Risk assessment involves four steps: hazard analysis, dose–response assessment, exposure assessment and risk characterization (Figure 1.3). These four steps will be detailed in this book for radioactive risk assessment of marine organisms.

The general principles of environmental risk assessment are shown in Figure 1.4. In the end, risk characterization consists of comparing the PEC (Predicted Effect Concentration) to the PNEC (Predicted No Effect Concentration) and risk is present when the ratio is equal to or greater than one.

In the environment, the assessment of adverse effects begins with the identification of hazards based on laboratory toxicity data for a single species tested and/or with data from different trophic levels, but the ecosystems are simplified as much as possible. The tests are standardized at the international level. Predicted No Effect Concentrations (PNEC) are estimated mainly by two methods (Figure 1.5): the scaling factor method and the statistical scaling method [AMI 17a]. The determination of the threshold value (PNEC) by extrapolation factors is based on toxicity data obtained by standardized laboratory tests. This method has two serious shortcomings, the extrapolation of laboratory data to the real impact in the field with in particular the problem of interactions (additivity, synergistic and antagonistic effects) and the extrapolation from the short-term to the long-term. The uncertainties on the threshold value obtained are mainly due to intra- and interspecific variations (biological variability) and to intra- and inter-laboratory variations of toxicity data. In order to take into account the different uncertainties, an extrapolation factor is applied. This extrapolation factor applied will be inversely proportional to the knowledge of the toxicity of the substance. In the marine environment, due to its high biodiversity and the small number of standardized ecotoxicity tests, the extrapolation factors applied to freshwater aquatic species are systematically increased by a factor of 10. Thus, if the PNEC is obtained with a single short-term lethal concentration value for three freshwater or saltwater species representative of three taxonomic groups (algae, crustaceans, fish) from three trophic levels, the maximum extrapolation factor is 10,000 [EC 96, AMI 17a].

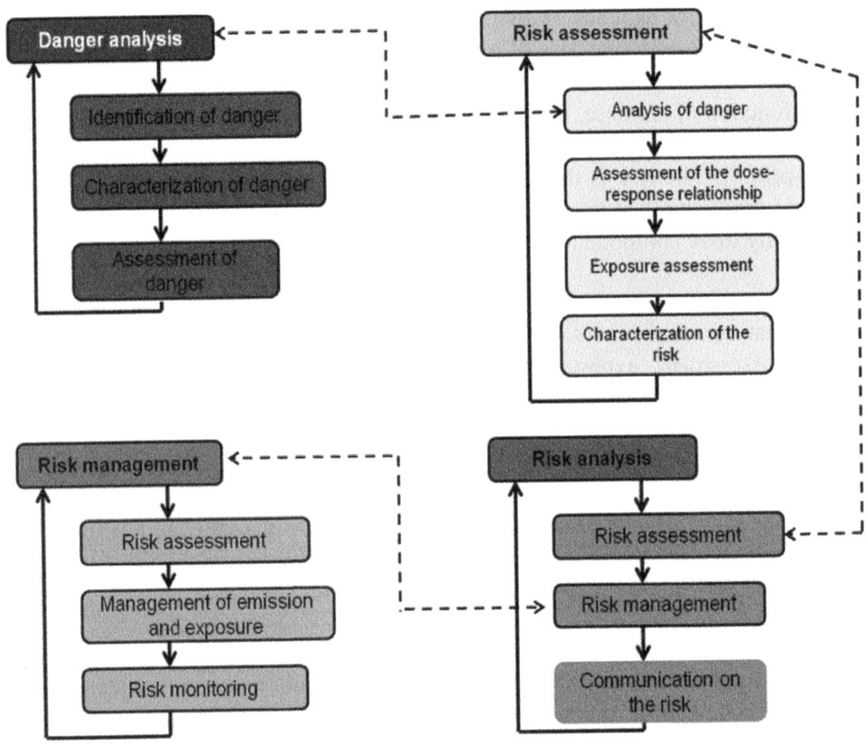

Figure 1.3. *Principle of risk analysis (modified from [DUF 01]). For a color version of this figure, seewww.iste.co.uk/amiard/radioecology.zip*

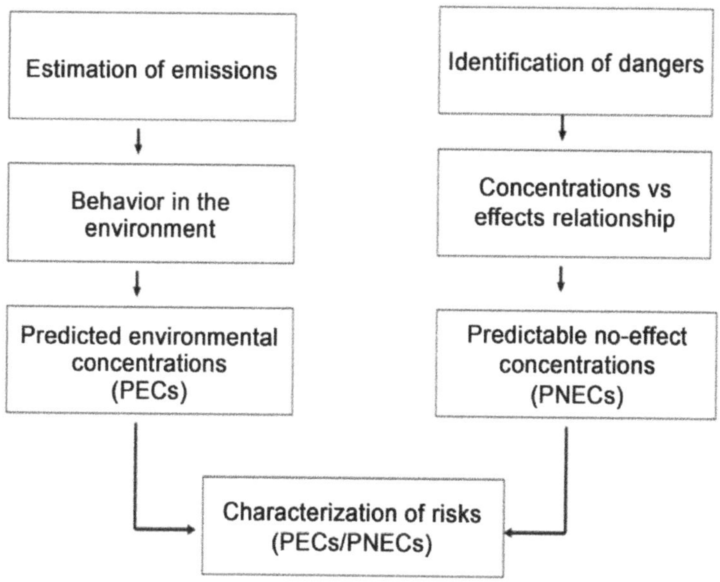

Figure 1.4. *Various steps of the risk assessment principle [AMI 17a]*

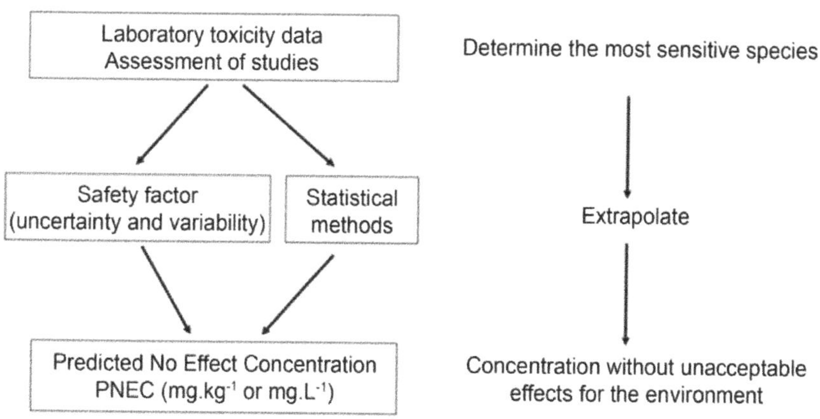

Figure 1.5. *Assessment of adverse effects (from [AMI 17a])*

The statistical extrapolation method is based on the species sensitivity distribution (SSD). The underlying assumptions [OEC 92] are that a species sensitivity distribution follows a known theoretical distribution function and the group of species tested in the laboratory is a random sample of this distribution. Depending on the percentage p (often 5%) of sensitive species that can be considered negligible, we estimate the corresponding PNEC (Figure 1.6).

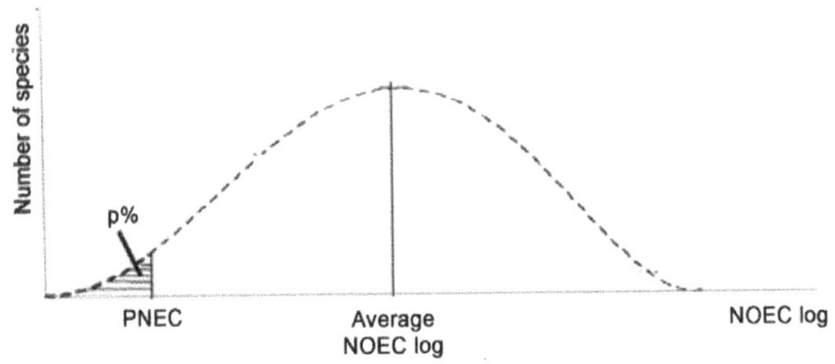

Figure 1.6. *Responses of various species of organisms to the NOEL for estimating a PNEC*

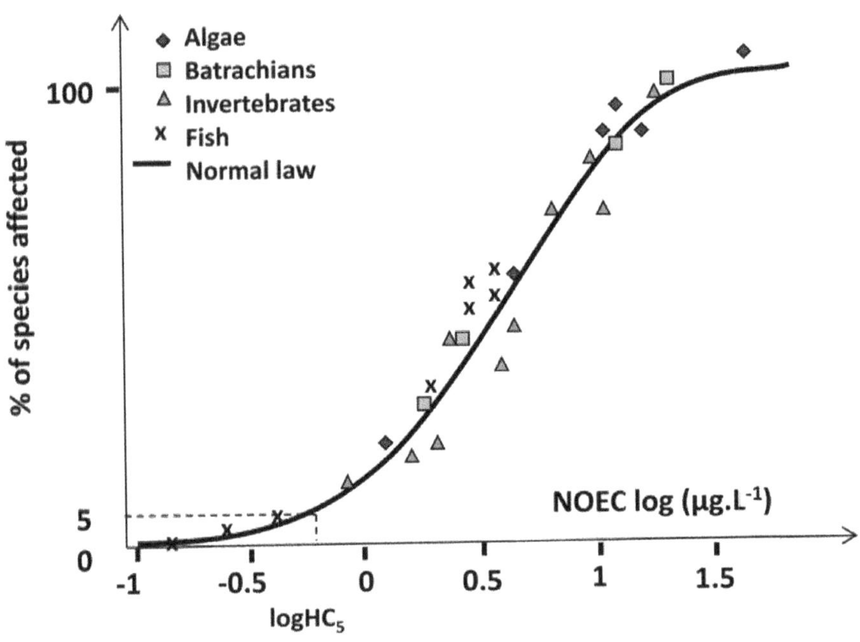

Figure 1.7. *Long-term toxicity of cadmium to freshwater organisms ($HC_5 = 0.5\ \mu g.L^{-1}$, IC90 % = [0.27; 0.79], PNEC = HC_5/FE and 1 $\leq$ FE $\leq$ 5). For a color version of this figure, see www.iste.co.uk/amiard/radioecology.zip*

In practice, this method can only be applied when a large number of data are available. The transformation of long-term toxicity data (NOEC, No Observed Effect Concentration) is carried out in log-logistic or log-normal form and at least eight species are required; the extrapolation method requires at least 10 NOECs, preferably 15 for at least eight taxonomic groups; the choice of outcome is usually the 5th percentile (HC5, Hazardous Concentration 5%) and the assessment factor (AF) from 1 to 5 to reflect the identified uncertainties. The formula is therefore PNEC = HC5/AF. An example is provided in Figure 1.7 for long-term ecotoxicity to freshwater organisms contaminated with cadmium [AMI 17a].

1.3. The particular case of the risk linked to ionizing radiation

When applying the characterization of the radioactive risk related to the presence of radionuclides in the marine environment and in marine organisms, the first step, the identification of the hazard, is obvious; radioactivity is dangerous for all forms of life. The hazard of a radionuclide depends primarily on three criteria: its radiotoxicity, its radiation type(s) and its half-life. The hazard of radionuclides can be related to their chemical toxicity or to their ionizing radiation, called radiotoxicity. For all radionuclides, with the exception of uranium[3], radiotoxicity is clearly more dangerous than chemical toxicity. Radiotoxicity varies greatly depending on the radionuclide.

Radioactivity corresponds to an unstable nucleus which, to regain its stability, spontaneously emits one or more particles. Radiation is mainly of three types: α radioactivity, where a helium nucleus is emitted; β radioactivity, where either an electron and an electron antineutrino (β-) or a positron and an electron neutrino (β+) are emitted; and γ radioactivity by which a nucleus loses its energy by high-energy electromagnetic radiation. To this must be added X-rays, which are of the same photonic nature as γ-emitters, but which are produced by electronic transitions, and neutrons, which are present during chain reactions in nuclear reactors or during the explosion of atomic bombs.

The two main properties of radiation are the power of penetration in matter and the ionizing power.

In a radioactive sample, the number of decays per unit of time is proportional to the number of unstable nuclei of the radionuclide, present in the sample. As a result, for a given sample, the number of unstable nuclei decreases progressively with time.

3 From an enrichment in 235U equal to 7%, it is the radioactive toxicity that becomes higher.

The period during which half of the unstable nuclei disappear is called the physical period (Tp) or physical half-life. It is a characteristic and unchangeable quantity of each radionuclide. Thus, radionuclides made up of very unstable nuclei will have a physical half-life of a few fractions of a second, while those made up of very stable nuclei will have a physical half-life of thousands of years. Thus, after four physical half-lives, the radioactivity of a sample is decreased by a factor of 16.

2

The Origins of Radionuclides in Marine Environments

2.1. Natural or anthropogenic origins of radionuclides in the marine environment

The presence of radionuclides in the marine environment has a double origin: a natural origin and an artificial origin linked to human activities. The natural radioactivity of the environment results from two distinct sources of radionuclides: (i) radionuclides born from the interaction of matter with radiation of cosmic origin (^{3}H, ^{7}Be, ^{14}C, ^{32}Si, ^{36}Cl, etc.) and (ii) the radionuclides present in the earth's crust since the origin of the earth (^{40}K, ^{87}Rb, ^{115}In, ^{138}La, ^{147}Sm, ^{176}Lu, ^{187}Re, etc.) and the descendants of the four radioactive families (^{238}U, ^{235}U, ^{232}Th and ^{237}Np) [EIS 73].

The anthropogenic origins of radioactive pollutants in the marine environment are various and varied. It is common to separate the intended releases of radionuclides into the environment (fallout from nuclear explosions and discharges of radioactive liquid and gaseous effluents into various environments) and the unintended origins of these pollutants (linked to incidents or accidents). Artificial radionuclides are mainly fission products and activation products as well as transuranics produced by neutron capture.

Some radionuclides have a dual origin, natural and anthropogenic. This is the case for tritium, which is produced by the interaction of cosmic rays with the atmosphere, and also by military nuclear tests and the civilian use of nuclear energy, and in the medium term even more so with the use of nuclear fusion (ITER). This is also the case for uranium and its descendants, especially during the first stages of its exploitation (mines).

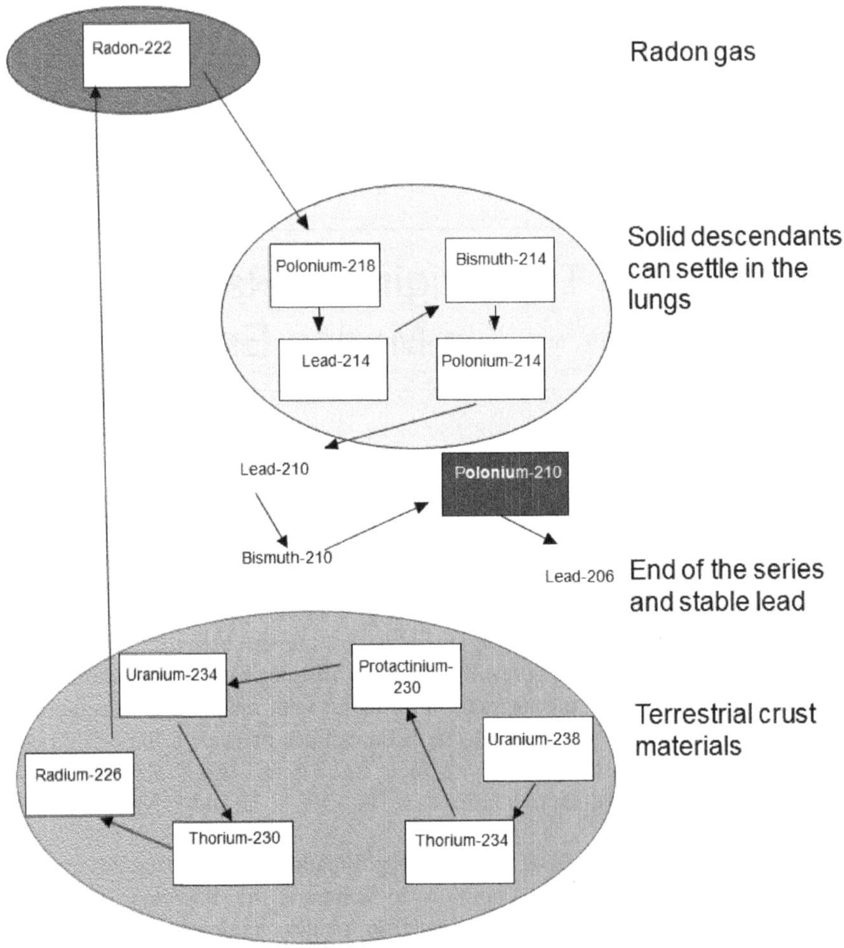

Figure 2.1. *The radioactive family of uranium-238 and its descendants. For a color version of this figure, see www.iste.co.uk/amiard/radioecology.zip*

Natural radionuclides are introduced into the hydrosphere primarily by leaching from the earth's crust. According to Arnold and Martell (1973 in [AMI 80]), in the ocean, ^{232}Th and ^{230}Th (which is a descendant of ^{238}U) precipitate very rapidly, uranium-238 a little less rapidly, while radium-226 (formed by decay of ^{230}Th, whose physical half-life is 1,600 years) can be resolubilized in waters close to the bottom. Some human activities will reintroduce into the environment various natural radionuclides present in the lithosphere. These will be, for example, phosphate plants, coal-fired power plants, oil exploitation or the first stages of the nuclear fuel cycle (mines, mills, etc.). Thus, a coal-fired power plant releases into the atmosphere 122.10^3 $Bq.h^{-1}$ of ^{226}Ra, 77.10^3 $Bq.h^{-1}$ of ^{228}Ra, 122.10^3 $Bq.h^{-1}$ of ^{232}Th, 298.10^3 $Bq.h^{-1}$ of ^{210}Pb and 298.10^3 $Bq.h^{-1}$ of ^{210}Po [KIR 77]. The offshore oil and gas industry is the largest non-nuclear sector contributor of radioactive substances to the marine environment. Almost all of the radionuclides released from this sector come from produced water (water extracted from the reservoir along with the oil and gas) and pipe descaling. A less significant source is the use of radioactive substances (e.g. tritium) as markers. Naturally occurring radionuclides in produced water include lead-210 (^{210}Pb), polonium-210 (^{210}Po) and radium-226 and 228 (^{226}Ra and ^{228}Ra) [PSP 09].

The fate of natural radionuclides that are descendants of uranium-238 involves various compartments of the environment according to their nature (gaseous, solid) and according to their speciation. Thus, radon is mainly in the atmosphere and polonium-210 in the hydrosphere (Figure 2.1).

There are seven major sources of anthropogenic radionuclides in the marine environment. These are military nuclear weapons programs, nuclear weapons testing, uranium mining and processing, nuclear power plants, commercial fuel reprocessing, ocean dumping of radioactive waste and nuclear accidents involving nuclear facilities or nuclear powered vessels [HU 10].

2.2. The natural origins of radionuclides

More than 60 radionuclides are known to occur naturally in the environment. These are the terrigenous radionuclides (^{40}K, ^{87}Rb) and those descending from the thorium series (^{228}Ra, ^{228}Th, ^{232}Th), actinium (^{227}Ac, ^{231}Pa, ^{235}U), uranium (^{210}Pb, ^{210}Po, ^{222}Rn, ^{226}Ra, ^{230}Th, ^{234}Th, ^{234}U, ^{238}U), as well as cosmogenic radionuclides (^{3}H, ^{7}Be, ^{10}Be, ^{14}C, ^{26}Al, ^{32}Si) [FUK 82]. Cosmic rays are 89% protons (hydrogen), 10% helium and 1% heavy elements [LAI 99]. The most abundant natural radionuclides in the ocean are potassium-40 with 15,000,000 PBq and uranium-238, 37,000 PBq [BUE 14].

The main characteristics of naturally occurring radionuclides in marine environments are listed in Tables 2.1 and 2.2.

Water produced by the offshore oil industry contains naturally occurring radionuclides [HOS 12]. As a result, high levels of radionuclides (mainly ^{226}Ra and

^{228}Ra) are discharged into the sea in connection with oil and gas production on the Norwegian continental shelf. The presence of added chemicals such as scale inhibitors in produced water has a marked influence on radium speciation [ERI 09].

Radionuclide	Physical half-life (years)	Concentration in surface water (Bq.m^{-3})	Concentration in sediments (Bq.kg^{-1})
Cosmogenic origin			
^{3}H	12.3	20–100	–
^{7}Be	0.146	1.1–3.4	–
^{14}C	5,730	5.5–6.7	–
^{32}If	172	0.2–3.3.10^{-3}	–
Land origin			
^{40}K	1.28.10^{9}	11,800–12,300	2–1,000
^{87}Rb	4.8.10^{10}	100	–
^{232}Th	1.34.10^{10}	0.4–29;10^{3}	12–50
^{228}Ra	5.76	0.8–8	–
^{228}Th	1.91	0.004–0.3	–
^{235}U	7.04.10^{8}	1.9	–
^{238}U	4.47.10^{9}	40–44	2.5–150
^{234}Th	0.066	0.6–6.8	–
^{234}U	2.45.10^{5}	47	–
^{230}Th	9.0.10^{4}	2–52.10^{-3}	–
^{226}Ra	1,617	0.8–8	10–100
^{210}Pb	22.3	1–4.5	100–280
^{210}Po	0.378	0.5–1.9	100–280

Table 2.1. *Concentrations in marine waters and sediments of some naturally occurring radionuclides (from [SZY 12, NIE 18a])*

Radionuclide	Physical period (year)	Distribution factor (Kd) between water and sediment	Bioconcentration factor (BCF)	Effective dose coefficient for adult ingestion ($Sv.Bq^{-1}$)
^{3}H	12.3	1.0	1.0	$1.8.10^{-11}$
^{14}C	5,700	1.10^{3}	2.10^{4}	$5.8.10^{-10}$
^{60}Co	5.27	1.10^{5}	7.10^{2}	$3.4.10^{-9}$
^{90}Sr	28.64	8.10^{0}	3.10^{0}	$2.8.10^{-8}$
^{99}Tc	213,000	1.10^{2}	8.10^{1}	$6.4.10^{-10}$
^{106}Ru	1.02	4.10^{4}	2.10^{1}	$7.0.10^{-9}$
^{125}Sb	2.77	2.10^{3}	6.10^{2}	$1.1.10^{-9}$
^{129}I	1.57.107	7.10^{1}	9.10^{0}	$1.1.10^{-7}$
^{131}I	0.0219			$2.2.10^{-8}$
^{134}Cs	2.06	4.10^{3}	1.10^{2}	$1.9.10^{-8}$
^{137}Cs	30.17			$1.3.10^{-8}$
^{238}Pu	87.74	1.10^{5}	1.10^{2}	$2.3.10^{-7}$
^{239}Pu	24,110			$2.5.10^{-7}$
^{241}Pu	14.35			$4.8.10^{-9}$
^{241}Am	432.2	2.10^{6}	1.10^{2}	$2.0.10^{-7}$

Table 2.2. *Characteristics of the main radionuclides introduced into the marine environment by human activities (from [NIE 18a])*

2.3. The military origins of anthropogenic radionuclides

Since World War II and the military use of nuclear energy, direct and indirect releases of radionuclides into the marine environment have been particularly significant, especially between 1945 and 1990. The impact on the marine environment is considerable due to the atmospheric fallout from the aerial testing of atomic bombs. Direct releases of radionuclides are also significant from military research and manufacturing centers for nuclear weapons, from poorly controlled military radioactive waste, and from releases related to the nuclear propulsion of ships and submarines.

2.3.1. *Radionuclides due to atmospheric fallout from atomic device explosions*

The input of long-lived radionuclides into the oceans was considerable, especially from 1945 to 1980 [EDG 80]. Between 1945 and 1965, 520 atomic bombs were tested, more than 100 of which were tested on islands or over/in large bodies of water. For the United States, these were the islands of Bikini, Eniwetok and Johnston; for the United Kingdom, Montebello Islands; for the United States and the United Kingdom, Christmas Island; for the USSR, the island of Novaya Zemlya; for France, the islands of Fangataufa and Mururoa; and for China, Lake Lob-Nor.

The principal sites of atomic bomb tests were located for the Americans in the coral atolls of the Marshall Islands at 110°N and 160–165 E. The result was a strong contamination of the marine environment [CAR 97]. The inventory of the quantities of radionuclides involved in the tests in French Polynesia (Mururoa and Fangataufa) is $1.7.10^5$ TBq [HU 10].

There are two main types of bombs, A-bombs or fission bombs and H-bombs or thermonuclear bombs (fission–fusion–fission).

2.3.1.1. *Atomic A-bombs*

The "classic" atomic (A) bombs involve the fission of a charge of uranium-235 or plutonium-239. The power of these bombs, which were very weak at the beginning (Hiroshima 16 ± 2 kilotons of TNT), now reaches more than 100 times a "unit bomb" (corresponding to 20 kt of the explosion of TNT or trinitrotoluene). The explosion of these A-bombs releases into the environment a large sum: the unreacted fuel (^{235}U or ^{239}Pu) (about 90%, or 10 kg or 222.10^{11} Bq for a unit bomb) and fission products (10%). The neutrons emitted during the explosion induce a significant radioactivity in the surrounding environment (activation products).

A "unit bomb" of 1 Mt releases about 1 kg of fission products, a complex mixture of about two hundred radionuclides corresponding to 35 different elements. The fission of heavy nuclei of ^{235}U or ^{239}Pu most frequently gives rise to two lighter nuclei with masses oscillating around two values: 90 and 140. The main fission products are ^{89}Sr, ^{90}Sr, ^{95}Zr, ^{103}Ru, ^{131}I, ^{137}Cs, ^{141}Ce, ^{144}Ba-La, ^{144}Ce and ^{144}Pr. One minute after the explosion, the radioactivity due to fission products and activation products is $3.03.10^{22}$ Bq for a "unit bomb". The decay of the radioactivity is rapid, but after one year, the residual radioactivity is still $4.1.10^{15}$ Bq.

Activation products typically account for only 25% of the radioactivity from fission products. The activation products formed will vary depending on the location of the explosion. In the case of an airborne explosion, there is formation of ^{3}H (< 37.10^9 Bq.Mt^{-1}), ^{41}A, ^{16}N, ^{19}O and especially ^{14}C ($1.26.10^{15}$ Bq.Mt^{-1}). If the explosion is underwater, it will be mainly ^{35}S, ^{38}Cl and ^{24}Na that will be produced and to a lesser degree ^{32}P, ^{51}Cr, ^{59}Ni, ^{64}Cu, ^{65}Zn, ^{99}Mo and ^{113}Sn. When the explosion is terrestrial or underground, one generally notes the formation of ^{45}Ca as well as ^{14}C, ^{24}Na, ^{28}Al, ^{31}Si, ^{32}P, ^{35}S, ^{38}Cl, ^{42}K, ^{54}Mn, ^{55}Fe, ^{65}Zn and ^{89}Sr (in [FON 60]).

2.3.1.2. *Atomic H-bombs*

H-bombs are composed of a nucleus of uranium-235 or plutonium-239 surrounded by a mass of lithium-deuterium or tritium, itself often contained in a uranium-238 envelope (3F bomb). During explosion, the fission–fusion–fission reactions follow one another extremely rapidly. The quantity of fission products formed is certainly proportional to the power of the device; on the contrary, the quantity of activation products must be much greater. The quantity of fissioned material is greater than for A-bombs.

Eight days after the explosion of an enriched uranium thermonuclear bomb (March 18, 1954), Kimura et al. (cited by [FLA 63]) estimated the following composition of the fallout: ^{237}U (20%), ^{144}Ce (16%), ^{106}Rh, ^{103}Ru, ^{129}Te, ^{131}I, ^{132}Te (15%), ^{147}Pm (9%), ^{147}Ne (9%), ^{91}Y (8%), ^{141}Ce (7%), ^{140}Ba (5%) and ^{95}Zr (5%). This composition, due to the large differences in the physical half-lives of the radionuclides, varies considerably over time. Thus, after four years, the main radionuclides are ^{144}Ce (28%), ^{147}Pm (24%), ^{90}Sr (19.6%) and ^{137}Cs (19.2%) (according to Siegel in [FLA 63]).

The radionuclides present in the fallout in significant quantities and dangerous because of a long or medium physical half-life, easy incorporation into the food chain, or significant retention in the human body, are ^{90}Sr (28.8 years – 37.10^{14} $Bq.Mt^{-1}$), ^{137}Cs (30.05 years – 37.10^{14} $Bq.Mt^{-1}$), ^{14}C (5,700 years – 81.10^{12} $Bq.Mt^{-1}$), as well as actinides

2.3.1.3. *The case of the Nagasaki explosion*

The explosion of a plutonium atomic bomb (10–15 kg) in Nagasaki, Japan, on August 9, 1945, led to the worldwide release of man-made radionuclides. This was a historic event. Part of the bomb was split, 1.2 kg of ^{239}Pu, releasing 21 kt of TNT energy with various fission products. The rest of the unspent fissile material, 13.8 kg ($3.49.10^{13}$ Bq) of $^{239,240}Pu$, was released into the atmosphere with the fission product, ^{137}Cs, 23.4 g ($7.44.10^{13}$ Bq). The fates of $^{239,240}Pu$ and ^{137}Cs were studied by analyzing local and global fallout. The highest concentration of $^{239,240}Pu$ was 64.5 $mBq.g^{-1}$ (181 $mBq.cm^{-2}$) while it was 188 $mBq.g^{-1}$ (526 $mBq.cm^{-2}$) for ^{137}Cs, both 2.8 km east of the hypocenter. The total amount of deposition in the local fallout region of 264 km^2 was 37.5 g ($9.48.10^{10}$ Bq) for $^{239,240}Pu$ and 3.14 mg ($5.88.10^{10}$ Bq) for ^{137}Cs. The Nagasaki explosion resulted in atmospheric fallout as far away as the Arctic, where deposition was about 0.16 µBq.cm^{-2} for ^{239,240}Pu and 20 µBq.cm^{-2} for ^{137}Cs in 1994 (Kudo et al. pp. 233–250 in [KUD 00]).

2.3.1.4. *The distribution of deposition between the northern and southern hemispheres*

From 1945 to 1980, there were 423 atmospheric tests[1] of atomic bombs (about 217 MT) releasing numerous radionuclides into the atmosphere (Hamilton et al. pp. 29–58 in [GUÉ 03]). These tests resulted in significant deposits to the oceans. Thus, for ^{90}Sr, the maximum cumulative deposits were 355 PBq in the Northern Hemisphere in 1966 and 107.3 PBq in 1972, 1973 and 1974 in the Southern Hemisphere (Hamilton et al. pp. 29–58 in [GUÉ 03]).

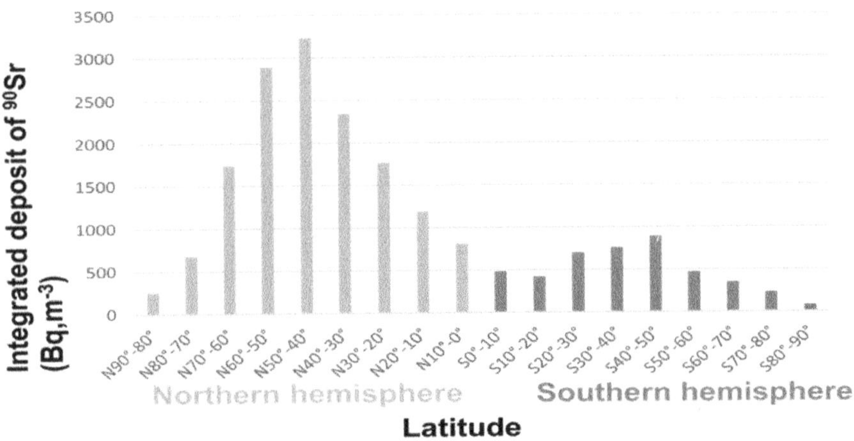

Figure 2.2. *Integrated 1995 deposits of strontium-90 corrected for radioactive decay (modified from [UNS 93]). For a color version of this figure, see www.iste.co.uk/amiard/radioecology.zip*

Because the currents that affect the stratosphere are mostly east-west and very weak in the latitudinal direction, the northern hemisphere, where most of the nuclear tests have taken place, is significantly more contaminated than the southern hemisphere. In each hemisphere, radioactive pollution is most intense between 25 degrees and 50 degrees of latitude. This is verified by the results of atmospheric fallout of most radionuclides. This is, for example, the case of the global deposition of strontium-90 (Figure 2.2) [NAS 71, UNS 93]. It is easy to find the atmospheric

1 According to the IRSN (2013), there were 543 atmospheric tests conducted by the United States, Russia, Great Britain, France and China between 1945 and 1980.

fallout of other radionuclides since the ratios between them remain fairly constant. These ratios are 1.5 between ^{137}Cs and ^{90}Sr, 9.4 between ^{55}Fe and ^{90}Sr, 0.017 between ^{239}Pu and ^{90}Sr, and 0.33 between ^{60}Co and ^{90}Sr [NAS 71].

Deposition of ^{137}Cs and $^{239,240}Pu$ into the Pacific Ocean from atomic bomb testing has been significantly greater in the Northern Hemisphere than in the Southern Hemisphere (Table 2.3) [HAM 03]. As a result, Pacific water concentrations are slightly higher in the Northern Hemisphere than in the South. In the South Pacific Ocean around the French Polynesian islands, the concentrations decreased from 1979 (about 6 Bq.m^{-3}) to 1994 (2.7 Bq.m^{-3}), with a turnover rate of 14 years for cesium-137 ([BOU 92]; Bourlat et al. pp. 75–93 in [GUÉ 03]), as well as for $^{39,240}Pu$ (0.62–0.32 Bq.m^{-3}) [BOU 95].

Similarly, atmospheric deposition of ^{210}Pb has been higher in the northern hemisphere than in the southern hemisphere. Moreover, deposition at the equator (10–15 mBq.cm^{-2}.yr^{-1}) is significantly higher than at the poles (< 5 mBq.cm^{-2}.yr^{-1}) [BAS 11].

The dispersion of radioactive aerosols will vary according to the power of the nuclear device, the location of the explosion, the weather conditions, and the nature and size of the debris.

The contamination of the ocean environment will be directly influenced by underwater explosions (< 0.1% of all explosions) and by explosions on the surface of islands (firing on barges). The distribution of radioactivity following an underwater explosion depends on the power of the device and the depth at which the explosion occurs. For explosions that take place at shallow depths, generally one-third to two-thirds of the debris is mixed near the explosion site (within a few hundred meters) and at the thermocline when it exists [NAS 71]. This distribution will be influenced by currents, turbulent diffusion and radioactive decay. As a 10 kt bomb volatilizes 7.10^{6} kg of seawater (1.3.10^{5} kg of chlorine, 7.7.10^{4} kg of sodium and potassium, etc.), the percentages of soluble and insoluble forms of radionuclides will vary according to the position of the explosion, underwater or on the surface [NAS 71].

Avargues [AVA 71] reports that in 1968, the total deposition of strontium-90 on the globe was 481.10^{15} Bq, of which 355.10^{15} was for the Northern Hemisphere (73.8%), or an average of 2,405.10^{6} Bq.km^{-2}. Estimating that the ratio $^{137}Cs/^{90}Sr$ is 1.6 in fallout, the cumulative global deposition of cesium-137 was about 740.10^{15} Bq in 1968, of which 555.10^{15} Bq was for the Northern Hemisphere. Similarly, the deposition of ^{55}Fe would have been 2,366.10^{15} Bq from 1962 to 1966 (including 1,900.10^{15} Bq for the Northern Hemisphere) [NAS 71].

Latitude (°)	Area (thousand km^2)	Deposit of ^{137}Cs (GBq.km^{-2})	Deposit of ^{239,240}Pu (GBq.km^{-2})	^{137}Cs inventory	^{239,240}Pu inventory
60–90 N	749	0.94	0.037	0.70	0.03
30–60 N	25,976	2.16	0.067	56	1.9
0–30 N	54,665	0.92	0.016	50	0.93
Subtotal				***107***	***2.83***
0–30 S	50,760	0.39	0.011	20	0.58
30–60 S	30,355	0.60	0.012	18	0.40
60–90 S	9,203	0.23	0.0024	2.1	0.024
Subtotal				***40***	***1.00***
Total				**147**	**3.83**

Table 2.3. *Deposits and inventories of ^{137}Cs and ^{239,240}Pu in the Pacific Ocean due to fallout from atomic bomb tests (modified from Hamilton et al. pp. 29–58 in [GUÉ 03])*

The difference in radionuclide concentrations between the northern and southern hemispheres is illustrated by two oceanographic expeditions carried out in 1988/1989 and 1991 by the Swedes (Holm et al. pp. 59–74 in [GUÉ 03]). Thus, for ^{137}Cs, surface waters had concentrations lower than 5 Bq.m^{-3} for the Southern Hemisphere and between 5 and 23 Bq.m^{-3} for the Northern Hemisphere. Similarly, for ^{239,240}Pu, concentrations in the Southern Hemisphere were less than 10 µBq.L^{-1} and often less than 5 µBq.L^{-1}, while in the Northern Hemisphere, they were often greater than 5 µBq.L^{-1} up to more than 15 µBq.L^{-1}. These authors estimate that the residence time in surface waters of plutonium is 8 years and about 100 years for cesium.

Before the use of nuclear energy, natural carbon-14 was in equilibrium in the biosphere at a concentration of 277.5 ± 99.9 mBq.g^{-1} C. The excess of carbon-14 due to thermonuclear bombs was estimated in January 1964 to be 63.4.10^{27} atoms or 155.10^{14} Bq (UNSCEAR, 1964 in [EIS 73]). The circulation of ^{14}C in the stratosphere, the troposphere, the biosphere and at the surface of the hydrosphere takes place in a few years but more slowly in the deep ocean. Thus, by the end of 1967, the tropospheric ^{14}C content had increased by 510% in the Northern Hemisphere and 390% in the Southern Hemisphere, resulting in a rapid increase in carbon-14 in the biosphere (Nydal, 1968 in [EIS 73]).

Taking the average production of $2,368.10^{15}$ Bq of iodine-131 per Mt of fission, the world production of iodine-131 due to nuclear explosions can be estimated in 1963 at 44.10^{18} Bq [EIS 73]. Iodine-131 from a bomb explosion of less than 100 kt will be distributed in the troposphere and will necessarily contaminate the aquatic environment.

The nuclear explosions that followed one another from 1947 to 1974 would have resulted in the introduction of $4,070.10^{12}$ Bq of plutonium into the immediate environment of the test sites, while $12,210.10^{12}$ Bq of plutonium would have been dispersed over the entire globe. In the fallout, ^{239}Pu would represent 58% of the total Pu, ^{240}Pu 39% and ^{238}Pu only 3% [WRE 74]. In the Northern Hemisphere where 80% of ^{238}Pu and ^{239}Pu have been deposited [HAR 74], this would result in a deposition of ^{239}Pu of 37 to 74.10^{6} Bq.km^{-2} in the temperate zone and of 0.74 to $1.48.10^{6}$ Bq.km^{-2} in the Arctic zone [AAR 70]. Thus, the oceanic zones would contain on average 259.10^{-4} Bq.L^{-1} of ^{239}Pu (Cherry and Shannon, 1974 in [AMI 80]). In the fallout, americium-241 would correspond to 25% of the activity of plutonium-239 [HAR 74].

The quantities of tritium produced by nuclear bombs are considerable since, according to Schell et al. [SCH 74], the quantities are 0.07 kg.Mt^{-1} for fission bombs and an average of 2 kg.Mt^{-1} for fusion bombs, that is, for all the tests from 1945 to 1962 (282 Mt for fission and 398 Mt for fusion), 816 kg of tritium. This is to be compared with the 200 g of tritium produced annually by cosmic rays, the total tank of natural tritium being estimated at 3.64 kg.

Tritium is a special case. It is deposited mainly on the continents and its deposition is minimal at the equator and increases linearly as one moves towards the poles (the northern hemisphere receiving more tritium than the southern hemisphere). Rainwater above marine stations for the years 1964 and 1967 contained, respectively, 7,240 and 1,206 Bq.L^{-1} TU (tritium unit, corresponding to 1 atom of tritium, i.e. for 10^{18} hydrogen atoms) at latitude 10°N and 72,372 and 12,062 Bq.L^{-1} at latitude 50°N. This corresponds to the deposition of about 23–230 ng of tritium per square meter and per millimeter of rain for the year 1964 at 10 and 50°N latitude, respectively [SCH 74].

For cesium-137, atmospheric deposition has been detailed by the IAEA [IAE 05] as a function of latitude and for the various oceans (Table 2.4).

Atmospheric deposition of cesium-137 and strontium-90 to the oceans varies strongly with latitude (Table 2.5).

Oceanic region	Pacific Ocean	Atlantic Ocean	Indian Ocean	Arctic Ocean	Total
N 90°–60°	2.1	16.3	0.0	7.4	25.8
N 60°–30°	114.6	91.7	0.0	0.0	206.2
N 30°–0°	105.5	49.3	21.3	0.0	175.8
S 0°–30°	42.6	16.5	23.4	0.0	82.4
S 30°–60°	41.3	24.6	36.0	0.0	101.9
S 60°–90°	4.8	2.7	3.4	0.0	10.9
Total	310.6	201.1	84.0	7.4	603.0

Table 2.4. *Inputs of ^{137}Cs to the various oceans from atmospheric deposition (in PBq) at various latitudes (1944–1980) (from [IAE 05])*

Latitude	^{90}Sr (PBq)	^{137}Cs (PBq)
N 90°–60°	16.1	25.8
N 60°–30°	128.9	114.6
N 30°–0°	109.9	105.3
S 0°–30°	51.5	42.6
S 30°–60+	63.7	41.3
S 60°–90°	6.8	4.8
Total	376.9	310.6

Table 2.5. *Global atmospheric deposition of ^{90}Sr and ^{137}Cs (in PBq) in the oceans from atmospheric atomic testing as a function of latitude (from [IAE 05])*

An atmospheric test of one Mt in the equatorial region disperses 35% of the debris in the lower stratosphere and 65% in the troposphere and in the polar region 99% remains in the lower stratosphere and only 1% in the troposphere [VAL 00].

For plutonium fallout, the case of Japan is well representative. The maximum deposition rate of ^{239,240}Pu was observed in June 1963, after the 1961–1962 period of atmospheric nuclear weapons testing between the USA and the former USSR. Annual deposition of ^{239,240}Pu decreased over the period 1963–1967 as a function of stratospheric residence time of nuclear debris. From 1968 to 1984, the level of ^{239,240}Pu deposition was primarily controlled by stratospheric fallout from atmospheric nuclear weapons testing in China. After 1985, the deposition of ^{239,240}Pu showed no decrease, which can be attributed to the resuspension of deposited plutonium (Hirose et al. pp. 251–266 in [KUD 00]).

2.3.2. *Direct releases of radionuclides from nuclear deterrence activities*

In the case of direct releases of radionuclides from military activities related to nuclear deterrence, information is neither in large supply nor easy to find. Discharges to the sea pose important legal problems [QUE 65].

In France, for the French Navy, the Cherbourg arsenal builds and dismantles nuclear submarines. Various radionuclide releases from this arsenal have been published by the GRNC [GRN 99b]. Liquid releases varied from 0.81 MBq (in 1981) to 3,011 MBq (in 1988), excluding tritium, and for tritium from 50 MBq (1996) to 1,790 MBq (1995). The liquid releases were due mainly to antimony-124 and 125.

In the case of nuclear reactors for naval propulsion, direct discharges into the sea were significant for the first generations of the reactors, and the current situation seems to be more environmentally friendly.

Nuclear-powered ships are equipped exclusively with pressurized light-water reactors (PWRs). The major difference between land-based reactors and ship-based reactors is that on the latter, there is not enough space for waste storage; therefore, most of it is discharged directly into the open sea (over the course of more than 12 miles for American ships). Moreover, with each start-up of the reactor, the volume of water contained in the reactor increases and the so-called expansion water goes directly into the marine environment. Since starting a submarine takes place several times a month, the average monthly loss of this expansion water is estimated at 1,900 L (in [NAS 71]). Avargues [AVA 64] estimates that at each startup $14.8.10^9$ Bq are lost. Other radioactive wastes from submarines are the demineralization resins used to decontaminate the primary circuit, the water protecting the reactor, the solid wastes from decontamination and maintenance, and the washing wastes. Pritchard et al. [PRI 59] found that the "*Savannah*" and the "*Nautilus*" released into the environment about 2,107 $TBq.yr^{-1}$ of liquid effluents containing fission products (^{90}Sr, ^{95}Zr-Nb, ^{106}Ru-Rh, ^{136}Cs, ^{144}Ce, ^{131}I) and activation products (^{60}Co, ^{55}Fe, ^{59}Fe, ^{182}Ta, ^{51}Cr, ^{64}Cu and ^{175}Hf). Ouvrard [OUV 72] estimated in 1972 that 300 PWR reactors of 60 MWe equipping submarines could eliminate 185.10^{12} $Bq.yr^{-1}$ by expansion water, 126 $TBq.yr^{-1}$ by losses and liquid wastes and 333.10^{14} $Bq.yr^{-1}$ due to ion exchange resins. Today, about 400 nuclear powered ships exist in the world, some of which have several reactors. However, it would seem that currently the discharges from expansion waters would be negligible for the new nuclear submarines, but all information is a "defense secret".

For French nuclear-powered vessels, two cases must be considered. In the case of submarines, nuclear effluents are discharged by dilution; there is temporary storage and then discharges of gaseous and liquid radionuclides directly into the sea (in the harbors of Brest, Cherbourg or Toulon). For the aircraft carrier *Charles de Gaulle*, gaseous releases are made by dilution through the chimney (nuclear ventilation at very low rates), while for liquid radionuclides, storage is achieved in an effluent tank and they are then transferred from the ship to an external tank in the port of Toulon to be treated at Cadarache. For some years now, there have been no more direct discharges into the sea [HUS 99, OUE 13].

2.3.3. *Indirect releases of radionuclides due to nuclear deterrence activities*

Among the military nuclear centers, the Hanford site in the United States is one of the oldest in the world, and its main activities (plutonium reactors and reprocessing plants) have resulted in significant direct discharges into the Columbia River. For example, Hanford's discharges from 1994 to 1972 were ^{24}Na (481,000 TBq), ^{32}P (8,510 TBq), ^{46}Sc (4,440 TBq), ^{51}Cr (266,400 TBq), ^{54}Mn (2,950,000 TBq), ^{65}Zn (18,130 TBq), ^{72}Ga (136,900 TBq), ^{76}As (92,500 TBq), ^{90}Y (16,650 TBq), ^{131}I (1,778 TBq) and ^{239}Np (233,100 TBq). In addition, the Hanford site released 12,400 TBq of ^{99}Tc and 1.7 TBq of ^{129}I, while the two European sites (Sellafield and La Hague) released 840 and 26 TBq, respectively, for these two radionuclides [HU 10].

Because the Hanford site is relatively close to the Pacific Ocean, a significant portion of the radionuclides arrive in the marine environment. The Columbia River plume has been tracked for 350 km from the mouth of the river by measuring its chromium-51 content in seawater [OST 65] (Figure 2.3).

Most of the ^{51}Cr and ^{124}Sb activities were associated with the soluble phase, while most of the ^{65}Zn and ^{46}Sc activities were associated with the particulate phase. The average transport rate of radionuclides after Vancouver, Washington was calculated. It was 55.5 GBq.d^{-1} for ^{124}Sb, 7,400 GBq.d^{-1} for ^{51}Cr, 407 GBq.d^{-1} for ^{46}Sc and 370 GBq.d^{-1} for ^{65}Zn. The corresponding oceanic equilibrium activities were 5,180 GBq for ^{124}Sb, 296,000 for ^{51}Cr, 48,100 GBq for ^{46}Sc and 129,500 GBq for ^{65}Zn [VAN 69].

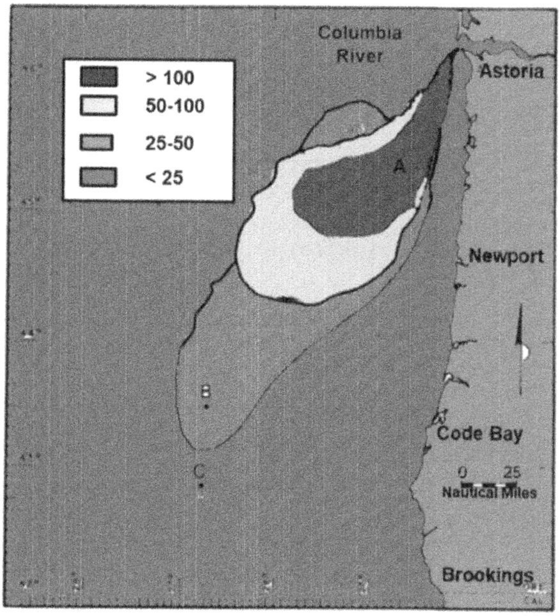

Figure 2.3. *Chromium-51 concentrations (expressed in Bq.L^{-1}) in surface marine waters from June 26 to July 1, 1965 in the Pacific Ocean at the mouth of the Columbia River (from [OST 65]). The values expressed by the authors in strokes per minute (spm, which in this case corresponds to one disintegration) for 100 L of seawater have been converted into pCi.L^{-1} by multiplying by 0.861 and then converting to Bq.L^{-1} (e.g. 100 cpm = 86.1 pCi.L^{-1} = 3.19 Bq.L^{-1}). For a color version of this figure, see www.iste.co.uk/amiard/radioecology.zip*

2.4. Civilian origins of anthropogenic radionuclides related to the nuclear fuel cycle

The use of nuclear fission as a source of energy requires a whole series of industrial operations that together form the nuclear fuel cycle. Each stage of this cycle produces industrial residues and gives rise, to varying degrees, to risks of radioactive pollution of the environment. The marine environment is not directly affected by all of these stages because the nuclear installations in contact with the ocean are in the minority. Figure 2.4 shows the various possible environmental impacts of human nuclear activity.

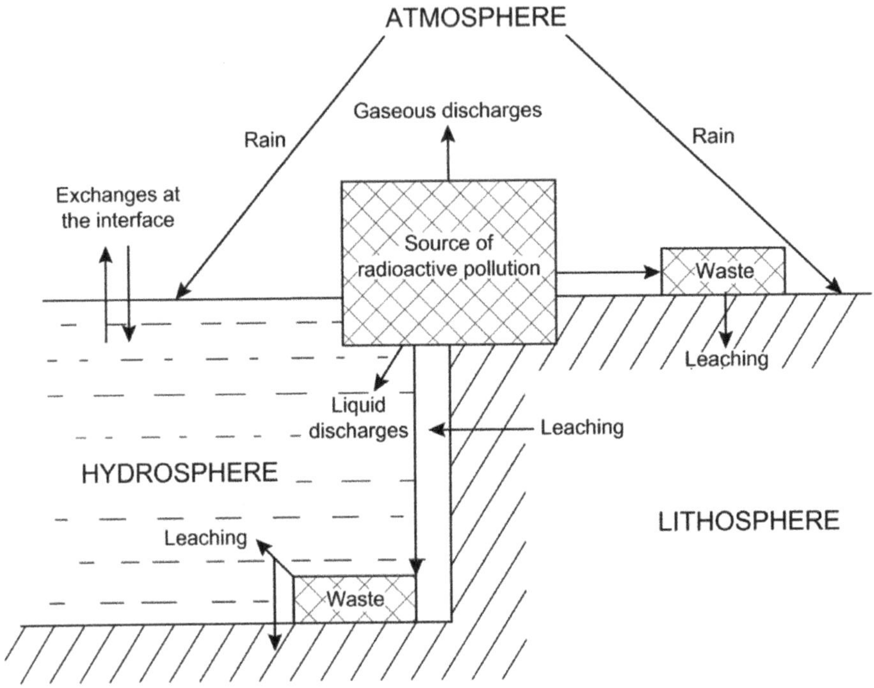

Figure 2.4. *Environmental impacts of a nuclear industrial activity (from [AMI 80])*

2.4.1. *Uranium mines and nuclear fuel fabrication plants*

Few mines and few manufacturing plants are close to the coast. Therefore, their radioactive impact on the marine environment will be only indirect through contaminated soil runoff, groundwater inflow and coastal river inputs.

To fuel a 1,000-MWe light-water reactor, 30 t of uranium enriched to 3.5% is needed annually, which corresponds to about 160 t of natural uranium[2]. Obtaining this 160 t of natural uranium implies the extraction of at least 64,000 t of ore (at 0.25%), which will provide as much solid waste called "waste rock". This waste rock contains ^{230}Th, ^{210}Pb, ^{210}Po [WRE 76] and about 26.10^6 $Bq.t^{-1}$ of radium-226 [CFD 75]. These radionuclides (^{210}Po and ^{226}Ra) will therefore be carried by rainwater to the waterways near the waste rock storage areas. In addition, a mill producing 40 t of uranium concentrated to 70–90% U_3O_8 per day will release 222.10^9 $Bq.d^{-1}$ of ^{226}Ra into the aquatic environment [WRE 76]. Moreover, the concentration of the ore leads to the formation of large quantities of liquid effluents

2 This implies a very low rejection rate of around 0.13%, which is rarely achieved.

estimated at 2,000 $m^3.t^{-1}$ containing at most $9.25.10^3$ $Bq.L^{-1}$ of radium-226 [JCA 59]. According to the CFDT [CFD 75], in 1970 Western Europe released 4,000,000 m^3 of liquid effluents containing 37.10^3 $Bq.m^{-3}$ of radium-226, while solid wastes are estimated at 300,000 t containing 14,800 Bq of radium-226. Because of the cost of fuel at the refining, conversion and enrichment stages, releases to the environment are low. The Capenhurst plant (Cheshire, near Liverpool) thus released in 1975 liquid effluents, representing a total activity of $0.67.10^9$ Bq [MIT 77].

Uranium hexafluoride (UF_6), enriched to 3.5% for use in light-water reactors (enriched to more than 90% or 95% for military needs), will be transformed into uranium oxide (UO_2). This operation will be carried out for French fuel either in Great Britain (BNFL at Springfields, near Preston, Lancashire) or in France at Romans (Drôme). In 1975, the Springfields plant released 80 TBq of β-emitters (excluding tritium) and 1.924 TBq of α-emitters into the aquatic environment [MIT 77].

2.4.2. *The use of nuclear fuel*

The main use of nuclear fuel is in power reactors. There are other, currently secondary uses, such as the manufacture of plutonium using so-called "plutonium-bearing" reactors, civil or military naval propulsion, or the propulsion of aerospace vehicles.

2.4.2.1. *Nuclear power reactors*

During the operation of an electronuclear reactor, the flow of neutrons emitted during fission reactions splits the uranium-235 nuclei, creating numerous fission products and activating various stable elements (or stable elements resulting from fission products, such as ^{133}Sb-^{133}Cs-^{134}Cs) contained in the fuel, the cladding, the coolant, the moderator, the auxiliary circuits, the cooling water, etc., giving rise to often radioactive activation products. Some of this radioactive waste will be released directly into the environment in the form of gaseous and liquid effluents. The data concerning the release of liquid and gaseous radioactive effluents from nuclear power plants are quite numerous. A distinction must be made between data resulting from theoretical calculations and data concerning actual discharges, either gross or in relation to the unit of net electrical energy produced. The qualitative composition of gaseous and liquid effluents will vary greatly from one type of reactor category to another, and also from one reactor to another. Thus, the actual liquid discharges from European power reactors show great fluctuations for the various reactors of the same reactor line. Moreover, the fluctuations are significant from one year to another for a given reactor. This is due to the fact that unusual operations may take place in certain years (intervention on the core or on the primary circuit, etc.) or that some incident may have disturbed the normal operation of the reactor.

More than 90% of the liquid discharges from light-water reactors are composed of four elements (Cs, Sr, Co and I) between 1972 and 1974 (Table 2.6). The same was true for the years 2002 and 2003 (Table 2.7). For natural uranium reactors, cesium, strontium and sulfur represent on average of nearly 75% of the liquid releases [AMI 13].

Radionuclides	BWR	PWR	UNGG and AGR
$^{58}Co + ^{60}Co$	27.3	19.0	0.3
$^{89}Sr + ^{90}Sr$	17.2	0.7	25.7
$^{134}Cs + ^{137}Cs$	45.4	50.0	36.8
^{131}I	4.7	21.0	–
^{35}S	–	–	11.0
Total	94.6	90.7	73.8

Table 2.6. *Composition of liquid effluents, excluding tritium (in percentages), from the various reactor types. Averages from 1972 to 1974*[3] *(modified from CCE, 1975 in [AMI 80])*

	2002	2002	2003	2003
Liquid discharges	900 MWe	1,300 MWe	900 MWe	1,300 MWe
Iodine ($GBq.yr^{-1}$)	0.01	0.014	0.009	0.008
^{14}C ($GBq.yr^{-1}$)	10.8	15.6	10.6	16.2
Tritium ($TBq.yr^{-1}$)	10.5	25.3	10.2	24.1
Other ($GBq.yr^{-1}$)	0.6	0.9	0.5	0.6

Table 2.7. *Average liquid discharges per reactor in 2002 and 2003 (modified from [EDF 04])*

Between 1985 and 2000, constant improvements were made to reduce releases from nuclear power plants by treatment at the source, development of specific operating procedures and improvement of installations according to requirements. In France, these actions have made it possible to reduce the activities released by a factor of 180 for the 1,300 MWe nuclear power plants and 100 for the 900 MWe nuclear power plants [AMI 13]. The effluents from the power plants are collected, sorted and treated according to different processes that allow the majority of them to be recycled and only a very small proportion to be released, in compliance with

3 At that time, EDF did not measure carbon-14 emissions, even though they were not negligible.

regulatory requirements. Thus, the average liquid activity released per reactor (excluding tritium and carbon-14) is 1 GBq per year, while the low-level waste sent annually to the Aube storage center is 3,000 GBq [SEY 11].

More recently, the liquid releases of tritium (expressed in $TBq.yr^{-1}$) are about 10 for a 900-MWe reactor, 25–30 for a 1,300-MWe and 30–35 for a 1,400-MWe. Annual releases of carbon-14 vary from 8 to 20 GBq per reactor, and those of iodine are less than 0.01 GBq per reactor. For the other fission and activation products (^{58}Co, ^{60}Co, ^{110m}Ag, ^{63}Ni), the annual releases are on average less than 0.5 GBq per reactor. A water sample makes it possible to control each hour the radioactivity released into the sea, which must be < 18 $Bq.L^{-1}$ for emitters and < 1,800 $Bq.L^{-1}$ (mid-discharge) and 900 $Bq.L^{-1}$ (daily) for tritium [EDF 20].

In France, in recent years, discharge limits have been significantly reduced for sites whose authorization orders have been renewed. For liquid discharges, the limits for radionuclides other than iodine, carbon-14 and tritium have been reduced from 375 to 15 $GBq.yr^{-1}$ for the 900 MW sites and from 550 to 12.5 $GBq.yr^{-1}$ for the 1,300 MW sites. On the contrary, tritium discharge authorizations have been greatly increased for several plants. For the two 1,300 MWe reactors at Penly, releases of tritium, carbon-14 and iodine remained the same between 2003 and 2019, respectively, 80,000, 190 and 0.1 $GBq.yr^{-1}$ [EDF 04, EDF 20]. For the EPR, the forecasts of discharges into the environment are decreasing, except for carbon-14 and tritium. Thus, radioactive effluents and radioactive waste would be reduced by 30% thanks to improved efficiency and increased combustion rates [AMI 13][4].

2.4.2.2. *Nuclear reactors producing materials*

A certain number of nuclear reactors have been used not to produce a maximum of electricity but to produce a maximum of plutonium-239 for military use. This is the case, for example, with the Hanford reactors in the United States and the plutonium-bearing reactors (G_1, G_2 and G_3) at Marcoule in France on the Rhône.

The nine Hanford reactors installed on the Columbia River, 560 km from the mouth, had only one water circuit that did not undergo any treatment after passing through the reactor core. Moreover, the quantitative releases from these reactors were considerable; from 1950 to 1960, the radioactivity released (which underestimates the radioactivity due to ^{51}Cr and ^{65}Zn) is estimated at more than 12.10^{15} Bq (in [NAS 71]). From 1960 to 1967, these reactors discharged into the Columbia River $15.69.10^{14}$ Bq of ^{32}P, 800.10^{14} Bq of ^{51}Cr and 40.10^{14} Bq of ^{65}Zn as well as a large quantity of neptunium-239 ($4.1.10^{14}$ to $9.6.10^{14}$ $Bq.yr^{-1}$) (in [NAS 71]). In total, the water that had been in the reactor for 4 hours contained

4 Only the EPRs installed in China are currently operating, but there is not enough time to assess whether these details are correct.

61 different radionuclides including activation and fission products. The Hanford site is the most polluted in the United States.

The Marcoule plant, through its activities, released plutonium into the environment, which was added to atmospheric fallout to contaminate the soils of the lower Rhone valley (50 Bq.m^{-2} in $^{239+240}$Pu and 1.5 Bq.m^{-2} in ^{238}Pu) and more strongly around Marcoule (120 Bq.m^{-2}). The erosion of the soils of the Rhône valley (about 100,000 km^{2}) introduced into the river from 1945 to 1998 about 92 GBq of ^{239}Pu [MET 10].

2.4.3. *Spent fuel reprocessing plants*

After an average stay of three years in a reactor, the nuclear fuel is encumbered with impurities that slow down the fission reactions. These impurities are mainly fission products (born from the breaking in two of the fissile uranium-235's nuclei) and activation products (born from the capture of a neutron by a stable isotope). In addition, the fuel still contains unfissioned uranium-235 and a new element, plutonium-239 (a fissile element born from the capture of a neutron by the element described as fertile uranium-238). Reprocessing of the fuel will consist of separating the waste (fission and activation products) from the elements that are still fissile (^{235}U and ^{239}Pu). Among the fission and activation products, some with civilian or military applications will also be separated from the waste.

This stage of the fuel cycle is by far the most polluting. Indeed, the uranium (or plutonium) rods, after a stay in the reactor, have been highly irradiated and contain a large quantity of radioactive products. The amount of radioactivity will depend on the rate of combustion (or irradiation) of the fuel rods. The composition will therefore depend on the nuclear reactor types. At the reprocessing plant, the first operation consists of cutting the cladding containing the fuel and dissolving it in boiling nitric acid. The reprocessing of 1,500 t.yr^{-1} of oxide fuel (i.e. the tonnage corresponding to about 50 1,000 MW units) involves the handling of large quantities of radioactive products (Table 2.8).

Worldwide, of these reprocessing plants, two of them are located directly on the coast. These are the British plant at Windscale, now Sellafield and the French plant of La Hague. The Sellafield site is located in Cumbria near the Elk estuary. The La Hague site is on the Channel coast west of Cherbourg and close to a powerful ocean current, the Raz Blanchard. In authorizing an increase in capacity for the La Hague site from 400 to 1,600 Mt U per year in the 1990s, French regulators did not accept an increase in the discharge limits for radioactive liquids and gases, nor did they allow subsequent license modifications to raise these limits [LEB 95].

The history of liquid and atmospheric releases from these nuclear complexes has been the subject of several synthesis publications with quantitative information, year by year, concerning the releases of various radionuclides. Thus, the releases between 1951 (the beginning of operations at the site) and 1992 have been recorded by Gray et al. [GRA 95]. The discharges from Sellafield were very significant, especially in the years 1970–1980. Thus, the maximum releases were more than 5,000 TBq for ^{137}Cs, 2,000–4,000 TBq for ^{3}H, more than 50 TBq for ^{239,240}Pu, 120 TBq for ^{241}Am, and about 2,700 TBq for ^{241}Pu ([JAC 00]; Mitchell et al. pp. 155–175 in [GUÉ 03]).

Radionuclide	Activity of one ton of irradiated fuel at 30,000 MWd.t^{-1} (Bq)*	Annual activity in the reprocessing plant for 1,500 t.yr^{-1} (Bq)
Fission products	163–185.10^{15}	244.10^{18}
Krypton (10.6 a)	311–407.10^{12}	610.10^{15}
Tritium (12 a)	14–26.10^{12}	41.10^{15}
Iodine 131 (8 days)**	81.10^{9}	122.10^{12}
Iodine 129 (1.6.10^{7} a)	1.4.10^{9}	204.10^{9}
Strontium 90 (28 a)	36.10^{14}	–
Actinides (including plutonium)	481.10^{13} (10 to 12 kg)	72.10^{11} –
Total	167.10^{15}	252.10^{18}

Table 2.8. *Activity of products processed annually by a reprocessing plant for irradiated fuel (modified from data collected in [EIS 73] and [CFD 75]). * except for plutonium in kg; ** radionuclide resulting from the fission of curium-224. Activity without taking into account decay*

For these two plants, the main radionuclides released were ^{137}Cs, ^{90}Sr and ^{239}Pu (Table 2.9)[5]. A significant fraction of the tritium (30–50%) and traces of all the other radionuclides (^{90}Sr, ^{137}Cs, ^{134}Cs, ^{60}Co, ^{106}Ru, transuranics) were released into the aquatic environment. In the previous scenario of reprocessing 1,500 t of irradiated fuel per year, liquid tritium releases were of 18.5.10^{15} Bq, not counting other radionuclides (Kahn, 1976 in [AMI 80]). This is fairly well verified since in 2010 for about 1,050 t reprocessed, the discharges to sea from La Hague were 9,950 TBq of tritium, 1.38 TBq of radioactive iodine, 7.34 TBq of carbon-14, 0.13 TBq of strontium, 1.08 TBq of cesium-137, 0.075 TBq of cesium-134, 1.03 TBq of ruthenium-106, 0.07 TBq of cobalt-60, 2.64 of other beta and gamma emitters, and 0.026 TBq of alpha emitters [ARE 11].

5 Since the 1990s, direct releases to the environment have decreased.

Plutonium has been released from the Sellafield (Windscale) site in the UK since 1952. The total cumulative release of plutonium corrected for decay until 1995 was 717 TBq, most of which resided in the marine sediments of the Irish Sea. Some of this plutonium is remobilized from the seabed and exported via the northern outlet of the Irish Sea, and is transported in solution to the Barents Sea and to Arctic waters ([KER 99]; Kershaw et al. pp. 304–328 in [KUD 00]). For plutonium, normalizing the concentrations to the amounts released, the releases are similar at both sites (about 20–40 µBq per GBq released). The distribution of Pu oxidation state is also similar, with a predominance of Pu (V, VI). A significant fraction of the Pu (III, IV) present in the filtered water is associated with colloids larger than 1 kDa. In each zone, tidal energy would appear to control the long-range dispersion of Pu in the filtered and particulate fractions [BOU 96].

Radionuclide	**Sellafield (until 1992)**		**La Hague (until 1985)**		**Atmospheric fallout over the North Atlantic**
^{137}Cs	41,000	*28,000*	940	*760*	76,000
^{90}Sr	6,200	*4,100*	755	*675*	51,000
$^{239,240}Pu$	592	*592*	3	*3*	1,300
^{99}Tc	794	*794*			

Table 2.9. *Releases (in TBq) from Sellafield and La Hague (in italics corrected decay in 1992 or 1985) (from [KER 95])*

For transuranics, the liquid releases at La Hague and Sellafield in 1999 and cumulatively since 1952 for Sellafield and 1966 for La Hague are presented in Table 2.10. The releases from Sellafield are much greater than those from La Hague. About 200 times more $^{239,240}Pu$ was introduced into the Irish Sea than into the English Channel (689.2.10^{12} vs. 3.4.10^{12} Bq) [GOU 04]. The releases also differ in their composition. The isotopic ratio $^{238}Pu/^{239,240}Pu$ of the releases increases steadily from year to year at the two sites because of changes in the characteristics (increase in the burnup rate) of the reprocessed irradiated fuels [GOU 04].

In addition to these three main radionuclides, others are released. Thus, the influence of the La Hague irradiated fuel processing plant could not be clearly demonstrated until 1974, on the one hand, on the overall fraction of ^{106}Ru, ^{95}Zr, ^{125}Sb, ^{144}Ce and, on the other hand, on the fractions below 28 µm of ^{106}Ru, ^{95}Zr. ^{125}Sb, ^{144}Ce, ^{65}Zn, ^{103}Ru, ^{110m}Ag, ^{134}Cs, ^{141}Ce and ^{151}Eu [GUE 76]. In addition, releases of ^{99}Tc and ^{129}I from the Sellafield reprocessing plant have increased significantly since the mid-1990s, in contrast to the general trend for most other radionuclides [KER 97].

	^{238}Pu	^{239,240}Pu	^{241}Pu	^{238}Pu/$^{239+240}$Pu	^{241}Am
La Hague releases (1999)	12.07	4.00	426.64	3.0	3.49
La Hague releases (1965–1999)	2,767	3,368	211,370	0.8	2,108
Sellafield releases (1999)	36.4	78.67	2,871	0.46	35
Sellafield releases (1952–1999)	50,477	689,157	20,588,420	0,073[a]	538,043

Table 2.10. *Discharges (in 10^9 Bq) from the COGEMA-La Hague and BNFL-Sellafield plants of different isotopes of plutonium, of ^{241}Am and isotopic ratio ^{238}Pu/$^{239+240}$Pu in activity for their periods of activity until 1999 (modified from [GOU 04]). A: this low value indicates that military plutonium was produced*

Seyve [SEY 11] provides the liquid releases from La Hague for the year 2002 for various radionuclides (Table 2.11) to which must be added 11,000 TBq of tritium, 1.42 TBq of ^{137}Cs and ^{90}Sr. More complete data are provided by the GRNC [GRN 99b] and by Zerbib et al. [ZER 09] for the year 2006.

^{106}Ru-Rh	1.6	^{90}SrY	0.9	Uranium	0.0046
^{125}Sb	0.51	^{99}Tc	0.14	^{238}Pu	0.0083
^{137}Cs	0.96	^{14}C	7.85	$^{239+240}$Pu	0.0048
^{60}Co	0.38	^{129}I	1.33	^{241}Am	0.014

Table 2.11. *Quantities (in TBq) of various radionuclides released in liquid form in 2002 from the La Hague reprocessing plant (modified from [SEY 11])*

Liquid discharges from European reprocessing plants for the years 1999 to 2003 are provided by Renshaw et al. [REN 11] for ^{3}H, ^{14}C, uranium and total releases from β/γ and α.

2.5. The origins of anthropogenic radionuclides linked to accidents

The recording of accidents of ships at sea involving radionuclides has been the subject of several censuses. They have taken place in all regions of the world.

The IAEA [IAE 01a, IAE 01b] has synthesized the studies concerning accidents at sea (Table 2.12). For the routes between Europe and Japan for plutonium transport, via South Africa, South America or the Panama Canal, the total risk of accident was estimated by Yamamoto et al. [YAM 00, YAM 01] to be $7.02.10^{-3}$, $6.88.10^{-3}$ and $5.19.10^{-3}$, respectively. The accident rates in the ports of Cherbourg and Paris were $3.47.10^{-4}$ and $4.13.10^{-4}$, respectively. Based on Lloyd's data for the period 1994–1997, the accident frequency for cargo vessels was $2.57.10^{-3}$. In addition, an analysis was performed on ship accidents recorded by the Marine Accident Investigation Branch (MAIO) for British ships for the period 1990–1996, the average fire frequency being 1.36. 10^{-2} per ship.year. For its part, the French Bureau Veritas, for the period 1978–1988, notes that this type of accident originates mainly from the engine room and premises and that, in a few cases, the duration of the fire recorded is more than one day [RAF 98, SCH 01].

For their part, Linsley et al. [LIN 05] list 23 accidents from 1962 to 2000. An update by the IAEA [IAE 15a] lists 30 accidents from 1950 to 2012 for a total potential activity of 48,055 TBq. The releases from the 1964 accident of the SNAP-9A generator of the Transit 5BN-3 satellite are estimated at 630 TBq and the American submarine *Thresher* sunk in 1963 had about 1,150 TBq (1984 value). The bombs lost by the B-52 bomber at Palomares in 1966 represented an activity of 1.37 TBq. The American submarine SNN-583 Scorpion sank in 1968 and lies at a depth of more than 3,000 m with an estimated activity of 1,150 TBq (2004 value). The two bombs of the Soviet submarine K-219 lost at a depth of 6,000 m in 1968, and found, represented an activity of 0.037 TBq. The activity due to the accident of the Soviet submarine K-8 with two reactors and bombs lying at 4,000 m is estimated at 9,250 TBq. The SNAP-27 generator of the Apollo 13 satellite contained 1.63 TBq sunk at a depth of 6,000 m (1970 accident). The 1978 accident of the Soviet nickel-based lightship represented 1.5 TBq. The Cosmos 1,402 satellite lost in 1983 had a generator whose core contained 1,000 TBq of ^{235}U, ^{90}Sr and ^{137}Cs. The containers lost by the French ship Mont-Louis in 1984 in the North Sea off Zeebrugge contained 6 TBq of uranium hexafluoride; they were later found. The Soviet submarine K-431 that crashed in 1985 in Chazhma Bay had a reactor core with 185 TBq. In 1986, the Soviet submarine K-219 sank at a depth of 5,000 m near Bermuda with two reactors of 9,250 TBq. In 1987, a Soviet helicopter crashed near the island of Sakkalin with a generator based on ^{90}Sr of 12,650 TBq. In 1989, a Soviet submarine lost 74 TBq of liquid radioactive waste. The Russian interplanetary station Mars 96 vanished into the Pacific Ocean off the west coast of Chile, losing 174 TBq. The French ship Carla lost in 1997 three packages containing 326 TBq of ^{137}Cs at 70 nautical miles from the Azores. As in 1987, in 1997 a Russian helicopter sank near Sakhalin Island with a 1,300 TBq ^{90}Sr generator; the whole package was recovered. In 2000, then in 2003, the submarines Kursk K-141 and K-159 sank near the Kola Peninsula, but their fuel was recovered (7,400 TBq for the second) [IAE 15a].

	Surface vessels	Submarines	Aircraft	Satellites	RTGs	Total
Accident with potential release	4 (1*)	13 (4*)	8 (5*)	6	2	33 (10*)
Non-release accident	1	2	0	1	0	4
Total	5 (1*)	15 (4*)	8 (5*)	7	2	37 (10*)

Table 2.12. *Accidents in marine environments involving radioactive material with potential radioactive releases (from [IAE 01a]). * Unconfirmed accident. RTG: Radioisotope Thermoelectric Generators*

The atmospheric burn of the SNAP 9A satellite dispersed in 1964 almost pure ^{238}Pu into the atmosphere over the Mozambique Channel at about 25°S latitude [CAR 97].

2.5.1. *The Thule accident*

In January 1968, a B52 plane carrying four nuclear weapons crashed on the sea ice about 12 km from the Thule air base in northwestern Greenland. The benthic marine environment in Bylot Bay, 180–230 m deep, was subsequently contaminated with about 1.4 TBq $^{239+240}$Pu (~ -0.5 kg) (Dahlgaard et al. pp. 15–30 in [KUD 00]).

2.5.2. *Releases due to submarine accidents*

The two American submarines (*Thresher* and *Scorpion*) contained 1,150 and 1,300 TBq, respectively, and the releases were estimated at 0.04 GBq for each of the submarines. For the first four sunken Soviet submarines, the quantities of radioactivity were, respectively, 9,250, 9,250, 3,590 and 1.2.10^{6} TBq. Only the submarine *Komsomolets* (K-278) caused radioactive releases estimated at less than 370 GBq [HU 10].

2.5.3. *The Chernobyl accident*

The marine environment was little affected by the Chernobyl accident. The seas closest to the accident reactor were the Black Sea (about 520 km) and the Baltic Sea (about 750 km). The primary pathways of contamination were mainly atmospheric fallout and to a lesser extent riverine inputs in the years following the accident. The atmospheric deposition of ^{137}Cs was 2.8 PBq on the Black Sea [ERE 95] and

3.0 PBq on the Baltic Sea [VAK 94]. The global atmospheric deposition on the oceans and seas of ^{137}Cs was approximately 15–20 PBq [AAR 98].

Atmospheric deposition of ^{137}Cs on the Black Sea was not uniform, varying from 14.8 to 503.2 mBq.L^{-1} in June to July 1986. In 1989, the mixing of the waters made the contamination of the waters more uniform between 40.7 and 77.7 mBq.L^{-1} [VAK 94].

The maximum fallout in the Baltic Sea occurred in the northeastern Gulf of Bothnia and the Gulf of Finland, resulting in a contamination of the water in ^{137}Cs of 930 mBq.L^{-1} [VAK 94].

Riverine inputs to the Black Sea were mainly from the Danube and Dnieper rivers and accounted for only 4% of the deposition of ^{137}Cs. Riverine inputs of ^{90}Sr were more significant and constituted 25% of the overall inputs [KAN 99]. For the Baltic Sea, riverine inputs accounted for 4 and 35% of total inputs of ^{137}Cs and ^{90}Sr, respectively [NIE 99]. This is due to the fact that the majority of ^{137}Cs was adsorbed on soils and on lake and river sediments. On the contrary, the deposition of ^{90}Sr occurred closer to the reactor than that of ^{137}Cs, this being probably due to a stronger volatilization of cesium [SMI 05].

The fluvial contribution of ^{90}Sr to the marine environment reached 25% of its total contribution during the first 10 years after the accident and continues to grow. Since 1987, the inventories of ^{137}Cs and ^{90}Sr in seawater have been decreasing, resulting from radioactive decay and elimination through the Bosphorus Strait into the Sea of Marmara. In 1994, the inventory of ^{90}Sr was approaching the pre-accident level and that of ^{137}Cs should reach its pre-accident level by 2025–2030 [KAN 99].

Before the Chernobyl disaster, the waters of the Baltic contained about 20–45 Bq.m^{-3} of ^{137}Cs. In June 1986, the concentrations of ^{134}Cs and ^{137}Cs reached, respectively, 5,900 and 1,100 Bq.m^{-3} in water and 1,500 and 2,500 Bq.kg^{-1} (dry weight) in plankton [MAT 04].

2.5.4. *The Fukushima accident*

The main synthesis of the impact of the Fukushima accident in 2011 was made by Povinec et al. [POV 13]. They estimated that the radionuclide releases from the Fukushima accident consisted mainly of atmospheric releases (^{131}I of about 150 PBq and ^{134}Cs and ^{137}Cs, each of about 15 PBq) and direct releases to coastal waters of about 4 PBq (for ^{134}Cs and ^{137}Cs as well), as well as ^{90}Sr with an estimated total activity of about 1 PBq. Other radionuclides were released secondarily to the ocean (Table 2.13).

Radionuclide	Amount released directly into the ocean (Bq)
^{125}Sb	$1.19.10^{11}$
^{144}Ce	$1.88.10^{12}$
^{54}Mn	$4.46.10^{10}$
^{60}Co	$1.01.10^{10}$

Table 2.13. *Quantities of radionuclides directly released into the ocean from the Fukushima accident (from [POV 13])*

Direct marine release fluxes resulting from the Fukushima Dai-ichi accident are estimated at 0.3 TBq.month^{-1} and river inputs through water were less than 0.2 TBq.month^{-1} and through sediment about 0.8 TBq.month^{-1}. At the end of 2012, ^{137}Cs inventories of the coastal area near Fukushima (0–200 m) were for ocean waters of 15 TBq and for the ocean sediment floor of 100 TBq and losses of ^{137}Cs by sediment mobility out of the area of 0.1 TBq.month^{-1} [BUE 14]. Various authors have estimated the input of ^{137}Cs to the North Pacific Ocean related to the Fukushima accident. The values vary from 10 to 50 PBq for atmospheric fallout and from 3.6 to 41 PBq for direct input to the ocean, for a total of 4–90 PBq [BUE 14].

In 2013 and 2014, the Fukushima plume spread over the Canadian continental shelf, and in 2015 and early 2016, it reached ^{137}Cs values of 6-8 Bq.m^{-3} in surface waters [SMI 17].

During the Fukushima accident, various iodine isotopes were released to the environment, especially iodine-129. Concentration levels and inorganic speciation of ^{129}I in seawater depth profiles collected off Fukushima were conducted in June 2011. Significantly high concentrations of ^{129}I in surface water were observed with the highest atomic ratio of $^{129}I/^{127}I$ of $2.2.10^{-9}$ in surface seawater 40 km off Fukushima. The amount of ^{129}I directly released from the Fukushima Dai-ichi nuclear power plant into the sea was estimated to be 2.35 GBq, and about 1.09 GBq of ^{129}I released into the atmosphere as a result of the accident was deposited in the sea off Fukushima. The total estimated release of ^{129}I from the Fukushima accident is 8.06 GBq (or 1.2 kg) [HOU 13].

A comparison of the releases of various radioactive iodine isotopes during nuclear accidents is provided in Table 2.14.

The inputs to the sea of ^{137}Cs due to the Fukushima accident were 15 PBq in mid-March 2011 via atmospheric fallout, 5 PBq in early April 2011 via direct discharges, 15–20 TBq.yr^{-1} via underground discharges and 10–12 TBq.yr^{-1} via inputs from rivers. The total radionuclide inputs to the ocean from the accident were

9–50 PBq, including 5–12 PBq from atmospheric fallout and 3.5–27 PBq from direct releases. The total radioactivity arriving in the North Pacific was 15.2–16.1 PBq [BUE 17].

Iodine	^{131}I	^{132}I	^{133}I	^{134}I	^{135}I
Chernobyl	629–1,760	1,000	900–2,500	30	250
Fukushima	90–380	168			
t_0 "EDF"	4,340	6,120	8,730	9,470	3,420

Table 2.14. *Various radioactive iodine isotopes released during nuclear accidents in PBq (10^{15} Bq) (from [ZER 16])*

2.6. Indirect anthropogenic origins due to rivers and groundwater

Several rivers carry radionuclides of anthropogenic origin. These are particularly the Ob in Russia, the Columbia in Oregon (USA) and the Rhône in France. For the first two rivers, the origin of the radioactivity is linked to military activities. In the case of the Rhone, military activities have become a minority unlike the years 1956–1986.

In the Rhone Delta, sedimentation rates vary from over 20 cm yr^{-1} near the river mouth to 0.2 cm yr^{-1} on the continental shelf. The delta area is a major sink for riverine inputs. The ^{137}Cs inventory over the study area (480 km^2) was estimated at 19.6 TBq in 1990, of which more than 40% was in the pro-delta (30 km^2), near the river mouth. This inventory appears to be well in excess (50%) of the direct and indirect inputs of ^{137}Cs from Chernobyl and fallout from weapons testing in this region. About 20% of the cumulative releases of ^{137}Cs from the various nuclear power plants and from the reprocessing plant of Marcoule were released between 1961 and 1990 [RAD 99, CHA 03].

Numerous studies have examined whether runoff from major Siberian rivers, particularly the Ob River, has the potential to transport radioactive contaminants from nuclear sites – SPA "Mayak" and Siberian Chemical Combine (SCC), located near these rivers, to remote locations, including the Kara Sea [STE 09].

Coastal waters contain elevated dissolved activities of four radium isotopes. These elevated activities result from desorption of radium from the surface of particles and the input of radium-enriched submarine groundwater. The long half-lives of ^{226}Ra (1,600 years) and ^{228}Ra (5.75 years) allow us to see that eddy diffusion controls their distributions in the coastal zone within 50 km of the shoreline [MOO 00]. Radium isotope fluxes and radium concentrations in coastal groundwater allow estimation of submarine groundwater discharge (SGD) fluxes to a large section of the South Atlantic. Total annual average SGD fluxes are three times greater than riverine fluxes. The highest radium and SGD fluxes occur in the summer off the Georgia coast of the United States. These SGD fluxes are not limited to the shoreline but occur throughout the continental shelf [MOO 10].

2.7. Anthropogenic origins of radioactive waste

An in-depth study of radioactive waste present at sea was carried out in Volume 5 of this series [AMI 21]. The IAEA carried out an inventory in 1991 [IAE 91]. This inventory shows that between 1946 and 1982, about 46 PBq of radioactive waste from research, medicine, the nuclear industry and military activities were packaged, generally in metal drums lined with a concrete or bitumen matrix, and disposed of at sea. This inventory includes unpackaged waste and liquid waste that was disposed of from 1950 to 1960. Beta-gamma emitters account for more than 98% of the total radioactivity in the waste, and tritium alone accounts for one-third of the total radioactivity stored in the Northeast Atlantic section of the sites. Other beta-gamma emitting radionuclides include ^{90}Sr, ^{137}Cs, ^{55}Fe, ^{58}Co, ^{60}Co, ^{125}I and ^{14}C. The waste also contained small quantities of alpha-emitting nuclides with isotopes of plutonium and americium representing 96% of the alpha-emitters present. An update was made in 1999 [IAE 99], for wastes immersed from 1946 to 1993 in the Atlantic, Pacific and Arctic oceans, then another in 2015 [IAE 15a].

In summary, the majority of radioactive waste has been dumped in the Atlantic and Arctic oceans (Table 2.15). To these must be added those from the nuclear reactors of Soviet and Russian submarines (Table 2.16). As early as 1995, Lynn et al. [LYN 95] listed six reactors containing their irradiated fuel and seven reactors without fuel that were immersed in the Kara Sea, mainly in Abrosimov Bay, for a total activity of $2.5.10^{15}$ Bq. The majority of the radioactivity consists of beta and gamma emitters (Table 2.17).

Type of waste	Ocean			Total	% of total activity
	Atlantic	**Pacific**	**Arctic**		
Reactors with nuclear fuel	0	0	$3.7.10^4$	$3.7.10^4$	43
Reactors without nuclear fuel	$1.2.10^3$	$1.7.10^2$	$1.4.10^2$	$1.5.10^3$	2
Low-level solid waste	$4.4.10^4$	$8.2.10^2$	$5.9.10^2$	$4.5.10^4$	54
Low-level liquid waste	$<1.10^{-3}$	$4.6.10^2$	$7.6.10^2$	$1.2.10^3$	1
Total	*$4.5.10^4$*	*$1.4.10^3$*	*$3.8.10^4$*	*$8.4.10^4$*	–
Percentage of total activity	53	2	45	–	100

Table 2.15. *Activities (TBq) of various types of submerged wastes in the Atlantic, Pacific and Arctic oceans (from [IAE 15a])*

Site	Year	Number of reactors		Total activity (TBq)
		Without fuel	With fuel	
Abrosimov Fjord	1965	3	3	45,300
	1966	2	–	500
Tsitvolka Fjord	1967	3	0.6	5,500
Novaya Zemlya	1972	–	1	29,600
Stepovoy Fjord	1981	–	2	7,400
Techeniye Fjord	1988	2	–	500
Total		10	6.6	88,800

Table 2.16. *Inventory of submerged reactors in the Kara Sea (from [IAE 15a]*

A much more heterogeneous mixture of liquids and solids containing a variety of radionuclides of low activity levels was packed in steel drums and sunk to the seafloor near the Farallon Islands off San Francisco, California, USA, between 1964 and 1994 [CAR 97].

In addition, liquid wastes released in the Arctic region represented 957.8 TBq. The largest releases were to the Barents Sea (522.6 TBq) and the Kara Sea (315 TBq) [AMI 21].

Ocean	Activity (TBq)				% of total activity
	Alpha	Beta-gamma (including tritium)	Tritium	Total	
Atlantic	675	44,587	15,569	45,262	53.43
Arctic	0	38,012	0	38,012	44.87
Pacific	0.03	1,446	0	1,446	1.70
Total				84,720	100.00

Table 2.17. *Alpha, beta-gamma and tritium activities in solid wastes (low and intermediate activity) dumped in the various oceans between 1946 and 1993 (from [IAE 15a])*

Sivintsev et al. [SIV 95] report that in October 1993, the contents of a TNT-27 tank were released directly into the Sea of Japan. This tank contained 1.10^{10} Bq of ^{137}Cs, $0.3.10^{10}$ Bq of ^{90}Sr and 0.074.109 Bq of ^{60}Co.

Discharges into Kola Bay by the Russian nuclear fleet varied in the 2000s on average from 10 to 130.10^{6} Bq.yr^{-1} [MAT 04].

2.8. Conclusion

The IAEA conducted a study to quantify the inputs to the oceans. The World Ocean Radioactivity Surveys (WOMARS) were conducted by the IAEA Marine Environment Laboratory in Monaco. Three anthropogenic radionuclides, ^{90}Sr, ^{137}Cs and $^{239+240}$Pu were chosen as the most representative of anthropogenic radioactivity in the marine environment. The results confirm that the main source of anthropogenic radionuclides in the marine environment is due to global atmospheric fallout. The total input of ^{137}Cs from global fallout was estimated to be 311 PBq for the Pacific Ocean, 201 PBq for the Atlantic Ocean, 84 PBq for the Indian Ocean and 7.4 PBq for the Arctic Ocean. For comparison, about 40 PBq of ^{137}Cs were released into the marine environment from reprocessing plants (including Sellafield and Cap de la Hague) until 2004 (Table 2.18). The Chernobyl accident contributed about 16 PBq of ^{137}Cs released to the sea, mainly in the Baltic and Black Seas, where the average concentrations in 2005 of ^{137}Cs in surface waters were estimated to be about 60 and 25 Bq.m^{-3}, respectively, while the global average concentration due to global fallout is about 2 Bq.m^{-3}. Mean residence times were similar for ^{90}Sr and ^{137}Cs in surface waters, 28 ± 3 years and only 13 ± 1 year for $^{239+240}$Pu [IAE 15a].

Hamilton [HAM 05] proposed an ocean inventory of fission products and transuranic elements from globally dispersed debris and local and regional deposits from atmospheric nuclear weapons testing. This inventory was repeated and completed for the four major anthropogenic radionuclides (Table 2.19). Inputs of ^{90}Sr and ^{137}Cs to the ocean from rivers increased by 6% and 1%, respectively, from the inventories reported in Table 2.18 [LIV 02].

The comparison of natural and anthropogenic origins can be made for tritium with a strong preponderance of atmospheric fallout from atomic testing. Similarly, in the case of cesium-137, the military origin is clearly greater than those of the two main nuclear accidents (Figure 2.5). In general, the sources of cesium-137 are mainly atmospheric fallout (950 PBq) and the various nuclear accidents at Three Mile Island (0.00004 PBq), Chernobyl (100 PBq) and Fukushima. For this last accident, the atmospheric and direct contributions were, respectively, 10–50 and 4–40 PBq [BUE 14].

Source	Overall contribution	Total input of ^{137}Cs	Total input of ^{137}Cs to the ocean
Atomic testing	200,000	1,000	600
Reprocessing plants	45		40
Chernobyl accident	5,000	85	15–20
Fukushima accident	500	20–40	15–20
SNAP-9A accident	0.638 (Pu)		
Palomarès accident	1.4 (Pu)		
Thule accident	0.03 (Pu)		
Underwater losses	1–2,000 (?)		
Other accidents	45		
Known submerged wastes	85		

Table 2.18. *Quantification (expressed as PBq) of anthropogenic sources (from [IAE 99; IAE 01a, HU 10])*

		^{3}H	^{90}Sr	^{137}Cs	$^{239,240}Pu$
Total contribution		186,000	622	948	10.9
Atmospheric fallout	Contributions to the ocean	112,693	377	603	6.6
	Inventory in the ocean	13,300	151	251	6.6
Direct discharges	Sellafield	39	6	41	0.6
	La Hague	87	1.1	1.0	0.1
Chernobyl accident	Total		10	85	0.06
	Inputs to the ocean		*	16	*

Table 2.19. *Major anthropogenic sources of radionuclides to the ocean (expressed in PBq) * negligible value (from [LIV 02])*

In the Baltic Sea, the sources of ^{137}Cs are 82, 14 and 4%, respectively, due to the Chernobyl accident, atmospheric fallout and civil use of nuclear energy [ZIE 18].

For example, radioactive contamination of the Arctic Ocean and adjacent seas (North Sea, Norway, Greenland, Barents, Kara, White, Laptev, Baltic) results from five primary sources which are explosion tests, reactor accidents, routine reactor discharges, international discharges and uncontrolled releases, and ocean dumping.

To these must be added four secondary sources which are (1) global atmospheric fallout, (2) river sediments after floods, floods and soil erosion, (3) natural sediments eroded by floods and finally (4) inputs from plants and organisms (excreta and post-mortem organic matter) [JOH 10].

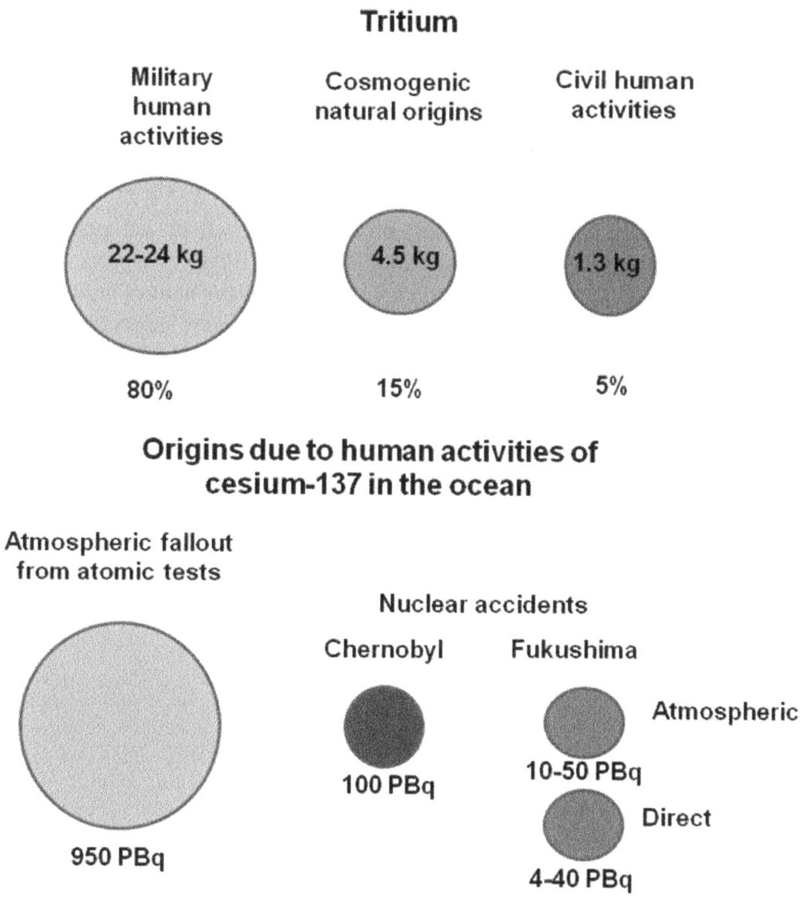

Figure 2.5. *The share of natural and human sources for tritium and the respective shares of human sources for cesium-137 (from [BUE 14, WHO 19]). For a color version of this figure, see www.iste.co.uk/amiard/radioecology.zip*

According to Aarkrog [AAR 03], the inputs to the world ocean in 2000 were 0.2 EBq of ^{90}Sr, 0.35 EBq of ^{137}Cs and 0.01 EBq of $^{239+240}$Pu. In the case of ^{137}Cs, 69% of its contribution to the ocean was due to atmospheric fallout, 21% to local fallout, 7% for direct releases from the Sellafield and La Hague reprocessing plants, and 3% for the Chernobyl accident.

The potential use of tritium as a fuel for fusion reactors in the near future could result in an additional source of tritium [JAI 11].

According to the synthesis by Guillemette [GUI 13], the global stock of natural iodine-129 is 7 GBq, the input to the ocean from nuclear weapons testing is 400 GBq, and the input from the reprocessing plants at La Hague (1966–2009) is 26,000 GBq and Sellafield (1952–2009) is 9,230 GBq. Thus, the reprocessing of spent fuel has multiplied the natural activity of iodine-129 by 5,000.

The vast majority of radionuclides produced by human activities (power generation, medicine, research, etc) are short-lived (half-lives <10 years) and the associated wastes can be managed by taking advantage of the natural rapid decay of isotopes. The main concern shared by past, present and future generations will be the longer-lived radionuclides (>10 years) (lanthanide fission products, ^{3}H, ^{90}Sr, ^{94}Nb, ^{99}Tc, ^{137}Cs, ^{226}Ra, ^{233,234,235,236,238}U, ^{237}Np, ^{239,240,241,242,244}Pu, ^{241,243}Am, ^{244,245,246,247}Cm, etc.), which will eventually decay into harmless isotopes as well, but only after hundreds or thousands of years. Thus, their potential for adverse effects on human health and our environment must be understood on a millennial scale [DEB 20].

3

The Fate of Radionuclides in the Marine Environment

3.1. The introduction of radionuclides into the marine environment

The sources of radionuclides in the marine environment raise several questions. Is the source of pollution natural or anthropogenic? Is the dispersion of the pollution local, regional or global? Is the source of pollution limited or diffuse? Is the source gaseous, liquid or solid? [AMI 17a].

3.1.1. *Diffuse or limited, chronic or intermittent introduction of radionuclides into the marine environment*

Generally, a natural source of radionuclides is rather diffuse and chronic, while an anthropogenic source will be limited and intermittent. The first case corresponds, for example, to the arrival of groundwater at sea or at the mouth of an estuary. The second case corresponds to the control of direct discharges of nuclear effluents into the sea, which will take into account various factors such as the tide to increase dilution factors and avoid concomitant discharges from several sources so as not to exceed any official threshold limits [AMI 13].

3.1.2. *Chemical speciation of radionuclides during their introduction into the environment*

Nuclear chemical speciation, or the chemical forms in which radionuclides are present, determines their chemical reactivity and mobility. The chemical speciation of radionuclides largely explains the migration of radionuclides in the environment, their transfer and toxicity to living organisms and their behavior in industrial processes. It is therefore essential to understand and predict this speciation in the various environments encountered in the nuclear industry, in the natural environment and in living organisms. This knowledge of speciation can be obtained by experimental approaches and by the development of predictive approaches based on thermodynamic models [ANS 03].

Existing methods and tools for the determination of speciation are relatively numerous (Gilbin et al. pp. 7–27 in [MEN 09]). Salbu et al. [SAL 01] propose a selection of methods to study the chemical speciation of radionuclides.

The behavior of radionuclides is almost identical to that of the corresponding stable elements. Thus, the mobility, the geochemistry and the speciation of a radioactive isotope of strontium are the same as those of a non-radioactive isotope of strontium. However, the decay of these radionuclides can give rise to other elements with their own geochemical characteristics in the environment, hence the complexity of this type of study [DES 91].

3.2. The behavior of radionuclides in the marine environment

The marine environment consists of two main physical entities: seawater and sediment. The sediment itself is composed of two sub-units, suspended particles and particles deposited on the ocean floor.

Radionuclides that arrive in the ocean either through industrial discharges or atmospheric fallout are dispersed by currents. They attach themselves to particles suspended in the water, which gradually settle to the bottom and are stored in the sediments, the radioactivity of which can be 100–10 million times higher than that of the seawater above. The radioactive contamination of sediments in place can occur directly from the supernatant water at the water–sediment interface or through precipitation or sedimentation phenomena leading to the deposition of contaminated particles. To understand the role of sediments in the biogeochemical cycle of radionuclides, it is therefore necessary to specify the sorption and desorption capacity of sedimentary particles according to their physical and chemical properties.

3.2.1. *Conservative and non-conservative radionuclides*

Szymczak [SZY 12] distinguishes four states of radionuclides, which explain their biogeochemical fate. Some radionuclides are conservative[1] (cesium, iodine,

1 A conservative element, or radionuclide, is an element that remains in its medium of introduction. In the case of the marine environment, the radionuclide remains soluble in the aqueous medium without absorbing onto the sediments.

radium, radon, tritium, uranium, strontium), or are bound to nutrients (carbon, cobalt, polonium, plutonium), or scavenged by particles (americium, lead, iodine, plutonium, thorium), or controlled by their redox potential (neptunium, iodine, plutonium, uranium).

The IAEA [IAE 11a] lists the nature and quantities of particles from the various atomic tests and from the various basic nuclear infrastructures (BNIs), as well as from the numerous nuclear accidents.

3.2.2. *The distribution coefficient K_d*

The importance of exchanges of radionuclides between the liquid and solid phases in aqueous media has long been known. Thus, for the marine environment, Duursma and Gross [NAS 71] evaluated the "decontamination" effect of water, resulting from the deposition of various sediments initially dispersed in radioactive water, as a function of the concentration of particles in the water and a distribution coefficient K_d [OLS 82].

The adsorption of a chemical substance on particulate material can indeed be represented by its distribution coefficient between the solid and liquid phase $K_d = C_s/C_e$, with C_s being the concentration of the pollutant adsorbed at the solid substrate and C_e being the concentration of the pollutant at equilibrium in the aqueous phase.

Distribution coefficients vary greatly from one radionuclide to another but are generally relatively high. Some examples are given in Table 3.1. The IAEA [IAE 85] has listed the K_d for the main radionuclides. Gil-Garci et al. [GIL 09] propose new estimates of K_d for a number of radionuclides, such as selenium, antimony and iodine.

A variant of K_d was introduced by several authors at the end of the 1980s. These are the fast distribution coefficients (FDCs) adapted to the study of short-term transfer processes. These FDCs correspond to the distribution coefficients obtained by means of radioactive markers for a tracer/particle contact time varying from 4 to 6 hours according to the method described by Mouchel [MOU 90].

Recently, a synthesis presents the different methods of estimating K_d and subsequent compilations of K_d data for various radionuclides (U, Ra, Th, ^{137}Cs and ^{60}Co) in soil and sediments ([KUM 20]).

Table 3.2 compiles average distribution coefficients for the major man-made radionuclides.

Radionuclide	K_d
^{90}Sr	15.8
^{137}Cs	$8.70.10^2$
^{106}Ru	$4.86.10^3$
^{144}Ce	$8.67.10^4$
^{239}Pu	$8.96.10^4$

Table 3.1. *Binding capacity of suspended silts to radionuclides released into Bombay Bay (modified from Pillai et al. in [AMI 80])*

Marine sediment	^{125}Sb	^{60}Co	$^{239,240}Pu$	^{137}Cs
K_d	2,000–4,000	3.10^5–5.10^7	1.10^4–1.10^5	1.10^3–4.10^3

Table 3.2. *Average distribution coefficients (K_d) between marine waters and sediments (Bq.kg^{-1} of dry sediment per Bq.L^{-1} of water) for some anthropogenic radionuclides (modified from [IAE 94])*

Many authors have drawn up an assessment of the distribution of various radionuclides between the main compartments of the environment (water, sediment, organisms) and in both experimental and natural environments; in the marine environment, the main exchanges take place between water and sediment. This justifies the frequently encountered expression of "sedimentary reservoir".

At Sellafield, the surface sediments retain the main radionuclides. Long-lived radionuclides (^{137}Cs, ^{238}Pu, $^{239,240}Pu$ and ^{241}Am) are deposited to depths of 18 cm. On the contrary, short-lived radionuclides (^{106}Ru, ^{144}Ce) remain more at the sediment surface. The K_d for cesium-137 in this system would be 10^4–10^5 (much higher than the average in other environments), but this radionuclide in reaching the deeper layers of the sediment (greater than 13 cm) is redissolved (about 77%). Conversely, transuranics remain strongly bound to the sediment [MAC 98].

Under natural conditions (La Hague coastline), Auffret et al. [AUF 71] indicate that if storms are combined with high tides, considerable quantities of fine particles from wave attack on the cliffs are washed out to sea, which should significantly reduce the concentration of ^{137}Cs in the water. In the area of the Windscale plant discharges, periods of bad weather are followed by phases of rapid decontamination of the water and accumulation of radionuclides in the sedimentary deposits. As a result, the areas where significant amounts of suspended particles are likely to pick up radionuclides and the sedimentation areas where these contaminated particles may accumulate will be the most contaminated aquatic areas.

3.2.3. *Factors influencing K_d*

Many factors influence the value of K_d, in particular those involved with sorption mechanisms and those interacting with horizontal or vertical migration.

3.2.3.1. *Sorption mechanisms*

The exchanges that occur between the liquid and solid phases depend on the physico-chemical characteristics of the radionuclides, on the one hand, and of the sediments, on the other hand.

3.2.3.1.1. The influence of radionuclide characteristics

For the same sediment, a very clear variation of the distribution coefficient K_d is observed for different radionuclides. Duursma (1969 in [NAS 71]) determined the following hierarchy with respect to their sorption capacity on marine sediments: ^{45}Ca < ^{90}Sr < U, Pu, ^{137}Cs < ^{86}Rb < ^{65}Zn <^{59}Fe, ^{95}Zr – Nb, ^{54}Mn < ^{106}Ru < ^{147}Pm.

For the same radionuclide, fixation by sediments varies according to the physico-chemical form involved in the exchange reaction. Thus, Guéguéniat [GUE 75] showed that the more ruthenium-106 was polymerized, the more it gave rise to significant fixation on marine sediments, the kinetics of fixation being itself influenced by the evolution of the equilibrium between the different physico-chemical forms following the initial sorption of the polymerized forms.

There is also indirect evidence of the influence of the physico-chemical form of radionuclides on the exchanges between water and sediment. In a reducing medium, the K_d factor decreases for ^{106}Ru, ^{59}Fe, ^{147}Pm, ^{54}Mn, ^{95}Zr/Nb and ^{144}Ce while the solubility of ^{90}Sr, ^{65}Zn and ^{60}Co is decreased. In many reactions, the modification of the physico-chemical behavior of these radionuclides results from the formation of complexes different from those encountered under normal oxygenation conditions rather than from a direct reduction of the element. However, ferric iron Fe^{3+} is transformed into ferrous iron Fe^{2+}, which is more soluble, while certain metals such as Zn or Ag can be precipitated in the form of sulfides (Duursma and Eisma in [NAS 71]).

3.2.3.1.2. Influence of sediment characteristics

For the same radionuclide, the sorption capacity of various sediments of various sediments depends on their physical (particle size analysis), chemical (carbonate content, organic matter, metal hydroxides, stable elements) and mineralogical (content and nature of clays) characteristics.

Particle size analysis

In general, the fine fraction of the sediment accumulates much larger amounts of radionuclides than the coarse fraction. This is illustrated in Table 3.3 for the marine environment. Two explanations are advanced by most authors. First, the percentage of aluminosilicate foliated minerals (which have the highest exchange and sorption capacities) is the highest among fine particles. Second, the selective sorption on fine particles is also related to the fact that the available surface area is considerable compared to the volume. In contrast, potassium-40, which exists naturally in the medium and is present within the particles, shows essentially constant concentrations regardless of particle size (Table 3.3).

Particle size	^{106}Ru	^{137}Cs	^{95}Zr	^{40}K
< 28 µm	84,878	4,699	13,667	11,671
28–50 µm	6,704	990	1,302	10,479
50–100 µm	5,984	277	792	13,830

Table 3.3. *Influence of particle size on the accumulation of various radionuclides by sediments (modified from [GUE 76, AMI 80])*

The organic matter content of the sediment

The organic matter content of the sediment has an effect on the sorption of radionuclides, but this effect can be positive or negative depending on the radionuclide in question and the environmental conditions (salinity, etc.). Most authors do not generally distinguish between inert organic matter – which can play a role in the contamination rate of sediments by complexing radionuclides – and living organic matter. However, Lowman et al. (*in* [NAS 71]) noted that bacteria show a marked propensity to associate with sedimentary particles and to accumulate trace elements. They thrive in particular in estuarine areas where nutrients and organic particles brought by rivers provide them with abundant surfaces available for their growth and division. In the marine environment, it seems that bacterial density is lower.

The carbonate content of the sediment

Data concerning this aspect seem very scattered and not very systematic, which would be necessary to understand the mechanisms of action of carbonates. However, the work of Pesret [PES 76] provides some elements of explanation, since this author has shown that fixation varies significantly according to the crystalline form of the carbonate (calcites fix more radionuclides than aragonites), to its purity (the experiments concerned pure calcite crystallized or obtained from lamellibranch shells, aragonite obtained from cuttlefish bone or shell sand, dolomite

(CO_3Ca – CO_3Mg)) and to the nature of the radionuclides (the K_d is high for ^{51}Cr, ^{54}Mn, ^{65}Zn, ^{109}Cd, ^{110m}Ag, lower for ^{60}Co, the fixation being practically nil for ^{125}Sb and ^{137}Cs).

The content of stable elements in the sediment

The influence of the content of stable elements on the association of radionuclides to the sedimentary compartment could result from two main processes, isotopic or ionic exchange phenomena and fixation through metal hydroxides or oxides.

Concerning isotopic or ionic exchange phenomena, experiments on freshwater or marine sediments have shown inverse variations in the sorption of ^{137}Cs and in the content of stable Cs or K in the experimental systems. A competitive effect of stable Sr or Ca on ^{90}Sr has also been observed [AMI 73, AMI 75a]. Duursma and Gross (in [NAS 71]) suggest that the low sorption of ^{90}Sr and ^{45}Ca on marine sediments could correspond to an exchange reaction with their stable isotopes. For these elements of simple physico-chemical forms, the phenomena of isotopic competition or competition with chemically related elements may be quantitatively important. For elements with more complex chemistries, the process becomes more complicated, as a physico-chemical form of the radionuclide is likely to compete only with the same physico-chemical form of the stable isotope. However, the physico-chemical behavior of radioactive materials recently introduced into the aquatic environment could differ significantly from that of their stable isotopes, natural constituents that have reached equilibrium. In this case, the influence of isotopic competition on sorption would be limited.

Sediment–seawater distribution coefficient (K_d) values in Japanese coastal areas were compared for the stable and radioactive elements of Sr and Cs. Between 1964 and 2010, the K_d values were 319 and 1,506 for ^{90}Sr and ^{137}Cs, respectively. The geometric mean K_d values of ^{90}Sr and ^{137}Cs between 2000 and 2010 were different being, respectively, $1.2.10^2$ and $5.8.10^2$. These authors found that the K_d of stable Sr (geometric mean 5.3) was much lower than the K_d of ^{90}Sr, while the K_d of stable Cs (geometric mean $1.7.10^3$) was higher than the K_d of ^{137}Cs. The result is that almost all of the stable Sr in the sediment had to be in the exchangeable fraction to account for the K_d of strontium-90 and only 7% of the stable Cs in the sediment had to be in the exchangeable fraction to fit the K_d of ^{137}Cs [UCH 17].

Salinity

The overall salinity can affect the exchanges between the solid and liquid phases. It seems that, in general, distribution coefficients are higher in fresh water than in seawater. Table 3.4 provides an example for ^{137}Cs. In addition, the transfer of the suspended particulate phase in waters of increasing salinity corresponds to desorption phenomena that are manifested by a decrease in the percentage of the particulate forms of certain radionuclides. Conversely, the mixing of seawater with fresh water causes a significant flocculation of the finest sedimentary material. The surface charges of the particles are affected by the dissolved salts. Furthermore, some freshwater soluble ions may precipitate upon mixing with seawater or co-precipitate with Fe and Mn hydroxides and subsequently be attached to the sedimentary material (Forster in [PRU 72]). Lowman et al. (1966 in [PRU 72]) showed that the rate of precipitation could be indicated by the mode of deposition in the coastal zone. For example, in the Columbia River, rapidly precipitating radionuclides (such as ^{54}Mn or ^{46}Sc) have essentially the same level of radioactivity in the inner and outer estuary sediments, whereas for ^{65}Zn and ^{60}Co, sediment radioactivity increases with distance from the coast.

	Total sediment	**Sediment without organic matter**
K_d in fresh water	6,750	1,450
K_d in seawater	80	270

Table 3.4. *Influence of organic matter content and salinity on the fixation of cesium-137 on the fine fraction (< 28 μm) of a sediment (modified from [AUF 71, AMI 80])*

Hydroxides and metal oxides

Regarding fixation through binding onto metal hydroxides or oxides, the colloidal phase of iron, aluminum and silica hydroxides and manganese dioxide is extremely important in the co-precipitation phenomena. In addition, Guéguéniat [GUE 75] reported that in seawater, metal hydroxides or oxides adsorbed on the surface of sediment particles can promote the fixation of certain metals. This author observed that the presence of weakly bound Mn (i.e. presumably Mn deposited on the surface of the particles) in the sediment, promoted the fixation of ^{106}Ru. From the study of 10 radionuclides, Duurma and Eisma (in [NAS 71]) concluded that the sediments with the highest iron concentrations had the highest distribution coefficients. For ^{106}Ru, Guéguéniat [GUE 75] specified that this effect is clear only for sediments in which the iron is in the form of hematite.

3.2.3.2. *Horizontal migrations*

Since radionuclides are frequently bound to the fine fraction of sediment, clay and hydroxides in particular, at the water–sediment interface, these microparticles are likely to be resuspended by currents or by the activity of benthic organisms. In their study of Lake Michigan, Edgington et al. [EDG 76] indicated that currents greater than or equal to 10 $cm.s^{-1}$ were sufficient to resuspend the flocculates and fine silt that make up the surface sediment. This phenomenon will therefore have consequences on the movement of sediments and on the dispersion of radionuclides.

During low-flow periods, radionuclides associated with the particle phase in the Columbia River may decrease by as much as 50% as spring floods resuspend sedimentary deposits (Nelson et al. 1964 in [PRU 72]). After an exceptional flood, Jennings (1966 in [PRU 72]) observed a scouring of previously present silt and fine sand at some points in the estuarine zone, associated with a marked decrease in surface radioactivity; on the contrary, other points with elevated surface radioactivity likely corresponded to areas of sediment re-deposition. In general, the estuarine zone can be said to be a "sediment trap", but large floods, especially when they coincide with a period of high tides, can carry abundant sedimentary material out of the estuary and onto the continental shelf. For a medium-sized river such as the Columbia, Gross (in [PRU 72]) estimates the mass of sediment deposited in the last 10 km of the estuary at 1.2 million tons per year, in the entrance channel at 2.5 $Mt.yr^{-1}$ and on the part of the continental shelf located within 10 km of the mouth at 4.5 $Mt.yr^{-1}$. Regarding the consequences on the distribution of radionuclides, the same author notes, however, that the inventory of radionuclides in the surface sediments of the continental shelf corresponds to only 11% of ^{51}Cr and 14% of ^{65}Zn brought with the river sediments during the previous nine months.

Radioactive effluents leaving the La Hague sea outfall at certain times of the tide were expected to be strongly diluted in the Raz Blanchard off the La Hague lighthouse and carried by the Coriolis effect towards the Pas de Calais. However, in studying the geographical distribution of sediments contaminated by radionuclides released by this plant, Guéguéniat et al. [GUE 76] observed that hydrolyzable elements (^{144}Ce, ^{106}Ru, ^{95}Zr) were essentially located in the sediments of the Nord-Cotentin region, with the sediments of Granville and the sediments of Brittany being much less contaminated. Soluble elements (^{125}Sb, ^{137}Cs) are more widely distributed around the point of discharge, but for ^{137}Cs, it should be noted that the respective contributions of atmospheric deposition and discharge are indistinguishable. Except in the latter case, the largest deposits are not located in the immediate vicinity of the point of discharge, but to the north-east of the Cotentin, in the port of Fermanville (Figure 3.1). This harbor, like all such harbors, is a quiet zone where the finest particles settle. Subsequent work [GUE 79] has demonstrated a re-concentration of particulate radionuclides ^{106}Ru and ^{144}Ce in the Normandy-Breton Gulf. The radioactivity stored in the Bay of Mont Saint-Michel in the first 10 cm of sediment would represent about 10% of the ^{144}Ce and 0.6% of the ^{106}Ru released by the La Hague plant. The duration of the transit between the discharge at the outfall and accumulation in the sediments of the Normandy-Breton Gulf can reach and even exceed two years. According to the authors, this could correspond to intermediate storage in an area that is not currently localized, with the contaminated particles subsequently being put back into transit under particular meteorological conditions (storms) and being associated with "gyres" [SAL 93].

These few examples indicate that the concept of a sedimentary reservoir or sink does not correspond to a static phenomenon but to a dynamic phenomenon influenced by a large number of factors, first by the internal movements of water bodies, as underlined by Deloffre and Lafite [DEL 09].

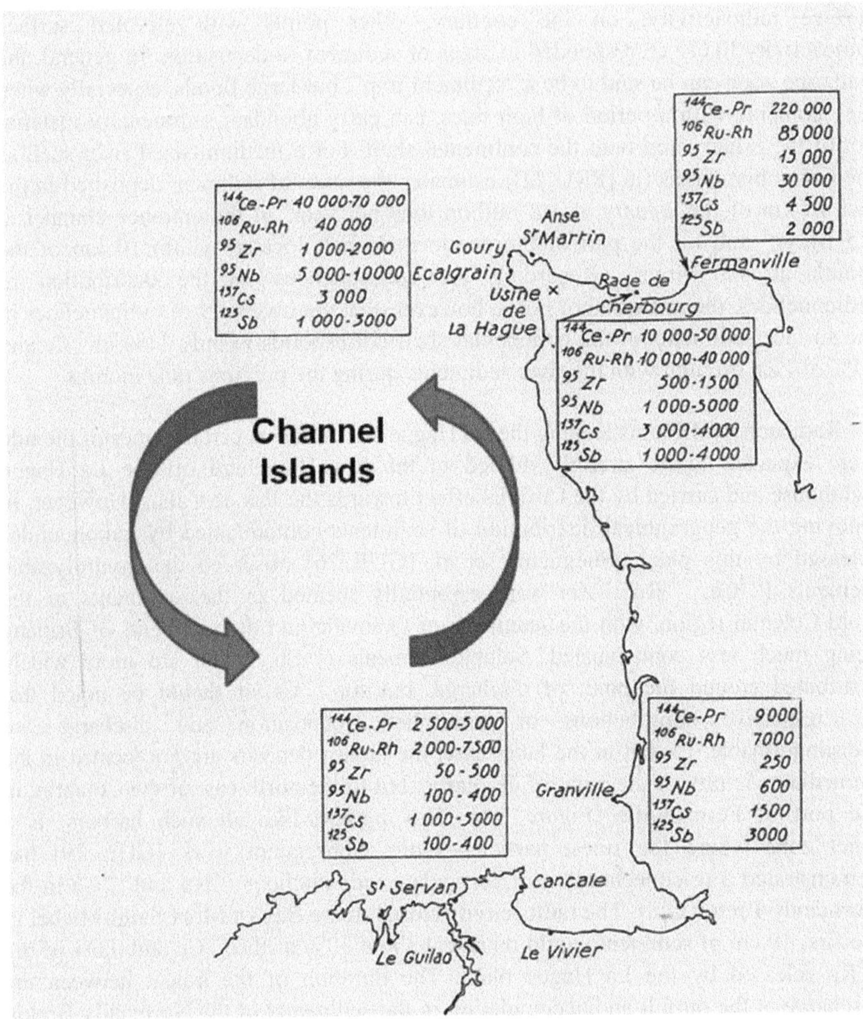

Figure 3.1. *Distribution of radionuclides released by the La Hague plant: radioactivity of the fine fraction < 28 μm (pCi.kg^{-1} dry weight) (modified from [GUE 76]). For a color version of this figure, see www.iste.co.uk/amiard/radioecology.zip*

3.2.3.3. *Vertical migrations and desorption phenomena*

Is the binding of radionuclides from the liquid to the solid phase a reversible or irreversible reaction? The question is often expressed in the following way: are sediments sinks or sources for radionuclides? This is a far-reaching question from a radiation protection perspective. Indeed, if the sorption of radionuclides onto sediments is definitive, this would mean that this reaction can be considered as a real "decontamination" of water and that the radionuclides associated with the sedimentary sink are no longer available to living organisms, nor for transfer along food chains.

In practice, this is a very difficult problem to study, which will be illustrated here with examples concerning plutonium. This choice is dictated by the relative danger of this radionuclide; but its behavior is extreme since it has a very strong affinity for sediments. Before 1964, samples taken in the North Atlantic off Cape Cod indicated that all the plutonium from nuclear explosions expected to fall at that latitude was associated with the sediment. But from 1973 to 1974, it became apparent that the overall inventory of plutonium in the sediments no longer corresponded to all the Pu in the fallout. This remobilization process had been masked as long as nuclear tests had been sufficiently numerous that losses were negligible in relation to inputs [BOW 75].

Based on the contamination of sediments, Hetherington et al. [HET 76] estimated that all of the plutonium released with the effluent from the Windscale (Sellafield) spent fuel reprocessing plant was incorporated into the sediment compartment. In fact, if molecular diffusion and desorption processes occur within the sediment, the quantities of plutonium involved are likely to be small compared to the quantities introduced by the discharges. At the same time, they are also likely to be less than the uncertainty in the estimate of Pu present in the sedimentary sink. Furthermore, the processes involved are slow and easily obscured by any reworking of the upper layers of the sediment as a function of currents or bioturbation. The effect of bioturbation on the fate of radionuclides in sediment particulate matrices will be developed in Chapter 4.

The plutonium emitted in the Channel and in the Irish Sea by the spent fuel reprocessing plants (La Hague and Windscale-Sellafield) associates mostly with the sediments. Boust [BOU 99] estimated the quantities of ^{238}Pu and ^{239,240}Pu stored in the sediments of the Channel at $1.05.10^{12}$ and $4.64.10^{12}$ Bq, respectively. Discharges from the Sellafield plant are responsible for a maximum contamination of the sediments of 8,000 $Bq.kg^{-1}$ for ^{137}Cs and 6,000 $Bq.kg^{-1}$ for ^{241}Am (Mitchell et al.

pp. 155–175 in [GUÉ 03]). However, this association is partially reversible. Gouzy [GOU 04] brought new data on the solid partitioning of plutonium and its solid–liquid partitioning in sediments. He demonstrated, following in situ and experimental studies, the role of reactive sulfides as a temporary phase (temporary sink) in anoxic sediments. Indeed, by oxidation, plutonium can be dissolved in pore waters and these waters can be carried by bio-irrigation to the surface. The plutonium is inserted into the carbonate matrix and adsorbed onto the carbonates in the form of surface complexes and the remobilizable activity of ^{238}Pu and ^{239,240}Pu is estimated, respectively, at 8.10^6 and 7.10^7 Bq (i.e. a fraction of approximately 10^{-5}) [GOU 05a, GOU 05b].

The rate of vertical migration within sediments depends on the possibilities of desorption of radionuclides from the contaminated particles into the pore water. The chemical composition of the pore water thus plays a decisive role. As mentioned above, in a reducing environment, the distribution coefficient K_d decreased for ^{106}Ru, ^{59}Fe, ^{147}Pm, ^{54}Mn, ^{95}Zr – Nb and ^{144}Ce [NAS 71]. Similarly, some organic-rich and highly reducing marine sediments accumulate very little Pu. On the contrary, ^{90}Sr and ^{60}Co appear to be less soluble in reducing media (in [AMI 80]).

This led Duursma and Gross [NAS 71] to predict the vertical movement of radionuclides in the sediments of anoxic systems. Such systems can occur in pore waters of sediments deposited in highly productive coastal areas.

As a result of atmospheric fallout, penetration into soils and sediments varies according to the radionuclide concerned, from 30 to 50 mm in 30 years for ^{137}Cs (30 years), from 0.7 to 10 mm in 53 days for ^{7}Be and from 10 to 30 mm in 20.2 years for ^{210}Pb [WAL 93].

3.2.3.4. *The influence of living organisms on the vertical migration of radionuclides in the marine environment*

The atmospheric fallout of plutonium and americium on the marine environment leads to heir rapid fixation on inert and living particles, then their filtration by various organisms, followed by excretion via feces and elimination by molts in crustaceans. Through remineralization processes, the organic parts of the feces and molts are degraded, leading to an increase in the concentration of these two radionuclides in these particles that sediment, carrying these transuranics to the seabed (Figure 3.2).

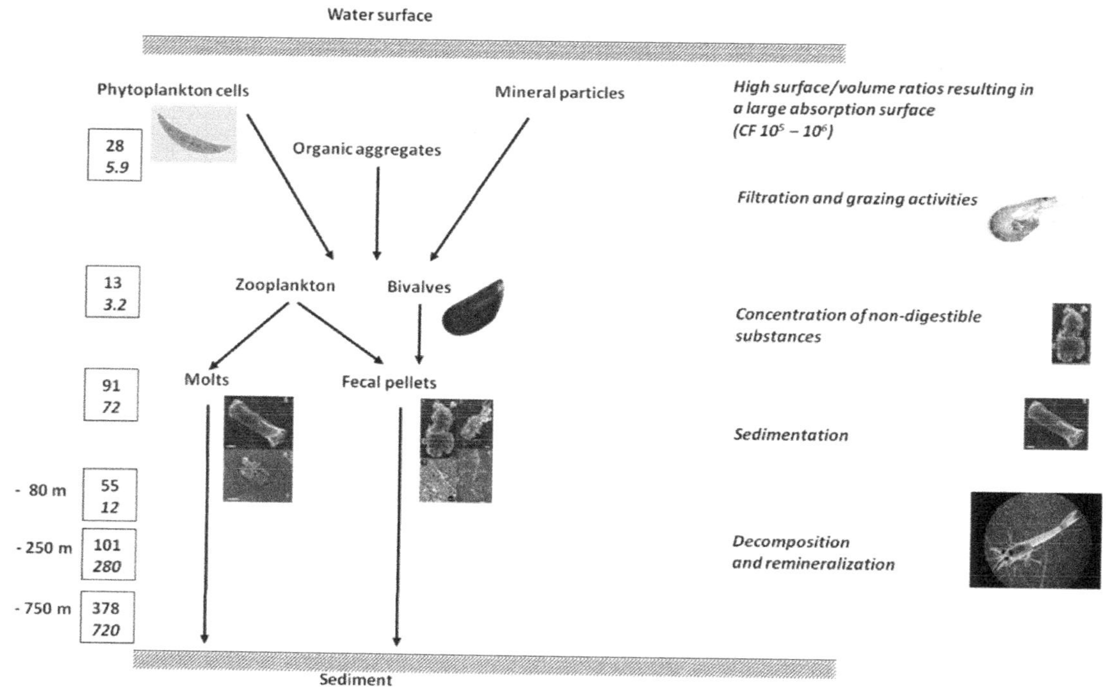

Figure 3.2. *Distribution of contaminants within the water body. Americium and plutonium values (in italics) in pCi.kg^{-1} ps*[2] *(modified from [FOW 89]). For a color version of this figure, see www.iste.co.uk/amiard/radioecology.zip*

2 The concentration can be reported on a dry weight (*dw*) or wet weight (*ww*) basis.

The concentrations of ^{210}Po and ^{210}Pb in fecal pellets of the zooplanktonic euphausiid *Meganyctiphanes norvegica* are reported. The activity ratio $^{210}Po/^{210}Pb$ is 2.2 ± 0.3. This reinforces the fact that there is a high probability that zooplanktonic fecal pellets play a significant role in the removal of these two nuclides from the surface layers of the ocean [BEA 78].

3.2.3.5. *Partial conclusions*

Any radioactive release introduced into a body of water disperses within the liquid phase. At the water interface – suspended particles and water – deposited sediment, exchanges occur between the liquid and the solid phase. Important sorption processes take place but, conversely, desorption reactions can be observed following a variation of the physico-chemical characteristics of the water (during a flood, because of a non-radioactive urban or industrial pollution). Based on experiments, Forster et al. [FOR 73] provide the distribution between these three phases of a few radionuclides chosen to be representative of very varied physico-chemical behaviors (Table 3.5). In some cases, waters are treated with antibiotics in order to eliminate the influence of bacteria on speciation. Exchanges also occur between the suspended particulate phase and the sediments that constitute the bed. These exchanges are likely to vary in space and time. Thus, the sedimentation process will be predominant in calm areas where the current is weak (such as the small port of Fermanville east of the La Hague reprocessing plant) and during low water periods, whereas resuspension phenomena will occur during a flood and an increase in flow velocity. These phenomena are well explained in Deloffre and Lafite [DEL 09].

Radionuclide	**Marine waters**		
	Soluble	**Suspension**	**Deposited**
^{60}Co	88–85	2–3	10–12
^{65}Zn	38–40	2–2	60–58
^{54}Mn	44–42	2–2	54–56
^{59}Fe	15–25	10–13	76–62
^{155}Eu	34–27	8–5	59–68
^{113}Sn	4–1	2–2	93–97
^{110}Ag	50–28	10–17	40–55

Table 3.5. *Distribution (in percentages) of radionuclides in marine waters according to three phases (soluble, suspended and deposited) (modified from [FOR 73, AMI 80]). The first number corresponds to natural waters, and the second number to waters treated with antibiotics*

Suspended solid flow and bottom scavenging are responsible for the downstream migration and dispersion of contaminated sediments. When they reach the estuarine zone, they are subject to significant fluctuations in the physico-chemical characteristics of the environment. This zone is characterized in particular by an increase in salinity, complex water movements, an abundance of particles in suspension (muddy plug phenomenon) and a significant "trapping" of sediments.

While the increase in salinity is responsible for desorption phenomena, the abundance of fine particles promotes the sorption of radionuclides coming from upstream or introduced with the effluents of local nuclear industries. The flocculation reactions of fine sediments, precipitation of dissolved salts and co-precipitation with hydroxides (Fe, Mn) further complicate the mechanisms of the equilibrium between the different phases.

Tidal currents are responsible for the resuspension of deposited sediment, and during periods of high water and high tide, the river's solid flow can carry large quantities of contaminated particles into the coastal zone, which will later settle in calm areas. It is not clear, however, that this corresponds to an ultimate and definitive deposition of radioactive pollutants, on the one hand, because of physical variations in the environment (storms causing resuspension of sediments, etc.), and on the other hand, because of the possible phenomena of desorption and migration within the marine sediment, mentioned above. It is, however, an environment in which the physico-chemical characteristics are relatively stable.

3.2.4. *Chemical speciation of radionuclides*

The chemical speciation of radionuclides frequently changes in the environment, especially in environments where salinity changes. Moreover, the chemical speciation of radionuclides is highly variable depending on the type of effluent.

3.2.4.1. *Speciation of radionuclides in estuaries*

Dange [DAN 02] using a speciation model (MOCO) was able to reproduce the geochemical behavior of trace elements and radionuclides during their estuarine transit. The model used predicts in particular the physico-chemical forms of Cd, Cs and Co and their radioactive analogues (^{109}Cd, ^{134}Cs and ^{57}Co). This author was able to predict the physico-chemical forms, including dissolved and particulate profiles of these metals, under which they are found in the main estuaries of the French coast (Seine, Loire, Gironde, Rhone). These last elements are present in estuarine waters under different dissolved species (free ions, mineral and organic complexes), colloidal, particulate (adsorbed, precipitated) or assimilated by different mechanisms in organisms.

3.2.4.2. *The change of the physico-chemical form of radionuclides from the river to the sea*

Experimentally, Forster et al. [FOR 73] determined the distribution of some radionuclides, representative of very varied physico-chemical behaviors, between the soluble phase and the particulate phase (Table 3.5). For the seven elements studied, the solubility increased from 1% for ^{113}Sn to 84% for ^{60}Co for increasing salinities. In situ, the main data are provided by work carried out on the Columbia River and summarized by Forster (in [PRU 72]).

The isotopes ^{238}Pu and $^{239,240}Pu$ introduced into the Rhône are mainly from low-level radioactive effluents discharged by the Marcoule reprocessing plant, located 120 km from the mouth of the Rhône. Given its very high affinity for the particulate phase and its long half-life, ^{238}Pu (87.74 a) is of particular interest as an anthropogenic tracer of the fate of solid inputs from the Rhône into the Mediterranean. In 2001, sedimentary Pu stocks were estimated at 92 ± 7 GBq in ^{238}Pu and 522 ± 44 GBq in $^{239,240}Pu$, for an area of about 500 km^2. The Rhône delta plays an essential role in the sequestration of plutonium, since it alone traps about 50% of these stocks over an area of 100 km^2. The assessment of the fate of the plutonium by Lansard [LAN 04a] indicates that at least 85% of the plutonium released by the Marcoule plant over the last 40 years has been trapped in the sediments opposite the mouth of the Rhône. Southeast swells are predominant in the erosion of deltaic sediments and in their export to the southwest. The remobilization of plutonium was understood through original experiments of resuspension in linear current channels. For the sediments of the Gulf of Lion and a given hydrodynamic stress (current intensity in particular), the potential fluxes of remobilization reach a maximum of 0.08 Bq.m^{-2}.h^{-1} in ^{238}Pu and 0.64 Bq.m^{-2}.h^{-1} in $^{239,240}Pu$ [LAN 04a, LAN 06].

3.2.4.3. *Radionuclides present in cationic form in effluents*

Cesium-137 and strontium-90 are conservative radionuclides that are characterized, in both seawater and freshwater, by the preponderance of the simple ionic forms Cs^+ and Sr^{2+}. Very low for strontium whatever the type of particles considered, it is important for cesium, in particular with regard to clays. In the Rhône, Bovard et al. [BOV 75] consider that cesium is mainly transported by the solid flow of the river. Nevertheless, during passage in brackish water and at the point of the outlet into the sea, the cesium would pass (at least partially) into the soluble state. This phenomenon can be compared to the existence of two types of binding in an aqueous medium, a binding involving low binding energies and a binding onto sites with a strong affinity for cesium where high binding energies exist. In freshwater, the two types of binding coexist. In seawater, binding to sites with high affinity becomes the priority mechanism because the cations in seawater compete with cesium and hinder binding phenomena involving low binding energies [AUF 71].

In the effluents from the Hanford reactors, ^{54}Mn and ^{65}Zn were entirely in soluble cationic form. Very quickly, upon introduction into the Columbia River, they associated with suspended particulate material. In the estuary, the percentage of particulate forms of zinc rapidly decreased from about 70% to about 40% and then to 5–9% in the coastal zone. Similar behavior was observed for ^{54}Mn and ^{60}Co

(Table 3.6). As with cesium, this is likely to be a desorption phenomenon resulting from competition with seawater cations.

	Percentage of particulate forms		
	^{54}Mn	^{60}Co	^{65}Zn
Columbia Estuary	51	84	72
Adjacent Coastal Zone	5	50	5–9

Table 3.6. *Percentages of particulate forms of various radionuclides in estuarine and marine waters (modified from [PRU 72, AMI 80])*

Experimentally, Bovard et al. [BOV 75] observed the reactions of ^{54}Mn and ^{60}Co with different types of water (distilled water, water from the Rhône, water from the Etang de Berre, water from the Mediterranean) on silica gel supports impregnated with ion exchange resins. They concluded that, upon discharge, these radionuclides complex with organic matter and are transported by the Rhône in the form of organic complexes of variable charge. At the time of contact with brackish water and seawater, anionic complexes appear in significant quantities.

3.2.4.4. *Radionuclides present in anionic form in effluents*

Chromium-51 and antimony-124 released in Hanford reactor effluents are found as complex anions $^{51}CrO_4^{2-}$ and $^{124}SbO^{3-}$. In the Columbia River, ^{124}Sb evolves rapidly to neutral forms while ^{51}Cr remains primarily in anionic soluble form. Downstream, the percentage of particulate chromium increases slightly and this increase appears to be little affected by passage through the estuarine zone [PRU 72].

3.2.4.5. *Particulate and colloidal radionuclides in effluents*

Scandium-46 is mainly found (70%) in particulate form in the effluents of the Hanford reactors. In the Columbia River, the particulate fraction increases downstream of the facilities and can reach 94% in the estuary. In seawater, this percentage decreases (16–57%) (Forster in [PRU 72].

The importance of colloids on the distribution of radionuclides (^{137}Cs, ^{106}Ru, ^{60}Co, ^{238}Pu and ^{239,240}Pu) has been quantified in the Rhône delta and the nearby marine waters by Eyrolle and Charmasson [EYR 04].

3.2.5. *The physico-chemical behavior of radionuclides in marine waters*

Marine waters are far from being homogeneous. It is traditional to distinguish between coastal waters and offshore waters. They are distinguished in particular by their particulate matter content. Natural particulate matter (organic or inorganic) has a high retention capacity for radionuclides. The distribution of biogenic or lithogenic particulate matter will therefore provide information on the dispersion of the pollutants linked to each.

The "green waters" of the littoral zone, enriched in mineral salts by seasonal exchanges with the bottom and by the contributions of coastal rivers, ensure important plant and animal production. Living particles in suspension and debris from excretion or post-mortem decomposition of organisms are therefore abundant in this zone. Oceanic "blue waters" are much less fertile and therefore have lower amounts of biogenic particles. The rivers carry considerable quantities of fine terrigenous particles to the sea. With each tidal cycle, the sea stirs up and resuspends the fine particles of the intertidal zone. This phenomenon is all the more significant when the tidal coefficients are high and the winds are strong. Consequently, coastal waters are still distinguished from oceanic waters by their richness in lithogenic particles. Finally, it should be noted that the coastal areas studied are mainly contaminated by radioactive effluents resulting from industrial uses of nuclear energy, while the radionuclides present in offshore waters come mainly from atmospheric fallout.

These few remarks make it possible to reconcile the apparently contradictory data provided by various authors. For example, in oceanic waters, Silker (1971 in [ROB 71]) determined that ruthenium is found essentially in unfilterable form. Similarly, Pillai et al. (1970 in [AMI 80]) did not observe any (short-term) precipitation of ^{106}Ru in seawater filtered to 0.22 µm. On the contrary, in coastal waters contaminated by Windscale (Sellafield) effluents, Templeton and Preston (1966 in [ROB 71]) observed that ruthenium is found essentially in particulate form. In water artificially enriched with suspended particles, Guéguéniat and Bovard (1969 in [PIR 70]) observed 90–95% in particulate form. This percentage drops to 25% for normal seawater and to 10–20% for filtered seawater.

In the marine environment, most authors note the clear dominance of particulate and colloidal forms of cerium. The same is true for aluminum and manganese, so Al_2O_3 and MnO_2 are even used experimentally in seawater to preconcentrate these stable elements or radionuclides present at trace levels in marine waters.

In oceanic waters or in filtered coastal waters (0.2 µm), manganese is essentially in soluble cationic form, simple (Mn^{2+}) or complex after solubilization by organic matter [ROB 71, FOR 73, MAR 74]. On the contrary, in coastal waters, particulate forms can be abundant, representing 58% of total manganese according to Slowey (1966 in [PIR 70]).

An ultrafiltration system was used to fractionate the low mass molecular (LMM) radionuclides in the effluent of the La Hague reprocessing plant [SAL 01]. Ruthenium-106 was mainly present in the low molecular masses (Figure 3.3). After three days of contact with seawater, about 20% of the ^{106}Ru was associated with colloids, while the low masses decreased and the neutral forms of ^{106}Ru increased. A similar trend was observed for ^{60}Co. Changes in effluent radionuclide speciation are essential to explain subsequent marine transport in the North Sea.

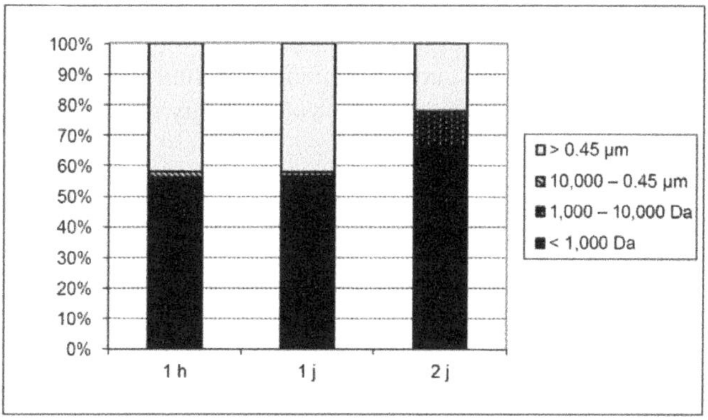

Figure 3.3. *Evolution over time of the forms of ^{106}Ru (size fractionation) of the effluent from the La Hague reprocessing plant mixed with seawater (modified from [SAL 01]). For a color version of this figure, see www.iste.co.uk/amiard/radioecology.zip*

In groundwater at Mayak, Russia, about 65% of $^{239,240}Pu$, 40% of ^{137}Cs, 10–15% of ^{90}Sr and ^{99}Tc are present in colloidal, pseudo-colloidal and particulate forms (Figure 3.4) [SAL 01].

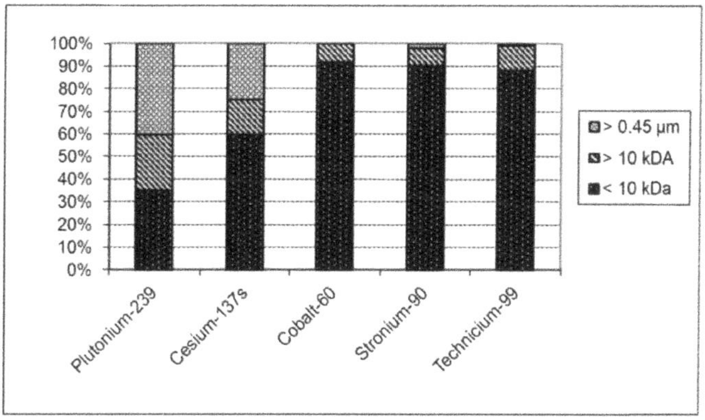

Figure 3.4. *Size fractionation of radionuclides in Mayak groundwater collected in 1994 5 km southeast of Karachai Lake (modified from [SAL 01])*

3.2.5.1. *Chemical speciation in seawater*

Factors such as solution phase chemistry (e.g. ionic strength and ligand concentrations), Eh and the nature of the mineral phases in the system have a critical effect on radionuclide speciation, controlling the distribution between solution and solid phases and thus migration [REN 11].

The relationship between ^{238}U concentration and salinity in the open ocean was re-evaluated using a larger sample set and a wider salinity range than previous open ocean work [OWE 11].

Data on the physico-chemical speciation of Pu in the Irish Sea and Western Mediterranean show that about 87 ± 6% of this element is in the oxidized state, as Pu (V). The ratio ^{241}Pu (IV)/$^{239.240}$Pu (IV) in filtered water from the northeast Irish Sea, near the source term of the Sellafield reprocessing plant, is significantly higher than the ratio ^{241}Pu (V)/^{239,240}Pu (V). Plutonium in oxidized form is truly water soluble. Overall, most of the Pu inventory in the Mediterranean water column is in the oxidized form, especially in open water, while the amounts in particulate or colloidal form are relatively small [MIT 95].

The oxidation state of plutonium varies with depth. Thus, in the North Pacific Ocean, plutonium was about equally divided between the reduced and oxidized forms, except near the bottom. There, the oxidized form was more abundant and constituted about 90% of the total plutonium [NEL 84].

Plutonium ^{239,240}Pu in pore water is characterized by a subsurface concentration maximum of about 0.168 Bq.kg^{-1} between 1 and 3 cm. There is a preferential downward transport of ^{137}Cs, on the contrary, for ^{239,240}Pu there is no diagenetic chemistry activity[3], and these isotopes are not significantly mobile in coastal sediments [SHO 84].

3.2.5.2. *Chemical speciation in the sedimentary matrix*

The availability of plutonium and americium for re-dissolution from offshore sediments is highly dependent on their physico-chemical forms. Sequential extraction studies have revealed that plutonium was mainly associated with strongly bound sesquioxides and organic complex fractions. Americium was mainly associated with the organic complex fraction, but a significant fraction was in the form of carbonate.

The measured distribution coefficient (K_d) values ^{239,240}Pu and ^{241}Am were both generally about 10^5 L.kg^{-1} [MCD 01].

---

3 Diagenesis refers to all the physico-chemical and biochemical processes by which sediments are transformed into sedimentary rocks.

The elements Ce, Co, Mn, Pb and Tl are concentrated in the solid phase by oxidative adsorption processes [LI 91].

Discharges of Cs, Pu and Am have been made into the Irish Sea for decades. Most of them are trapped in the mud of the Irish Sea. Sequential extractions have been performed to estimate their fate. Depending on the sediment samples, cesium-137 is between 15 and 40% in exchangeable form and 10 and 40% in reducible form. For plutonium, the exchangeable fraction is minimal and 30–50% is reducible, and for americium, more than 90% is in residual form. The result is that about half of the cesium is in remobilizable forms, the plutonium is very moderately remobilizable, and the americium is apparently permanently trapped [RAY 20].

Plutonium speciation in estuarine particles has been little studied. Solid partitioning of plutonium under anoxic conditions has been studied in the intertidal sediments of the Ravenglass estuary in the northeast Irish Sea. A significant proportion of the plutonium in these sediments is associated with relatively mobile geochemical phases. The results are consistent with the relatively high level of plutonium remobilization now known to occur throughout the northeast Irish Sea [LUC 04].

3.3. Dispersion of radionuclides in marine waters

Mainly due to potassium-40, the natural radioactivity of seawater is about 13,000 $Bq.m^{-3}$. To this must be added anthropogenic radioactivity, of which cesium-137 is one of the principal radionuclides. The oceans are relatively homogeneously contaminated by fallout from atmospheric testing of nuclear weapons. Shallow seas and estuaries are more heavily contaminated due to discharges from spent fuel reprocessing plants at Sellafield (UK) for the Irish Sea and La Hague (France) for the English Channel. These discharges have been decreasing significantly for more than 30 years.

3.3.1. *The passage of radionuclides through estuaries*

Estuaries are characterized by two types of water movement: one is oscillatory that is linked to the tide and the other is a non-tidal circulation. Indeed, the less dense river water tends to flow into the sea forming a layer of fresh water on the surface, separated from the denser seawater, which rises in the lower layers of the estuary. This interface can be destroyed by tidal currents because the turbulence due to the currents can cause more or less mixing. Based on the characteristics of the non-tidal circulation, vertical stratification, and the extent of lateral homogeneity, Pritchard [PRI 58] distinguishes three main hydrological types of estuaries.

The first type is called "salt wedge" or highly stratified. Stratification is maximal, the freshwater/seawater interface is very clear and vertical exchanges are practically non-existent. This mode of circulation is promoted by low tides, high river flows (cases of the Rhone, Mississippi, Orinoco deltas) or by a low width/depth ratio (case of fjords). The second type is called partially mixed. Compared to the previous type, vertical exchanges are increased. The freshwater/seawater interface is replaced by a more or less marked salinity gradient. In addition, in high latitude estuaries, the influence of the Coriolis effect induces a lateral salinity gradient. This is the case of the Gironde, the Loire and the Columbia rivers. The third type is called homogeneous. There is turbulent and rapid mixing and homogeneity is quickly achieved.

In fact, it is rare that a given estuary can be attributed to one of these three hydrological types at any given time. Two main factors must be considered, the river flow whose increase promotes stratification and tidal coefficients.

In a partially mixed estuary, which is the type of estuary most frequently encountered in nature, the dispersion of a soluble element discharged into an estuary is subject to three concurrent hydrodynamic processes: tidal oscillatory motion, non-tidal residual circulation and turbulent diffusion.

If an effluent of density equal to that of the receiving water is discharged at the beginning of the flow at a point near the bottom, the flow (tide) and bottom transport (residual circulation) will carry it upstream. Since the tidal oscillatory motions are the most important, the ebb tide will carry the effluent back downstream but against the bottom transport, so that the latter point will be upstream of the initial discharge point and the cycle will resume. The resulting movement will therefore be upstream. At the same time, turbulent diffusion will cause the contaminated cloud to disperse in all three dimensions, and thus, some of the cloud may reach surface waters. If, on the contrary, the effluent is discharged into surface waters, it will form an elongated horizontal slick under the action of prevailing tidal currents. During the ebb tide, the slick will drift downstream and upstream during the flood tide. Since vertical and horizontal diffusion contribute to the dispersion of the effluent throughout the contaminated slick, the radionuclide concentration will decrease as the distance from the source increases. In these surface waters, transport from residual circulation is directed downstream, and effluent movement resulting from this transport and tidal movements is also directed downstream. The diluted effluent will therefore be carried seaward. In the coastal zone, advection and turbulent diffusion phenomena may bring a fraction of the pollutant back to the near-bottom waters, from where it can again enter the lower layers of the estuary.

The physical characteristics of the effluent (density related to temperature and salinity) as well as the technique of introduction and the depth at which it is introduced, may initially modify the general pattern presented above. If the density of the effluent is lower than the density of the outfall water and the discharge is into surface water, the initial vertical mixing is inhibited and the effluent will be rapidly discharged to the sea. On the contrary, if a dense effluent is discharged near the bottom, vertical mixing and subsequent discharge will be slowed. If the density of the effluent is low but the discharge is near the bottom, the contaminated slick will rise to the surface by diluting in the overlying water encountered, the level of dilution achieved depending on the depth. If a dense effluent is discharged to surface waters, a similar dilution phenomenon will occur but in the opposite direction, as the contaminated slick descends into the deep waters. Diffusion can be enhanced if the discharge is made through a multi-pore diffuser.

As for the rivers, let us finally point out the existence in estuaries of bank effects, eddies being frequently associated with the existence of small bays or points.

Trace metals and radionuclides transiting through estuaries undergo significant changes (Figure 3.5). This is due to the fact that in these transitional waters, the mixing of waters of different nature is at the origin of strong variations of physico-chemical parameters. In particular, the distribution between the particulate and dissolved phase as well as the chemical forms, are affected. There are three types of behavior of trace elements and radionuclides in the dissolved state that have been classically identified. The first type is called "conservative" where the concentration of the element decreases as a function of the dilution of freshwater in marine waters. The other two types of behavior are called "non-conservative" and "reactive"; the concentration in water depends on exchange reactions with the particles present (desorption or adsorption). This results in a higher or lower concentration in the estuary than in the river (Figure 3.6). Here, the term adsorption encompasses all the phenomena responsible for the transition of metals from the dissolved to the particulate state.

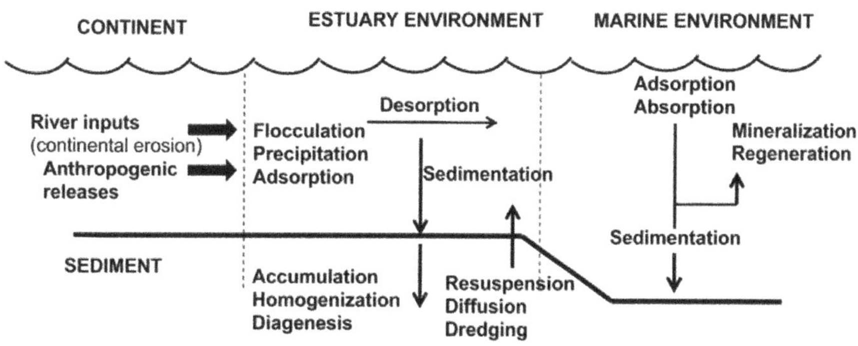

Figure 3.5. *Diagram of physico-chemical and biochemical processes affecting the behavior and dissolved/particle distribution of metallic elements and radionuclides at the continent/ocean and continent/hydrosphere interface (modified from [OLS 82, DAN 02])*

Several examples of dispersion of various radionuclides in sites with variable salinity such as estuaries or fjords exist. For example, vertical profiles of dissolved and particulate ^{210}Po and ^{210}Pb have been measured in the redox transition zone in the Framvaren Fjord, Norway. Both radionuclides are efficiently sequestered by particles in surface waters. Their peaks coincide with those of microbial biomass suggesting a strong biological influence on the behavior of these radionuclides [SWA 99].

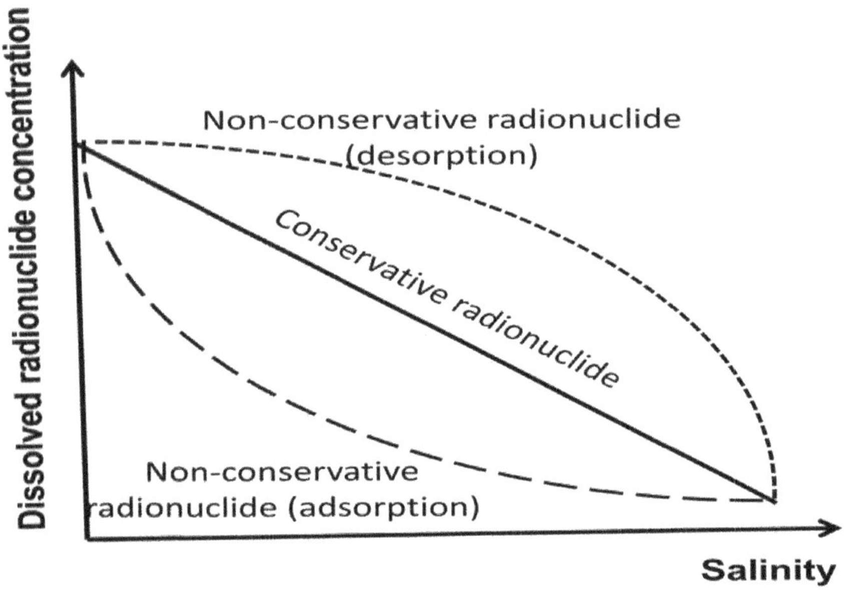

Figure 3.6. *The various behaviors of elements and radionuclides in estuaries (modified from [BOU 99, ROM 01])*

The dispersion of dissolved and particulate radionuclides ^{226}Ra, ^{210}Pb and ^{210}Po was monitored in the Tagus River, Portugal, the estuary and the nearby maritime system. The ^{226}Ra in the river particles dissolves in the estuary and contributes to the increase of dissolved ^{226}Ra concentrations in the estuarine water. In the estuary, dissolved ^{210}Pb and ^{210}Po from river flow and atmospheric deposition are scavenged by suspended matter, which in turn becomes enriched in these radionuclides relative to riverine particles. As a result of these processes, estuarine water flowing into the coastal sea contains increased concentrations of dissolved ^{226}Ra, but is depleted in dissolved ^{210}Pb and ^{210}Po. The average residence times of ^{210}Po and ^{210}Pb in the estuarine system were 18 and 30 days, respectively [CAR 97].

The distribution of plutonium in the Seine estuary shows that sorption–desorption processes occur in specific salinity ranges, Pu(V) of marine origin, from the La Hague plant discharges with an isotopic ratio (IR) $^{238}Pu/^{239,240}Pu$ of about 0.7. On the contrary, plutonium of fluvial origin from atmospheric fallout in the watershed is not very reactive to changes in salinity in the estuary, but 45% of the plutonium from sources within the watershed, with an IR > 1.7, flocculates as early as 0.5 g L^{-1} [GAR 97]. Plutonium isotopes behave in a non-conservative manner with an excess of dissolved ^{238}Pu (IR > 1) in low salinity waters (0.5 and 1.5 $g.L^{-1}$) and depletion (IR <1) in the intermediate salinity zone (8 to 15 $g.L^{-1}$) relative to conservative dilution [GAR 96].

3.3.2. *Dispersion in marine waters*

The oceanic dispersion of pollutants, and in particular radionuclides, will be different depending on whether they are in the soluble or particulate phase (Figure 3.7). This dispersion is governed by several factors (Pritchard et al. in [NAS 71]). The simplest case consists of a short-term, well-localized release of an effluent of density equal to that of the receiving environment and containing radionuclides in soluble form. The local currents will be responsible for a general movement of the contaminated water table, which is an overall drift, while the small-scale turbulent movements will produce a diffusion of the pollutant. Diffusion is responsible for an increase in the volume of the water table and a decrease in the concentration of radionuclides. Subsequently, movements affecting bodies of water with a volume greater than that of the contaminated slick will produce a new drift, while movements affecting smaller bodies of water will cause dispersion by turbulent diffusion. In the marine environment, for a 100-m slick, it is the swell that ensures diffusion while the tide generates the drift. These phenomena can play in the horizontal direction and in the vertical direction, but in the latter case, the scale may be limited by the proximity of the bottom or by the presence of a thermocline.

For the particulate fraction, the general movement of the contaminated water mass is superimposed by the particles' own movements, with episodes of deposition and resuspension governed by the system's hydrodynamics.

In the marine environment, it is common to distinguish a wide variety of currents. In the coastal zone, we have tidal currents, swell currents, estuary currents, drift currents and slope currents, to which are added the great oceanic currents of the open sea. Moreover, the ocean circulation of pollutants will depend on both regional currents, such as the Gulf Stream or the Labrador Current, or general currents that affect the entire ocean and which in theory allow a polluting substance to cross the world ocean in 1,000–1,600 years.

The phenomenon of thermal stratification occurs in the marine environment. With respect to the fate of radionuclides, marine hydrographers have long considered the thermocline as a barrier that would limit the penetration of radioactive pollutants from fallout into deep waters and limit the contamination of surface waters where marine life is richest when the watertightness of submerged waste containers fails. In fact, this barrier role is reduced by the importance of biological exchanges (migrations of living animals, sedimentation of phytoplankton, fecal pellets, molts and zooplankton corpses, etc.) between the deep and surface waters (Chapter 4).

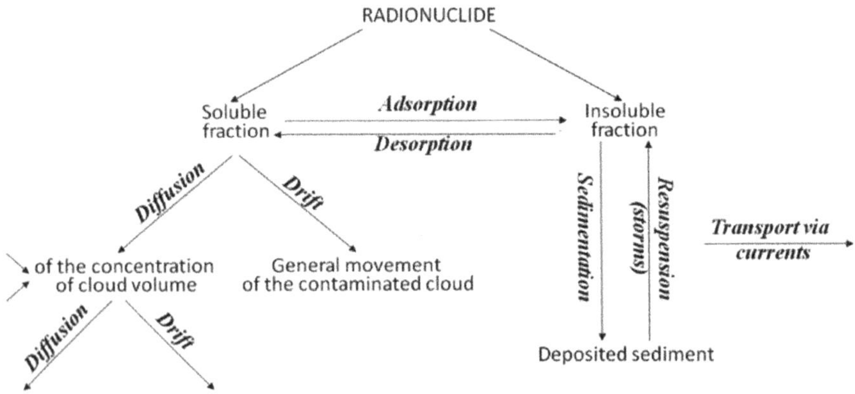

Figure 3.7. *Oceanic dispersion of pollutants*

In the coastal zone, the movement of water presents great variations in space and time. While tidal currents are generally weak offshore, they are very significant near the coastline where they correspond to the transfer of water masses from large maritime bays that the sea periodically uncovers and covers. Thus, in the Raz Blanchard, a strait between the island of Alderney and the west coast of the Cotentin, from Cap de la Hague to the Nez de Jobourg (the area where the discharges from the reprocessing plant for irradiated fuels at La Hague are carried out), the maximum current speeds reach 10 knots in the swell and 7 knots in the ebb.

The swell originates from the strong winds that accompany strong atmospheric depressions and can propagate over great distances. In general, in the wave breaking zone, a current whose direction depends on the direction of the swell develops parallel to the coast (longshore swell).

The swell, coming from distant atmospheric depression zones, is usually overlaid by waves formed by the action of local winds. Outside the wave breaking zone, these local winds and the salinity gradient linked to the inflow of continental water dominate the non-tidal currents. This explains why the direction of the flow in the surf zone can be opposite to the direction of the currents outside the surf zone.

Estuary currents are the outflow of water from a river. The extension and location of low salinity water is controlled by climatic variations that influence both the river flow and the wind regime. Local winds can have a direct or indirect action. In shallow waters, surface water transport directly induced by the wind is directed slightly to the right of the wind (Northern hemisphere). Onshore winds therefore tend to transport surface water offshore. During the summer thermal stratification period, warm surface water is carried offshore while cooler deep water is carried inshore. According to Pritchard et al. [NAS 71], the resulting distribution of densities will produce a current parallel to the coastline and directed in such a way that, in the Northern Hemisphere, an observer at sea looking in the direction of the current will have the coastline on his right. This theory seems to be confirmed by in situ observations made in the coastal zone adjacent to the Columbia estuary or by the "coastal river" extension of the Seine in the Pas de Calais strait.

As depth increases rapidly near the coast, wind-driven drift currents and the resulting slope currents give rise to near-shore water movements that can affect the deep layers. In the Northern Hemisphere, the winds coming from the right looking out to sea are ebb winds that cause deep water upwelling. This happens, for example, on the French Riviera when the Mistral wind sets in. The winds coming from the left or inflow winds conversely cause a descent of surface waters or cascading.

Turbulent currents are caused by the friction of other currents either on the coastline or on the bathymetric lines. In the Raz Blanchard, the speed of the tidal currents is such that the slightest relief on the bottom causes violent eddies even in calm seas.

In coastal areas, water mass movements are governed mainly by tidal phenomena and by the interplay of ebb and flow. The resulting dynamics are represented by so-called "residual" currents which are constant. The resulting dynamics of the water masses is permanent on a yearly scale, allowing the dispersion of marine discharges from nuclear facilities to be modeled by dilution factors (*DF*) of their discharges. This dilution factor will be the average ratio between the water activity (background deducted) and the radioactivity flux released by the plant.

In the case of a coastal discharge, all of the hydrodynamic phenomena lead, after a fairly long time, to the establishment of a general low-level concentration in the vicinity of the source, on top of which is superimposed a cloud of high concentrations corresponding to recent discharges. The steady state depends on the rate of exchange between coastal and oceanic waters. On the east coast of the United States (Cape Cod to Cape Hatteras), Ketchum and Keen (1955 in [NAS 71]) estimated the average residence time of waters in the coastal zone at 1.5 years. Tritium, one of the principal radionuclides present in liquid discharges from the La Hague plant, takes 110–152 days to travel from La Hague to Pas-de-Calais [BAI 95].

Hydrodynamic models for the simulation of radionuclide dispersion in the marine environment are numerous [BAI 99, VIN 01, BAI 05a, BAI 05b, BAI 05c].

The major ocean currents caused by quasi-permanent weather and climate conditions are relatively better known than coastal currents, and the physical oceanography literature provides maps of surface, intermediate and deep currents for different ocean areas. Thus, large-scale ocean circulation may play an important role in the long-term distribution of radionuclides introduced into the marine environment by coastal discharges.

Given its surface area, the world ocean is also largely affected by deposition and two-thirds of all deposition arrives at the ocean surface. Pritchard et al. (in [NAS 71]) describe the fallout as a large planar source (atmosphere–sea interface). According to these authors, the spatial distribution of the radionuclides thus introduced is determined mainly by four phenomena: vertical diffusion, vertical advection, descent into the intermediate and deep waters and horizontal movement of the radioactive material with the intermediate and deep water masses.

For oceanic waters, most of the anthropogenic contribution of radionuclides is through the atmosphere. As a result, for conservative radionuclides such as tritium, cesium-137 and strontium-90, the highest concentrations remain at the surface, with contamination reaching the ocean floor only very slowly. Conversely, for non-conservative radionuclides such as transuranics ($^{239,240}Pu$, ^{241}Am), surface residence times are much shorter than for conservative radionuclides. This is explained by their affinity to the particulate fraction [VIN 01].

3.3.2.1. *The global dispersion of radionuclides*

Due to the general circulation of the world ocean, a "theoretical" radionuclide deposited in the Bay of Biscay, for example, will cross the various oceans and return to its starting point between 1,000 and 1,600 years later.

In a comparison of radionuclide concentrations in the marine waters of the three major oceans, Bourlat et al. (pp. 75–93 in [GUÉ 03]) note that for cesium-137, the waters of the Pacific, and especially of the west Pacific, are more contaminated than those of the Atlantic and Indian Oceans. For strontium-90, the findings are relatively similar. In the Atlantic Ocean, radionuclides tend to sink to the bottom, as is well documented by Nyffeler et al. (pp. 1–28 in [GUÉ 03]) between 1972 and 1981 for tritium. The depth distributions of radionuclides (^{3}H, ^{90}Sr, ^{137}Cs and ^{239,240}Pu) are well correlated. In the North Atlantic, currents send the discharges from Sellafield and La Hague to the Arctic. Thus, the radionuclides released from Sellafield take about 10 years to reach the Arctic and are relatively diluted (Table 3.7). The dilution is expressed as a transfer factor (TF), which represents the concentration in water in Bq.m^{-3} obtained with an annual release of 1 TBq (Table 3.7). In units of weight, this TF is expressed in ng.m^{-3} per annual ton (Nyffeler et al. pp. 1–28 in [GUÉ 03]).

	Transport time (years)	**TF (ng.m^{-3}/t.a^{-1})**
Kattegat	4	15
N-W current of Norway	3–4	10–50
S-SE Barents Sea Current	5–6	5–10
Kara Sea	5–6	5–10
West Spitsbergen	5	5
Arctic Ocean	6–10	1–10
East Greenland Current	7–10	1–2
North Atlantic Faroe Islands Current	14–17	>> 0.3

Table 3.7. *Transfer factors (TF = Bq.m^{-3}.(PBq.yr^{-1})$^{-1}$ = ng.m^{-3}.(t.yr^{-1})$^{-1}$ from Sellafield to various surface waters (modified from Nyffeler et al. pp. 1–28 in [GUÉ 03])*

The average residence time in the world ocean is 27.5 ± 4.6, 28.6 ± 2.1 and 12.7 ± 1.4 years, respectively, for ^{90}Sr, ^{137}Cs and ^{239,240}Pu. For the various oceans, the average residence times are presented in Table 3.8 [IAE 05].

Ocean	**^{90}Sr**	**^{137}Cs**	**^{239,240}Pu**
Pacific	15.9 ± 2.4	15.9 ± 4.3	7.4 ± 1.1
Atlantic	22.3 ± 2.4	20.4 ± 3.0	9.1 ± 0.5
Indian	–	20.3 ± 1.8	9.0 ± 2.9
World Ocean	19.1 ± 3.2	19.8 ± 1.5	8.8 ± 1.0

Table 3.8. *Mean residence times (expressed in years) of three radionuclides in surface waters of various oceans (from [IAE 05])*

The observed residence time for ^{90}Sr and ^{137}Cs in the surface waters of the Northeast Atlantic is 20 years, which corresponds to an average residence time of 80–100 years for stable elements. The observed ^{99}Tc/^{90}Sr ratio (3.10^{-3}) in the Azores samples is 10 times higher than the theoretical fission yield decay corrected to 1992 [DAH 95b].

3.3.2.2. *Dispersion in the Atlantic Ocean*

The principal contributions of radionuclides to the Atlantic Ocean at the end of the 1990s were ^{239}Pu (50.8%) and ^{241}Am (39.1%), and ^{137}Cs (27.5%), ^{106}Ru (21.3%) and ^{241}Pu (13.9%). The main sources are the spent fuel reprocessing plants and in particular Sellafield (Table 3.9) (Nyffeler et al. pp. 1–28 in [GUÉ 03]).

Between 1949 and 1982, the releases of radionuclides into the Northeast Atlantic were 666, 26,150 and 15,500 TBq for α and β emitters and tritium, respectively (Nyffeler et al. pp. 1–28 in [GUÉ 03]).

Type of installation	**α emitters (%)**	**β emitters (%)**	**Tritium (%)**
Reprocessing	97.4	98.6	69.7
Power plants	0.2	0.5	19.6
Research	0.01	0.02	10.8
Other	2.4	0.9	0
Sites	**α emitters (%)**	**β emitters (%)**	**Tritium (%)**
Sellafield	95.2	86.9	52.9
La Hague	0.52	5.3	16.4
Dounreay	1.8	6.7	0.33
	Total (PBq)	**Total (PBq)**	**Total (PBq)**
	1.4	150	46

Table 3.9. *Relative contributions of various sites to total radionuclide discharges in the Atlantic (modified from Nyffeler et al. pp. 1–28 in [GUÉ 03])*

3.3.2.3. *Dispersion of releases from the Sellafield and La Hague reprocessing plants*

^{134}Cs, ^{137}Cs, ^{90}Sr and ^{99}Tc, which behave conservatively in seawater, have been widely used as tracers of water movement through the North Sea, Norwegian Coastal Current, Barents Sea, Greenland Sea, Fram Strait, Eurasian Basin, East Greenland Current and Denmark Strait overflow. It was estimated that 22% of the ^{137}Cs release from Sellafield passed into the Barents Sea en route to the Nansen Basin via the Bjomoya-Fugloya stretch and 13% via the Fram Strait. This is equivalent to 14 PBq of ^{137}Cs. In the late 1970s and early 1980s, because of the substantial decrease in discharges of reprocessing waste from the Sellafield plant (United Kingdom), the principal radionuclides were ^{134}Cs, ^{137}Cs, ^{90}Sr, ^{99}Tc and Pu, while the discharges from La Hague (France) contained a greater proportion of ^{129}I and ^{125}Sb [KER 95][4].

The flux of ^{137}Cs to the Barents Sea in 1994 was estimated to be 300–400 TBq, of which 100–200 TBq was from atmospheric fallout. About 120–180 TBq of additional ^{137}Cs, other than atmospheric fallout, were transported through the Fram Strait into the polar ocean [KER 97].

The estimated transfer factor for Sellafield releases to this stream is 52 $Bq.m^{-3}$ per $PBq.a^{-1}$ with transit times from Sellafield to the Pentland Firth and Lowestoft of about 9 and 24 months, respectively. These transit times, derived from ^{99}Tc data, are significantly shorter than previous estimates based on ^{137}Cs and ^{90}Sr data from the 1970s and 1980s [MCC 02].

Transport of European coastal contaminants to the Barents Sea, Kara Sea, Arctic Ocean and East Greenland was quantified using a transfer factor calculated as the quotient between observed environmental concentrations and an average discharge rate t years earlier, where t is the transport time. A transfer value of 1–2 $ng.m^{-3}.t^{-1}$ per year was found for Sellafield discharges to polar waters in the East Greenland Current [DAH 95a].

Antimony-125 is entirely associated with the water mass; its uptake in biological constituents, suspensions and sediments is low. It is also a specific marker for the La Hague plant, with releases two to four times more than Sellafield (releases to the Irish Sea). Cesium-137 is a specific marker of the Sellafield spent fuel processing plant with discharges 100–1,000 times higher than at La Hague. Therefore, the monitoring of Sb/Cs ratios can allow differentiation between the effects of Sellafield and La Hague whenever interference is possible. ^{137}Cs is also one of the principal radionuclides introduced into the environment by atmospheric fallout following nuclear tests in the atmosphere, and it follows that ^{137}Cs is measurable in all natural waters. Guéguéniat [GUE 86] estimates that waters with an activity level above 4 $mBq.L^{-1}$ in ^{137}Cs are subject to industrial influence. Only a few water bodies in the Channel are not under industrial influence.

The distribution of plutonium, discharged into the sea by the spent fuel reprocessing plant at La Hague, in the waters of the English Channel is different from that of conservative elements such as antimony [GAR 97].

---

4 The option taken at La Hague is to release about 95% of the iodine-129 to the sea, or about 1.2 $TBq.yr^{-1}$. This translates into the release of almost all of the tritium, carbon-14 and iodine-129 activity into the sea.

With the radionuclides ^{125}Sb and ^{99}Tc, transit times were estimated between Cape de la Hague and the Straits of Dover between July 1987 and January 1992 at 110–152 days and at 170–250 days for the entry into the Baltic Sea. The seawater flux at the Straits of Dover is estimated at 97,000–195,000 $m^3.s^{-1}$ between January and July 1988 [BAI 95].

A monthly release at La Hague of 10,000 GBq in the central part of the Channel is subject to the Blanchard current and is responsible for radioactive contamination, in $Bq.m^{-3}$, at Goury from 68 to 89, at Cherbourg from 50 to 70, in the Strait of Dover from 21 to 45, at the entrance to the Norwegian Sea from 5 and about 2.2 at the entrance of the Barents Sea [GUÉ 03] (pp. 121–154). The transport of radionuclides can be rapid in the central part of the eastern Channel or much slower if the radionuclides are in the water bodies near the coast in what is called the "coastal river". According to [GUÉ 03] (pp. 121–154), the concentrations of ^{125}Sb, ^{90}Sr, ^{137}Cs and ^{99}Tc in the Norwegian Sea are 0.32, 0.67, 0.22 and 0.015 $Bq.m^{-3}$ and in the Barents Sea 0.33, 0.33, 0.06 and 0.002 $Bq.m^{-3}$. Discharges from La Hague would take 17 months to reach the Norwegian Sea and 36 months to enter the Barents Sea.

Neptunium-237 was released in larger quantities than the plutoniums at Sellafield in the 1990s. Releases varied from 0.03 to 0.67 $TBq.yr^{-1}$ for ^{237}Np and from 0.15 to 58.0 TBq for plutoniums between 1978 and 1996. Because of these high releases, it is necessary to have data on this radioelement. It is clearly more mobile (widely distributed in the Irish Sea) but less bioavailable than other transuranics [ASS 99].

3.3.2.4. *Dispersion of radionuclides in the northern seas*

The sources of radioactive contamination of the Arctic Ocean are multiple (atmospheric fallout, releases from nuclear facilities, waste releases). Thus, the expulsion by rivers is estimated for ^{90}Sr at 2,368 TBq between 1961 and 1994, including 1,355 TBq for the Kara Sea alone. The waste discharges to the Kara Sea are presented in Table 3.10.

	^{137}Cs	^{90}Sr	$^{239,240}Pu$
Tsivolky Fjord	493	447	11
Fjord Abrosimov	352	319	1.6
Stepovogo Fjord	189	185	<0.1
Novaya Zemlya	69	74	0.4

Table 3.10. *Inventory of major radioactive discharges to the sea from Kara from 1993 to 1995 (in TBq corrected for decay) (modified from Strand et al. pp. 95–119 in [GUÉ 03])*

The current regime in the Arctic Ocean and adjacent seas has been studied using radioactive tracers, in particular iodine-129 released by the Sellafield plant and especially the La Hague plant. Karcher et al. [KAR 12] present numerous maps and demonstrate a change in the current regime of the northern seas. The residence time of iodine-129 in the North Atlantic Ocean and the Arctic Ocean is $1.6.10^7$ years [KAR 12] and that of ^{137}Cs is 30.2 years [SMI 11a].

The evolution of the distribution of cesium-137 in surface waters in the various seas of the North Atlantic (Irish, North, Baltic, Norwegian and Barents seas) is strong. Between 1976 and 1980, the Irish and North Seas had the highest concentrations, between 1981 and 1985 the concentrations of ^{137}Cs in the Norwegian Sea began to increase, then between 1986 and 1990, the maximum concentrations were observed in the center of the Baltic, following the Chernobyl accident, and finally from 1991 to 1995, the concentrations of cesium decreased in all these seas [POV 03, NIE 18b].

3.3.2.5. *Dispersion in the Mediterranean Sea*

Vertical fluxes of ^{239,240}Pu and ^{241}Am and temporal changes in their inventories in the northwestern Mediterranean Sea have been measured. The submarine maxima in 1990 were between 100 and 400 m for ^{239,240}Pu and between 100–200 m and 800 m for ^{241}Am. The total inventory of ^{239,240}Pu had not changed much between 1976 and 1990 while ^{241}Am had decreased by about 24%. Upper water column residence times (0–200 m) based on direct flux measurements ranged from 20 to 30 years for ^{239,240}Pu and 5 to 10 years for ^{241}Am [FOW 00].

	^{137}Cs (PBq)	^{239,240}Pu (TBq)
Full input in 1986	12.0	190
Fallout from Chernobyl	2.5	0.02
Atmospheric fallout (1986–1996)	0.3	7.5
Marcoule plant until 1995	0.03	0.3
Black Sea (1986–1996)	0.3	
Exchanges with the Atlantic (1986–1996)	=	-6.3
Total in 1996	15.1	191.5

Table 3.11. *Inputs of ^{137}Cs and ^{239,240}Pu in the Mediterranean waters by various routes until 1996 (modified from Papucci et al. pp. 177–197 in [GUÉ 03])*

More than 90% of the ^{137}Cs deposited at the surface of the Mediterranean Sea was still present in the water column in 1995, while the remaining 10% was located in the surface sediments (Table 3.11). The surface concentrations were only one-third of the concentrations present in the 1970s, while the deep waters had concentrations 30% higher than in previous years (Papucci et al. pp. 177–197 in [GUÉ 03]).

3.3.2.6. *Dispersion in the Black Sea*

The Black Sea represents a particular ocean basin with a large flow of freshwater, large amounts of deep anoxic water and has received significant fallout from the Chernobyl accident. This resulted in high concentrations of ^{137}Cs in 1990, such that only the Baltic Sea and the Irish Sea had higher concentrations, respectively, 52, 125 and 55 $Bq.m^{-3}$ (Buesseler and Livingston, pp. 199–217 in [GUÉ 03]). The Baltic Sea also received high atmospheric fallout of ^{137}Cs from Chernobyl. The inventory of this radionuclide in the waters of the Baltic was 325 TBq in 1985; it rose to 4,300–5,000 TBq in 1986 and has been decreasing steadily since then (Nies and Nielsen, pp. 219–231 in [GUÉ 03]).

3.3.2.7. *Dispersion of radionuclides after the Fukushima accident*

Historically, the maximum release of ^{137}Cs into the sea following the Fukushima accident was observed on April 6, 2011 with a concentration in water of 10^8 $Bq.m^{-3}$, while the background was less than 2.0 $Bq.m^{-3}$ [KAE 17]. After the Fukushima accident, the dispersion of radioactive cesium in the North Pacific and adjacent seas has been carefully monitored. The radioactive cesium drifted from Fukushima first eastward as surface water and spread to the eastern shores of the North Pacific in 2014. Southward intrusion of groundwater was also observed as subtropical mode water and central mode water [KAE 17]. Between June and August 2015, the most contaminated surface water in $^{13}7Cs$ (10 $Bq.m^{-3}$) from the Fukushima accident was located at 50°N 150°W, forming a large concentric zone [SMI 17].

3.4. Dispersion of radionuclides with marine sedimentary particles

Average sediment dispersal velocities range from a few kilometers to a few tens of kilometers per year. They are about two orders of magnitude lower than the residual tidal currents responsible for the dispersion of conservative elements (1,000 $km.yr^{-1}$), suggesting that sediment movements occur only significantly during violent events [BOU 97].

Concentrations of artificial (^{60}Co, ^{137}Cs, ^{238}Pu and ^{239,240}Pu) and natural (^{40}K, ^{212}Pb and ^{214}Pb, daughter nuclides of the ^{232}Th and ^{238}U series) radionuclides in the bottom sediments of the English Channel have been reported. Average sediment velocities ranged from a few kilometers to a few tens of kilometers per year. Sedimentary inventories of artificial radionuclides show that a significant portion of the ^{60}Co and Pu isotope input was immobilized in the Channel seafloor while most of the ^{137}Cs input was removed by water mass circulation [BOU 99].

In the Baltic Sea and Skagerak, the presence of ^{137}Cs is predominantly due to the Chernobyl accident. Sediment accumulation rates were strongly dependent on sediment type and ranged from 0.05 to 1.8 cm.yr^{-1} [PET 06]. In another study, the sediment accumulation rates showed a wide range (60–6,160 g.m^{-2}.yr^{-1}). The total amounts of ^{137}Cs activity were generally considerably higher at points in the Bothnian Sea than in the Gulf of Finland and Bay of Bothnia [MAT 06].

Colloidal fractions are of great importance for the transport of artificial radionuclides ^{106}Ru, ^{60}Co, ^{238}Pu and ^{239,240}Pu. In the dissolved phase (<450 nm), colloidal compounds explained about 40% of the transport for both Co and Ru, 60% for ^{238}Pu and ^{239,240}Pu, but have no significance on ^{137}Cs flux [EYR 04].

3.5. Conclusion

The behavior of radionuclides in the marine environment will depend on many factors. The most important factor is the nature of the radionuclide itself according to whether it will be conservative, that is, whether it will remain soluble in the water mass, or not. Then, it is the chemical speciation of the radionuclide, that is, the physico-chemical forms, which it has at the time of its introduction into the receiving environment. This factor is itself dependent on the origin of the radionuclide, natural or anthropogenic, and in the latter case, on its mode of formation such as an atomic test, a chronic discharge from a nuclear reactor, a discharge from a reprocessing plant or even following a nuclear accident. The evolution over time of its speciation will depend on the various environments it encounters, its adsorption onto sedimentary particles or its absorption by a living organism.

The dispersion of the radionuclide in the marine environment will be conditioned by its conservative or non-conservative behavior. If it is a conservative radionuclide, its mobility will be rapid and it will disperse over large distances. In the case of a non-conservative radionuclide, it will remain more localized near its point of introduction and its eventual remobilization will be subject to environmental variations such as a physical or chemical change in the environment (agitation, input from another body of water, etc.).

In the case of a nuclear accident, it is essential to assess the input of radionuclides into the marine environment, to understand their dispersion in seawater, and to estimate their transfer to biota and sediments [LAG 12].

4

The Fate of Radionuclides in the Marine Biosphere

4.1. Introduction

One of the objectives of radioecology is to estimate the radionuclide contamination of living organisms, excluding humans. To do this, one of the first concepts proposed was that of the concentration factor (CF). Subsequently, radioecologists have tried to quantify the distribution of radionuclides in organisms and their elimination. Indeed, in addition to the information that they provide on the metabolism of elements, organotropism studies have a double interest. From the point of view of human protection, it is important to determine, by experimenting on edible species, whether or not a given radioactive pollutant is concentrated in the consumable parts of the organism. Furthermore, with regard to environmental protection, the impact of a possible radioactive contamination on animal species depends on the absorbed dose of radiation. This dose can vary considerably depending on the pollutant's organotropism.

The radionuclides incorporated in living organisms will be eliminated more or less quickly and more or less completely by these organisms. Any elimination will be especially evident when individuals are in an unpolluted or less polluted environment. This elimination will result from the combination of various processes such as urinary, gill or fecal excretion, passage across membranes, elimination with the exuvia during the molt of crustaceans, etc. Elimination studies of radionuclides thus provide information on the self-purification power of polluted organisms, on the biological periods and the rate of radionuclide renewal. In addition, combined with organotrophy studies, they provide an overview of the elimination mechanisms of radionuclides and indicate the critical organs where they are stored.

Researchers quickly realized that bioaccumulation and elimination were dependent on a number of natural ecological (climatic factors, salinity) and biological (size/age/weight complex, molting cycle) factors. Finally, radioecologists have attempted to quantify the trophic transfer of radionuclides.

4.2. Penetration pathways of radionuclides in living organisms

Individual radionuclides behave like elements. Their crossing of biological barriers, external biological membranes such as the skin or the digestive tract, and internal membranes, follows various patterns according to the taxonomic groups of organisms concerned [AMI 13].

4.2.1. *Exposure pathways of flora and fauna*

In the marine environment, for aquatic flora, the only exposure pathway is almost always the water column because they generally do not have a sea grass system. For marine animals, the vectors are water, food and sediment. Only the first two concern pelagic species. For benthic species, i.e. those living in contact with sediment, the latter plays the dual role of substratum and, through its organic fraction, possibly of food. Moreover, this vector is not negligible since the sediment is the major reservoir of most radionuclides.

4.2.2. *Transmembrane passages*

Penetration occurs through various processes [AMI 17a]. When a pollutant enters an organism, whatever its route of entry, oral, gill or skin, it first reaches the general circulation (blood or its equivalent). Then, from this internal liquid, it is distributed to the different tissues and organs before eventually being excreted.

Movement through the body involves the passage across multiple membrane barriers. Transport across membranes is varied. Membranes are generally assemblies of phospholipids and proteins in molecular bilayers whose protein hydrophilic poles are oriented towards the outside and hydrophobic poles inside the membrane. Only lipophilic molecules can therefore cross the membrane by simple diffusion, according to the concentration gradient, from the most concentrated medium to the most dilute medium. Hydrophilic molecules will have to pass through pores or channels or with the help of various types of transporters.

Transmembrane passage leads to neutral water-soluble molecules, ionized or not, and to a lesser degree liposoluble and insoluble molecules, depending on their size. Transport can be passive, and therefore without energy expenditure, but with a facilitated diffusion. The facilitation is due to two types of transport proteins, protein channels which form pores and transport proteins (transporters).

Radionuclides being elements will be concerned mainly by facilitated diffusion and by transporters. The problem may be different when the radionuclide is incorporated in an organic molecule, in which case passive diffusion is possible.

Radionuclides of metals, like their stable counterparts, can use membrane transporters to cross the lipid barrier. These transporters are dedicated proteins embedded in the membrane. There are a large number of transporters such as ZIP (Zinc-Iron Permease family) transporters, OPT (OligoPeptide Transporters), magnesium transporters, Nramp (Natural resistance-associated macrophage proteins), CDF (Cation Diffusion Facilitator), co-transporters, ion channels, P_{1B}-ATPase and ATP-binding cassette transporters (Catty et al. pp. 65–82, *in*: [MEN 09]).

For radionuclides, some of the transmembrane passages occur via facilitated transport through ion channels (*in:* [BLU 01]). Ionic competition between isotopes of the same element or between isotopes of chemically close elements (such as cesium and potassium, strontium and calcium, etc.) explains a good part of the bioaccumulation of radionuclides (*in:* [BLU 01]).

In marine mammals, transfers through the epithelial (skin and digestive tract) and cerebral (hemato-encephalic and hemato-liquid) barriers will vary greatly depending on the radionuclides. Percutaneous passage may be greatly facilitated if the skin is damaged. Skin passage is relatively low for uranium (0.4%), cesium (1.3%) and cadmium (0.6%). The intestinal passage of uranium in mammals is 0.06–2.8% and in humans 1.4–5%. The intestinal passage of cadmium is similar (5%), but that of plutonium is one hundred times less than that of uranium. On the other hand, that of cesium is 80% and that of tritium in the form of tritiated water is rapid and total (Dublineau et al. pp.133–151, *in*: [MEN 09]).

The blood–brain barrier (BBB) is not absolutely impermeable to the passage of stable metals and their radionuclides. Thus, this passage has been noted for uranium [LEM 03]. The hemato-liquid barrier constituted by the choroid plexuses also allows the passage of various metals and their radionuclides. This is particularly true when these are in organic forms, such as methylmercury (Dublineau et al. pp.133–151, *in:* [MEN 09]).

4.2.3. *Body distribution*

After passing through the membrane barriers, the pollutant is distributed throughout the body and reaches the sites of its biological or toxic action. This cellular and tissue distribution is mainly carried out from the blood in vertebrates or its equivalent in invertebrates; this liquid thus carries the pollutant to all the organs.

4.2.3.1. *Blood transport and its equivalent*

Blood transport of radionuclides in vertebrates (or its equivalent in invertebrates) takes place in two forms, a free form, which corresponds to the fraction of the water-soluble pollutant dissolved in the aqueous phase of the blood, and a bound form, which results from the binding of the non-water-soluble fraction to certain plasma constituents. The free form mainly concerns water-soluble chemical pollutants, i.e. neutral or ionizable substances (ionized or not), i.e. the ionized forms of fat-soluble acid or basic molecules. The bound form concerns especially the liposoluble or insoluble fractions in the blood which can only be transported fixed on certain blood constituents, proteins and/or blood cells. In vertebrates, the proteins involved in this binding are mainly albumin and secondarily globulins and fibrinogen. For radionuclides such as cesium, it is the elements of the blood (red blood cells, white blood cells and blood platelets) that ensure the transport. The transporter-pollutant bonds are usually of low energy (hydrogen bonds, hydrophobic, electrostatic). They are therefore immediately reversible and non-specific since several pollutants are transported by the same type of carrier.

4.2.3.2. *Cellular and tissue distribution*

From the blood and in their free form, the pollutants are then distributed to the various organs. This distribution, or organotropism (see section 4.4), obeys the laws that govern membrane passage. It occurs mainly by passive diffusion and by facilitated diffusion at the level of the basal laminae.

4.3. Radionuclide bioaccumulation mechanisms

Two modes of contamination are involved in the accumulation of radioactive isotopes by living organisms, adsorption and absorption. The available surface area determines the importance of the role played by the adsorption phenomenon. The distribution of radionuclides in organisms and the influence of the surface area/volume ratio on accumulation will therefore be relevant elements for evaluating the part played by adsorption in an observed contamination. Adsorption is a passive phenomenon. Furthermore, when ecological factors, such as the absence or presence of light or the presence of inhibitors of cellular respiration, influence the radioactive contamination of an organism, it is possible to conclude that the accumulation of the radionuclide considered is, at least in part, an active phenomenon, requiring the consumption of energy by the cell.

4.3.1. *Accumulation mechanisms in plants*

With regard to contamination mechanisms, it is necessary to distinguish, first of all, for aquatic plants, between algae, which can only take up radionuclides in open water, and aquatic phanerogams, equipped with roots or rhizomes, which can also be contaminated from the sediment and interstitial water.

While these mechanisms have rarely been considered in a systematic way, a certain number of elements of interpretation are provided for various radionuclides by classical radioecology experiments.

4.3.1.1. *Contamination by adsorption*

The authors generally admit that a phenomenon of fast contamination indicates a preponderant role of adsorption. It seems that this is generally the case for the elements of group I which are easily hydrolyzable. Polikarpov (1960, 1961 *in* [POL 66]) showed that in green, brown and red algae and in phanerogams, the state of equilibrium between the organisms and their environment was reached in a few hours to a few days in the case of contamination by ^{91}Y, ^{143}Pr, ^{144}Ce–^{144}Pr. This result is confirmed by Rice [RIC 56] for yttrium-90 in the unicellular alga *Carteria* sp.

For plutonium, Folsom et al. [FOL 75] report that the phenomenon of fixation is very rapid since in certain coastal zones where the algal populations are very dense (30 cm^2 of algal surface corresponding to 1 L of ambient seawater), the algae can retain up to 10% per day of the Pu present in the water that bathes them.

For ruthenium-106, Jones (1960, *in* [POL 66]) showed that unicellular algae, the diatom *Phaedactylum tricornutum,* were contaminated in a very comparable way to fine sedimentary particles, the amount of ^{106}Ru extracted from seawater being proportional to the concentration of the algal cells.

The contamination dynamics of a sediment and the macroalga *Ulva pertusa* are variable according to the radionuclides. Thus, the equilibrium state is quickly reached by the alga for ^{95}Zr–Nb and ^{106}Ru–Rh as well as for a radionuclide of group III, cobalt-60, whose stable isotope, a constituent of vitamin B_{12}, is essential to life. On the other hand, for ^{137}Cs, the equilibrium state is not reached in 14 days [NAK 77].

The surface area/volume ratio conditions the quantitative variations of adsorption in different species. The surface available for adsorption is constituted by the external surface of aquatic plants as well as by the internal surface corresponding to the extracellular spaces. There are few such precise data, but the morphology of algae already provides information on the importance of the available surfaces in relation to the volume. Amiard-Triquet and Amiard [AMI 80], in order to compare the contamination of algae according to their morphology, collected a few examples selected so that the capacity of accumulation of the different species with respect to each radionuclide was successfully determined under homogeneous conditions. Generally speaking, the species which, with equal weight, present the highest surface areas are also the most contaminated. This is the case for phytoplankton and enteromorph macroalgae for ^{54}Mn, *Sargassum* for ^{60}Co, *Polisiphonia* and coralline algae for ^{144}Ce, coralline algae for ^{95}Zr, ^{106}Ru and plutonium. For 110m silver, the larger of the two diatom species is the less contaminated. On the other hand, another explanation must probably be sought to account for the high contamination of *Fucus vesiculosus* by ^{65}Zn. For plutonium, we generally observe a higher contamination of unicellular algae (CF = $7.1.10^3$–5.10^5 depending on the species and the determination conditions) than of multicellular algae (CF = 85–4.850). Two exceptions can be noted for plutonium, a CF of 660 for dinoflagellates and an average CF of 20,000 (with a maximum of 100,000) for *Sargassum* (*in* [AMI 79c]).

Using autoradiography on kelp (*Laminaria japonica*), Nakahara et al. (1975) conclude that cobalt-60 is five times more concentrated in the cortical zone than in the internal tissues. However, for this species, it is the growth zones that have the highest concentrations of ^{60}Co. In contrast, in *Eisenia bicyclis*, *Sargassum thungergii* and *Chondrus ocellatus*, ^{60}Co accumulates mainly in the older parts of the alga. This distribution is interpreted by the authors as a response to the structural conditions of the algal surface as well as to the additional effects of adhesion of microorganisms and organic or inorganic particles promoting metal binding.

Organisms buried in the sediment are sometimes more contaminated and sometimes less contaminated than those immersed in open water.

4.3.1.2. *Contamination by absorption*

Numerous experiments have shown the influence of metabolic factors on the accumulation of various radionuclides by aquatic plants, supporting the hypothesis of active transport. Thus, various works have shown a depressive effect of inhibitors of cellular respiration on the contamination of *Ascophyllum nodosum* and marine phytoplankton by ^{239}Pu in colloidal form [ZLO 70, ZLO 71a, ZLO 71b]. Polikarpov (1963, *in* [POL 66]) studied the action of light on the radioactive contamination of *Ulva* sp. and established CFs that indicate active transport for cobalt-60 and cesium-137.

The unicellular alga *Nitzschia closterium* accumulates 80% of the ^{65}Zn available in the experimental medium (27 μCi/l) in 1 h and almost all of it in 24 h. In the dark, non-dividing cells accumulate only 87% in 24 h (Chipman et al. 1958, *in* [POL 66]). High levels of phosphates or phosphates and nitrates promote cesium accumulation by the unicellular alga *Chlamydomonas reinhardii* and the multicellular alga *Fucus vesiculosus* (Gertz, 1973 *in* [VAN 75b]). The light energy and the available nutrients that condition the rate of photosynthesis thus influence the radioactive contamination of aquatic plants. While in macrophytes, it can be assumed that metabolic processes are involved in this phenomenon, for unicellular algae, it is not possible to decide whether the important factor is the formation of new protoplasm or the increase in surface area available for adsorption, related to cell division. Williams (1960, 1970, *in* [VAN 75b]), however, indicates that living cells accumulate more cesium than dead cells. It is interesting to note that in *Carteria* sp. the accumulation of strontium-90 depends directly on the mitotic index.

4.3.2. *Accumulation mechanisms in animals*

The accumulation of radionuclides in animals occurs by two main entry pathways which are the direct entry at the level of external integuments and osmoregulatory organs (gills of crustaceans and teleost fish, skin of amphibians, rectal glands of selachian fish or nasal or lingual glands of birds) and by entry in the alimentary tract. These entry routes function with varied kinetics and can have a regulation mechanism. In addition, organisms can acquire a certain tolerance to radionuclides present in their environment (see section 4.2.2.4).

4.3.2.1. *Accumulation kinetics*

Accumulation kinetics depend on the ability of radionuclides to cross biological membranes. They will also vary greatly depending on the route of entry.

In the case of contamination by ingestion, one of the most important parameters is the capacity of the radionuclide to cross the barrier of the digestive tract. In the case of mammals, this is expressed by the gastrointestinal absorption rate f_1 or f_A [ICR 07]. This will be discussed in section 4.7.4.

4.3.2.2. *Some examples of bioaccumulation*

Carvalho [CAR 18] made a brief annotated review of radionuclide concentration processes and CFs in marine organisms for key groups of radionuclides such as fission products, activation products, transuranic elements and naturally occurring radionuclides. For their part, Olsen and Vives i Batlle [OLS 03] propose a model of technetium-99 biokinetics in lobsters (*Homarus gammarus*).

4.3.2.3. *Regulation phenomena*

When variations in the chemical composition of the environment occur, many animals are able to maintain their internal chemical composition at a level compatible with the normal course of vital functions. This capacity varies from one species to another, and such regulation becomes increasingly common and more precise towards the top of the zoological scale. However, the regulatory capacity of organisms is limited and is only effective for moderate variations of concentrations in the environment. Furthermore, it appears that this capacity is manifested to varying degrees depending on the physiological function of the elements (Figure 4.1).

4.3.2.3.1. Unregulated elements

Elements that do not normally exist in the natural environment or have physiological functions are not likely to be regulated. In this case (Figure 4.1(A)), when the concentration of their radioactive isotope increases in the environment, it increases proportionally in the organisms and, as a result, the CF remains constant.

The literature provides a few data on the metabolism of group I elements that have no specific role in the organism. The work on the contamination of some marine invertebrates by ^{106}Ru and ^{51}Cr introduced into the water analyzed by Nassogne [NAS 70] seems to indicate that the accumulation of these radionuclides is not regulated. This is probably what happens for the 110m silver present in the food for a crab *Carcinus maenas* where the contamination of this crab follows that of the food that is provided experimentally [AMI 78a].

4.3.2.3.2. Co-regulated elements

The elements of group II (Cs, Sr) have no defined physiological functions but are close in their chemical characteristics to major ions (K, Ca), whose biological accumulation is regulated. Always present in the medium at negligible concentrations compared to potassium and calcium, cesium and strontium are not regulated independently but co-regulated with the major ions [BER 68]. Various works by Bryan (*in*: [NAS 70]) have shown that variations in the concentration of cesium (stable or radioactive) in the environment do not lead to changes in the CF of this element in many marine invertebrates, indicating that there is no regulation in this case. On the other hand, if the concentration of mineral salts in the water decreases or if the concentration of K alone decreases, an increase in the CFs of ^{137}Cs and K occurs (Bryan, 1964, 1965, *in:* [NAS 70, AMI 75c]). Analyzing numerous data, Bernhard [BER 68] reports that a similar situation exists for Sr and Ca.

This means that the concentration in the organism of a co-regulated element is proportional to its concentration in the environment as long as the latter concentration remains negligible compared to the concentrations for which the corresponding major ion is regulated.

4.3.2.3.3. Regulated elements

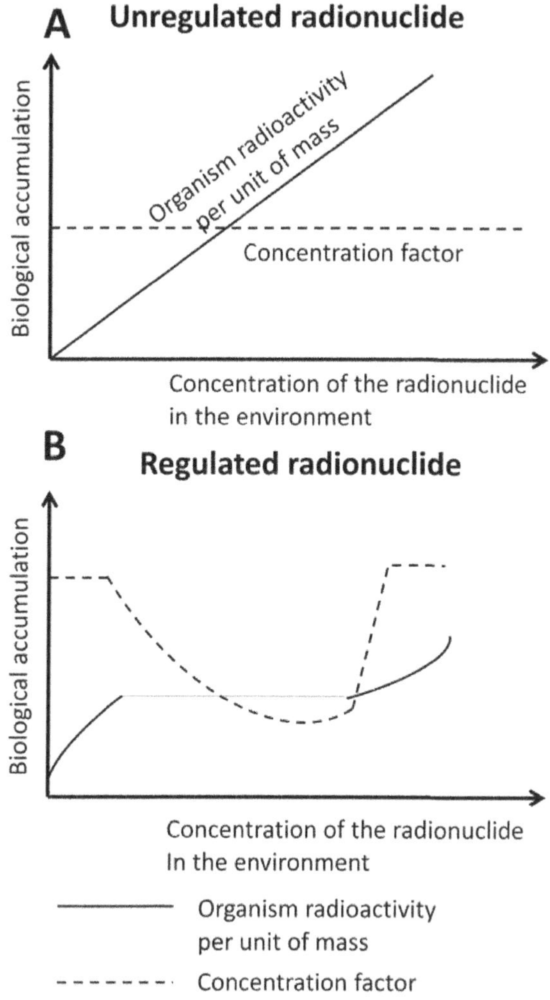

Figure 4.1. *Schematic representation of the respective variations of the concentration factor and the concentration of radionuclides in the body as a function of the concentration in the environment (modified from [BER 70, AMI 80])*

For regulated elements, Bernhard [BER 70] proposes the theoretical pattern that we have included in Figure 4.1(B). When the concentration in the medium of an element indispensable to life is lower than the optimal concentrations for the operation of vital functions, the organisms accumulate all the quantity available and there is no regulation. At average concentrations in the environment, the regulation phenomenon intervenes to maintain a constant concentration in the organism while the CF decreases. When environmental overloads increase further, a toxic effect may occur, resulting in, among other things, a disruption of regulatory mechanisms.

The existence of this mechanism had already been suggested in some marine animals (the bivalve *Scrobicularia*, crabs, plaice) [AMI 80]. A similar regulation probably also exists in the case of manganese.

4.3.2.4. *Phenomenon of tolerance and resistance*

The terms resistance and tolerance appear frequently in the literature, but despite an attempt to clarify the terminology by Lotts and Stewart [LOT 95], there is no consensus on the definitions. According to these authors, acclimation is defined as a physiological or biochemical adjustment by an organism to an alteration in environmental conditions that increases its chances of survival; resistance refers to an organism capable of surviving for a limited period of time in an environment that could potentially exert a lethal effect; and tolerance refers to an organism capable of surviving indefinitely under particular environmental conditions as a result of genetic adaptation [AMI 11].

The acquisition of tolerance to pollutants, especially metallic ones, is a general phenomenon. It has been recognized in particular for phytoplanktonic organisms, macrophytic algae, polychaetes, bivalves and crustaceans [AMI 08a, AMI 09, AMI 11].

This tolerance can probably be explained by an amplification of chemical detoxification mechanisms. This may be related to an increase in the synthesis of ligands that specifically bind the pollutant. Some species have thus duplicated the gene controlling the synthesis of metallothionein. However, this chemical detoxification does not protect the individual from the effects of ionizing radiation. Tolerance can also be the result of the genetic selection of certain individuals in a stressed population. Some works, such as those of Theodorakis (see Chapter 7), show genetic adaptation in very radioactive environments.

The fact that species or populations are tolerant or acquire tolerance to environmental pollutants will have a beneficial effect on the conservation of biodiversity. However, the picture is not entirely idyllic, as tolerant organisms, depending on the detoxification strategies adopted, may become highly contaminated links in food chains, favoring the prey–predator transfer of pollutants and possibly their toxicity. Moreover, an individual's energetic allocation is primarily devoted to ensuring the basic metabolism (its survival), then enabling its growth and allowing its reproduction. Any energy that is allocated to ensure the defense of the individual against a chemical stress will therefore necessarily be used to the detriment of its growth and reproduction.

Tolerance to a chemical stress quite frequently leads to a high concentration in the tolerant individual. This is particularly true for radionuclides since the two main detoxification mechanisms (see section 4.6) result in the storage of these elements (biomineralization or binding to metallothioneins).

4.4. Influence of ecological factors on radioactive contamination

A number of biotic or abiotic factors have a significant impact on the biological accumulation of radionuclides. The complete list of these variables is long, such as temperature, light, salinity, physico-chemical form and concentration of stable and radioactive isotopes, amount of food available, age, size, sex of organisms, state of sexual maturity, growth rate, molt and inter-molt cycle in crustaceans. The mode of action of these factors can be very complex. Thus, light will have a direct effect on the energy available to primary producers and therefore on their metabolism. Light will also have several indirect ecological effects through the development of aquatic plants, by increasing the amount of food in the system and the surface area available for adsorption, by putting in circulation in the waste products of metabolism. In addition, in the natural environment, certain factors exert their influence in a concomitant way. This phenomenon is particularly evident in the seasonal variations of the contamination of organisms.

Some unknown or uncontrolled parameters may play a significant role in the radioactive contamination of aquatic organisms. For example, Kane et al. (1969, *in:* [PAT 75]) observed no significant difference in the elimination of zinc-65 by the crab *Pachygrapsus marmoratus* maintained in seawater taken directly in situ and overloaded or not with stable zinc (+ 0.1 or + 1.0 mg $Zn.L^{-1}$). On the other hand, elimination is considerably slowed when crabs are placed in decontamination in seawater from the laboratory tank, without the authors being able to identify the precise cause of this phenomenon.

Conversely, flora and fauna participate in the radionuclide cycle in aquatic environments. Indeed, the biochemical composition of the environment can be modified by the introduction of excretion products or organic detritus (exuvia, corpses, etc.). Animals also play a role in the distribution of radionuclides through their behavior: vertical or horizontal migrations, filtration of suspended particles, foraging, etc.

4.4.1. *Influence of abiotic factors on radioactive contamination of aquatic organisms*

The main abiotic factors influencing the radioactive contamination of aquatic organisms are physico-chemical forms of radionuclides, isotopic discrimination, isotopic and chemical competition, salinity, temperature and vectors of contamination.

4.4.1.1. *Physico-chemical forms of radionuclides and their stable isotopes*

In order to compare the biological accumulation of different physico-chemical forms of radionuclides, Fraizier [FRA 74] defined the relative contaminating power (Cp). This contaminating power is equal to the ratio of the CF due to contamination by physico-chemical form A to the CF due to contamination by physico-chemical form B. For ruthenium-106, it shows that the contaminating power of polymerized forms is much more significant than that of isolated soluble forms towards various plant and animal species. A more detailed analysis on plaice indicates that this is the case only for the external organs, the internal organs accumulating only small amounts of soluble ruthenium. In *Mytilus galloprovincialis*, for the same radionuclide, chloride complexes (essentially neutral with small anionic and cationic fractions) are 10 times more concentrated than nitrate complexes (essentially in anionic form, with smaller cationic and neutral forms). An identical but less pronounced phenomenon was observed for another bivalve *Tapes japonica* (*in*: [ROB 71]).

The experimental study of the accumulation of ^{237}Pu by a freshwater diatom (*Asterionella formosa*) showed a variation in the CF from 1,700 to 100,000 depending on the physico-chemical form of the radionuclide. In contrast, in the polychaete *Nereis diversicolor*, the CFs are identical (CF ~ 200) after immersion in water contaminated by $^{237}Pu^{4+}$ or $^{237}Pu^{6+}$ (*in:* [AMI 79c]).

For iron-55, the specific activity found in salmon is significantly higher than in seawater, corresponding to a CF of 10^3–10^4. The migrations carried out by this species in variously contaminated areas are not sufficient to explain this difference, which is probably due to the fact that the iron-55 recently introduced with fallout is probably in a more assimilable or adsorbable physico-chemical form than the stable iron in equilibrium in the environment. This hypothesis is supported by the fact that diatoms (*Phaeodactylum tricornutum*) accumulate much more of newly formed ferric hydroxide than aged ferric hydroxide in seawater (*in:* [ROB 71]). In mussels and blennies, contamination is clearly differentiated according to the physico-chemical form of iron. The relative contaminating power (Cp) of the insoluble forms is particularly high in the external organs and the digestive tract, suggesting the intervention of a passive accumulation phenomenon. In contrast, for most internal organs of fish, soluble forms are the most assimilable (*in:* [AMI 80]).

In many organisms, cobalt bound as cobalamin is more available than ionic cobalt (*in:* [AMI 80]). Laboratory experiments and in situ observations reported by Robertson [ROB 71] indicate that the biological accumulation of zinc-65 differs from that of stable zinc, probably, in many cases, because of different physico-chemical forms.

Contamination of the environment (natural or experimental) by tritiated water leads to a concentration of tritium in the free water of the tissues close to that of ambient water. On the other hand, the accumulation of tritium from tritiated organic molecules leads to much higher incorporation rates (Kirchmann, 1975, *in:* [AMI 10]). However, the ratio of specific activity in organisms to the specific activity of the environmental medium can become much higher than 1 when the source of tritium is food or complex tritiated compounds [AMI 10].

Metal ions can be grouped into three classes based on their ability to form ionic forms or covalent complexes with various ligands. Class A has low affinity with most ligands (e.g. Cs). Class B has strong affinities with most ligands (e.g. Ag). The transition metals such as Co form the C class.

4.4.1.2. *Isotopic discrimination*

For the lightest elements, the relative variation in atomic mass between stable and radioactive isotopes of the same element may be sufficiently large that isotopic discrimination becomes likely. Thus, the atomic masses of deuterium and tritium are, respectively, double and triple those of hydrogen.

Isotopic discrimination by organisms of various groups (plants, mammals, insects) has been abundantly demonstrated for the isotopes of the lightest elements (H, C, O, N). It seems that it can be generalized to aquatic organisms. Steemann-Nielsen (*in:* [SOU 71]) estimates that marine phytoplankton discriminates between the isotopes ^{12}C and ^{14}C to the disadvantage of the latter and that, as a result, the measurement of ocean primary productivity by the ^{14}C method must be increased by 5 to 6%. In general, in aquatic environments, this phenomenon does not appear to be very significant for tritium in the form of tritiated water and when there is discrimination, it is against tritium [BRU 73, VAN 75b].

Isotopic discrimination will depend on the position occupied in the biological molecules by the radioactive isotope (*in:* [AMI 80]). It can be practically null or on the contrary very strong if, for example, the radioisotope occupies a specific active enzymatic site, if the intramolecular bond is more or less easily broken, if the chemical reaction is more or less facilitated or if the isotopic exchange rates are different.

Isotopic discrimination for heavy elements has not been demonstrated and if it exists, its importance to the radioecologist is likely to be negligible compared to other factors.

4.4.1.3. *Isotopic and chemical competition, salinity*

When isotopic discrimination becomes negligible, the biological accumulation of a radionuclide is conditioned by the abundance of other stable or radioactive isotopes of the same element present in the natural environment in the same physico-chemical form. Under these conditions, an increase in the concentration of the stable isotope in the environment will correspond to a lower contamination of living organisms for the same concentration of the radioactive pollutant; this is the isotopic competition phenomenon. However, this is not always a linear relationship because the organism will sometimes respond to the overall increase in the element (stable and radioactive isotopes) with regulation.

For cesium and strontium, the concentrations in natural waters are very low compared to those of the chemically related elements, K and Ca. For radiocesium and radiostrontium, chemical competition will therefore be added to isotopic competition both in seawater and in environments of variable salinity (*in:* [AMI 80]). It is necessary to recall on this subject the theory of Bernhard [BER 68] according to which Sr and Cs are co-regulated with the corresponding major elements.

An increase of salinity corresponds to an increased competition between the increasingly concentrated mineral salts and the radionuclides. This explains why the CFs are generally higher for freshwater organisms than for marine organisms (*in:* [AMI 80]). It should also be noted that mineral metabolism varies with salinity. Thus, in marine environments, alkaline earths enter fish through the digestive tract or the skin (at the level of the fins), whereas in brackish environments, the gills would be the main site of absorption (*in:* [AMI 78a]).

4.4.1.4. *Temperature*

The effect of temperature on the metabolic rate is related to the thermal dependence of the enzymatic reactions that catalyze vital processes (Cossins and Bowler, 1987, *in:* [BLU 01]).

An increase in temperature very generally corresponds to an increase in reaction rates and exchange rates in poikilothermic organisms. A literature review shows that this phenomenon is real in the case of the biological accumulation of radionuclides and their excretion. It is possible to quantify the increase in radio-contamination or decontamination kinetics using the thermal coefficient Q_{10}, corresponding to a temperature rise of 10°C [AMI 81]. For 95% of the experimental data, the thermal coefficient Q_{10} is less than 5 (in $\log_{10}$, 0.7). In most cases where it is greater than 5, it corresponds to a lethal or sublethal temperature and therefore to intense physiological disorganization of the cells or organisms.

The accumulation of various radionuclides by the brown macroalga *Fucus vesiculosus* (L.) was compared at 2°C and 12°C. CFs in the thallus were temperature-dependent and decreased significantly at 2°C for ^{110m}Ag, ^{109}Cd, ^{60}Co, ^{54}Mn and ^{65}Zn (CFs of 439, 51, 101, 92 and 162, respectively) but not for ^{241}Am, ^{133}Ba, ^{134}Cs, ^{152}Eu, and organic ^{57}Co associated with cobalamin (CFs of 329, 210, 3.3, 470 and 214, respectively) [BOI 97].

Seasonal variations of radio-contamination of aquatic organisms confirm the general tendency for contamination to increase with rising temperature (except in fresh water) and for decontamination to accelerate. However, we must not forget that the seasonal change in temperature is accompanied by variations in the physiological characteristics of the organisms (fasting, molting rhythm in crustaceans, sexual maturity), in the available food, in the light. The parameters that vary are more numerous in rivers than in the marine environment.

However, our statement must be qualified because great differences exist between radionuclides and species, and even between different populations of the same species. Thus, using a biokinetic and bioenergetic approach, Baines and Fisher [BAI 08] modeled the bioaccumulation of various metals (Ag, Am, Cd, Co, Se and Zn) by the dietary route or directly from water in blue mussels, *Mytilus edulis*, acclimated at 2 and 12°C. Metal accumulation from the aqueous phase was not affected by temperature. However, for Ag, Am and Zn from food, bioaccumulation was higher by a factor of 5.7, 5.3 and 2, respectively, at 2°C than at 12°C, largely because these three metals were assimilated into food more efficiently at the lower temperature. In contrast, the bioaccumulation of Cd, Co and Se via the food pathway was not significantly affected by temperature although the removal constant for these metals was up to 4-fold lower at 2°C than at 12°C. Total bioaccumulation of Ag, Am and Zn was up to 1.7- to 3.6-fold higher at 2°C than at 12 C°. It is therefore necessary to consider the variability regarding the influence of temperature on metal bioaccumulation by taking into account the adaptation of management strategies developed by individuals living in temperate seas or polar zones.

Mussels (*Mytilus edulis*) continuously exposed to effluents from different polonium-210-rich phosphorus plants have a bioaccumulation of this radionuclide that is doubled in winter compared to mussels caught in summer [PEN 97].

In tropical species, retention capacities appear to be greater. This is well illustrated by the bioaccumulation of ^{134}Cs in five tropical marine species, three bivalves (the oysters *Isognomon isognomum* and *Malleus regula* and the clam *Gafrarium pectinatum*), one decapod crustacean (the prawn *Penaeus stylirostris*) and one alga (*Lobophora variegata*) [MET 16].

4.4.1.5. *Contamination vectors*

Radionuclides introduced into the aquatic environment are mainly transferred to biota by two contamination vectors: water and food. Various studies have shown that the mode of contamination has a very great influence on the distribution of radionuclides in the organism as well as on their subsequent elimination. Thus, in the planktonic crustacean *Meganyctiphanes norvegica,* particulate radionuclides with no known physiological functions (^{141}Ce, ^{238}Pu and ^{106}Ru) accumulate mainly in the exuvia after contamination by water, whereas they are rapidly eliminated with the feces after contamination by food. For ^{110m}Ag, contamination by water affects mainly the external organs, the shell of the lamellibranch *Scrobicularia plana*, exoskeleton and gills of the crab *Carcinus maenas*. In these two species, after contamination by food, ^{110m}Ag is mainly accumulated by the digestive gland or hepatopancreas, respectively. Subsequently, the external organs for which contamination by adsorption predominates will eliminate silver-110m more rapidly than the digestive gland or hepatopancreas which plays the role of a storage organ.

For cobalt-60, the water vector is mainly responsible for external contamination in marine organisms. These include the shell of the mollusk *Scrobicularia plana*, and the exoskeleton of the crustaceans *Crangon crangon* and *Carcinus maenas*. Ingestion of contaminated food is responsible for a preferential accumulation of cobalt-60 in the kidneys and liver of vertebrates and in similar organs of invertebrates (digestive gland, hepatopancreas). This influence of the contamination vector on the organotropism of cobalt-60 has the further consequence of determining the relative importance of the different compartments during elimination [AMI 78a, VAN 75a].

In addition, it has been frequently noted that the slower the elimination of various radionuclides, the longer the contamination phase and the longer time until the steady state was reached.

The route and duration of contamination will have a great deal to do with the total radiation dose to contaminated organisms (see Chapter 6).

The nature of the food available affects the retention of associated radionuclides by predators. In plaice, the retention of zinc-65 is greater after the ingestion of labeled starch or gelatin capsules than after the consumption of contaminated shrimp, mussels or annelids (Kevern 1966; Pentreath 1973, *in:* [AMI 76a]). More generally, the availability of radionuclides to the predator will depend on their location in the prey and the degree of breakdown of the food during digestive transit.

4.4.2. *Biological factors in the contamination of organisms*

Each species, depending on its biological and physiological characteristics, has a particular behavior towards radionuclides. Among the morphological characteristics, one of the most important is the surface area/volume ratio which determines the adsorption capacity of the organism or organ. Moreover, at the physiological level, regulation mechanisms are increasingly common and more precise for the species located towards the top of the systematic scale. Finally, within the same species, radio-contamination can vary considerably during the life cycle.

4.4.2.1. *Age and size of animals*

The smallest and youngest individuals usually have the highest levels of radioactive contamination. This is a common phenomenon for many species and many radionuclides. Two hypotheses have often been put forward to explain this age effect. Frequently, the surface-area-to-volume ratio is greater in younger individuals, promoting a high surface contamination for particulate or colloidal radionuclides. Metabolism is more intense in young individuals than in adults, and this may play a role in the accumulation of radionuclides whose stable isotopes have physiological functions or are close to such elements (groups II and III) (*in:* [AMI 75a, AMI 76a, AMI 78a]).

As organisms grow, the concentration of the radionuclide in the body per unit weight decreases.

A broad synthesis of allometric relationships of polonium-210 concentrations in marine organisms was performed with nearly 400 data sets covering nine taxa of marine organisms. Data for each taxon were fitted to the allometric equation $\log Q = \log a + b \log M$, where Q is the concentration of ^{210}Po (mBq.g^{-1} dry mass) and M is the dry mass per individual (g). For ^{210}Po, eight of the nine taxa, the slope was negative and for six, the correlation was statistically significant. For ^{210}Pb, seven of these taxa, b was again negative, and for four, the correlation was statistically significant. The weighted average of the nine slopes of ^{210}Po was -0.24 ± 0.05 and the eight slopes of ^{210}Pb was -0.22 ± 0.05 [CHE 91].

An allometric scaling approach was used to calculate biological elimination half-lives and CFs for radionuclides in marine biota [VIV 09].

4.4.2.2. *Sex and sexual maturity*

These two factors, sex and sexual maturity, have been little studied but some examples indicate that they are not negligible. In the annelid *Arenicola marina*, radiocesium contamination of adults results in greater contamination of genital products for spermatozoa than for ova [AMI 75c]. An in situ study of the distribution of plutonium in the organs of the crab *Cancer pagurus* showed that the contamination of certain internal organs (hepatopancreas) was more intense in females than in males [GUA 76]. In *Cancer irroratus* and *Carcinus maenas*, an increase in the concentration of trace elements (Fe, Cu, Zn, Mg, Mn and Co) in the ovary occurs during ovogenesis. The concentrations measured in the testes of *Carcinus maenas* are always lower than those determined for the ovary [MAR 75].

The reproductive cells are therefore likely to accumulate various radionuclides and consequently to undergo irradiation whose effects on the reproduction of the species are currently poorly evaluated.

4.4.2.3. *The molt and intermolt cycle*

Many radionuclides adsorb in significant quantities on the external organs of crustaceans. Under these conditions, the molt and the intermolt cycle will have a definite effect on the accumulation and biological elimination of radionuclides. A review of this subject has been carried out [AMI 79a].

Since accumulation depends on the contact time between the organism and the pollutant, a high frequency of molts, such as that observed in small planktonic species, may limit external contamination, and the state of equilibrium between the organism and the environment may, in some cases, never be reached. While Fontaine [FON 60] reports that after molting, the soft tegument is more permeable to radioactive pollutants than the highly calcified carapace of the earlier stages, several authors have shown that in various species of crayfish, the exoskeleton accumulated much more cobalt-60 before molting than during the early stages of post-molt. The intermolt cycle includes some stages (A_1 -A_2, B_1 -B_2 post-molt stages, D_3 -D_4 pre-molt stages) during which the animal fasts, which may limit the concentration of radionuclides that are accumulated mainly by the dietary route.

More generally, it can be said that the absorption and accumulation of trace elements and their radioactive isotopes is influenced by their metabolic role, their physico-chemical form, the function of the different organs and the structure of the external surfaces.

In crustaceans, decontamination is largely determined by exuviation, which, by eliminating the radionuclides fixed on or in the exoskeleton, constitutes a true detoxification process (*in:* [AMI 80]). Thus, for the fiddler crab, *Uca pugnax*, the average percentage of the total amount of metal eliminated during molt with the exuvia for a highly contaminated site varies according to the metal from 12% for Cu, 76% for Pb and 22% for Zn. The average percentage of the total amount removed during the molting process for a relatively clean site also varied for each metal and was significantly lower for this site, respectively, 3%, 56% and 8% for Cu, Pb and Zn [BER 07]. This is also the case for radionuclides such as technetium-99 [SMI 98; OLS 03].

Molting plays an important role in the decontamination of crabs after immersion in silver-110m contaminated seawater due to the high contamination of exuviae [AMI 78c].

The absorption and retention of technetium from seawater by juvenile lobsters, and the effects of molting on these processes, were studied using ^{95m}Tc, body CFs of over 2,000 being measured in some lobsters. Absorption was highest immediately before a molt; after molting, there was a decrease in accumulation. The biological half-life was calculated to be 53 days. The molted exoskeletons of these lobsters contained about 0.6% of the total day zero body activity [SWI 85].

Cresswell et al. [CRE 21] point out that ecdysis (molting) has significant effects on the absorption and loss kinetics of radionuclides. When molting occurred during exposure, it caused significant increases in ^{85}Sr bioaccumulation and ^{134}Cs removal under constant aqueous exposure.

4.4.3. *Influence of living organisms on the cycle of radionuclides in the aquatic environment*

In the photosynthetic zone of aquatic environments, trace elements and their radioactive isotopes can be incorporated into phytoplankton biomass and transferred into food webs. The catabolism, excretion, death, decomposition of organisms and regeneration of trace elements, on the other hand, involve the entire depth of the water layers. In most aquatic ecosystems, phytoplankton and zooplankton represent the highest biomass, the largest surface area available for adsorption and the highest CFs of radionuclides. In areas of intense biological activity where the utilization rate of trace elements and their radioactive isotopes is very high, a significant fraction of the radionuclides introduced into the water may be bound to the biological material. This raises the question of the importance of the biological cycle of radionuclides in relation to the geochemical cycle.

This phenomenon has been well studied in the case of plutonium. Accumulation by shrimp was relatively slow, and the degree of absorption was strongly influenced by molting. The sinking of molts at depth contained large fractions of the plutonium content of shrimp. In mussels, CFs were higher than in shrimp, with most of the accumulated isotope (>80%) in the shell. In worms, which readily accumulated plutonium-237 in the +4 or +6 state, reaching CFs of about 200, retention studies indicated a relatively slow loss of plutonium-237 from all animals studied. Plutonium half-lives were on the order of two years in mussels and 1.5 months in shrimp [FOW 76].

4.4.3.1. *Influence of organisms on the distribution of radionuclides within sediments*

Burrowing species are well represented among polychaete annelids, mollusks (scaphopods, lamellibranchs, some gastropods), as well as include many holothurians, enteropneusts, some sea urchins and some crustaceans. These burrowing activities are often correlated with a mode of feeding (mud ingestion and other forms of deposit feeding) which also leads to a reworking of the sediment. Some research allows an evaluation of the importance of this phenomenon.

Thus, Rhoads [RHO 67], working on the Atlantic coast of the United States, indicates that the quantity of sediment ingested by the polychaete annelid *Clymenella torquata* is estimated at 274 $mL.individual^{-1}.yr^{-1}$ and at 400 $mL.individual^{-1}.yr^{-1}$ in *Pectinaria gouldii*. When the density of animal populations is high, this leads to a very high level of biogenic reworking of the sediment. In the intertidal zone, this reworking can affect the sediment over a depth of about 30 cm with a maximum in the first 10 cm. In the subtidal zone[1], biological activity is limited to the first 10 cm.

Amoureux [AMO 63] notes that the presence of a population of the lugworm *Arenicola marina* (polychaete annelid) with a density of one individual per dm^2 leads to a rapid descent (1 dm in 15 days) of the sub-surface layer of the sediment and the establishment of a vertical gradient in particle size. The role of the lugworm in the distribution of radionuclides within the sediment is therefore easily explained. Thus, Amiard-Triquet [AMI 74, AMI 75c] showed that in an experimental environment, in the absence of living organisms, ^{144}Ce, ^{59}Fe and ^{60}Co introduced into the water fixed rapidly to the sediment surface. After the introduction of the lugworms, the author observed a progressive descent of the most contaminated layer and a homogenization of the rate of radioactivity within the different layers. Figure 4.2 represents this process for ^{144}Ce but the phenomenon is very similar for the other two radionuclides. Similarly, Cross (1968) demonstrated the role of a burrowing amphipod (*Anonyx* sp.) in the distribution of zinc-65 within the sediment.

1 The subtidal zone is located in the littoral environment below the variations of the water level due to the tides, and consequently always submerged.

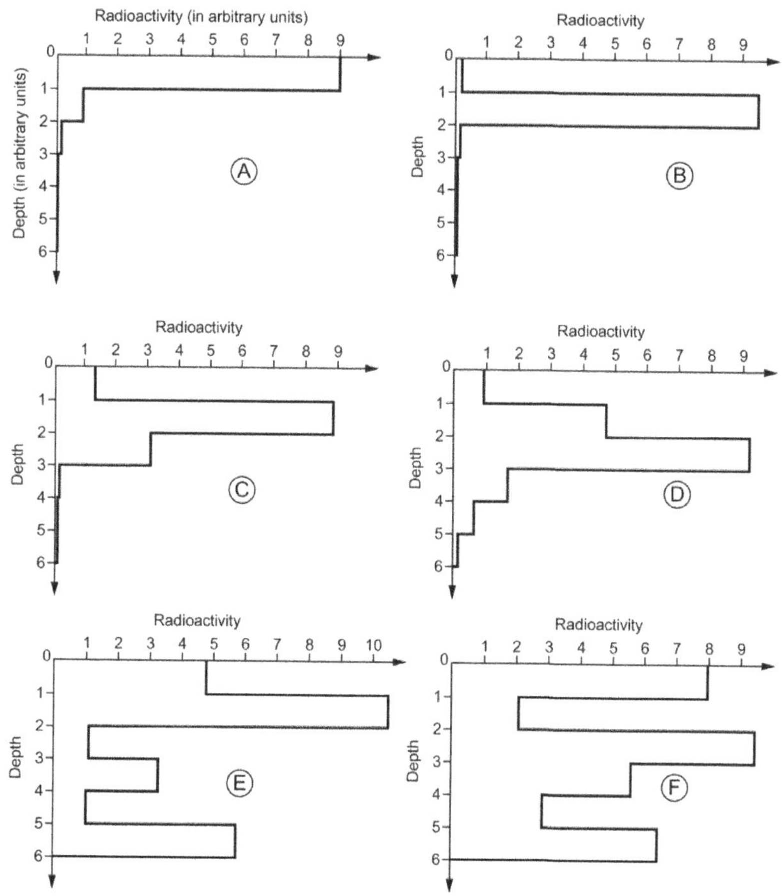

Figure 4.2. *Process of sediment contamination by cerium-144 in the presence of Arenicola(modified from [AMI 74]) (A–F: time after sediment contamination 0, 1, 2, 5, 12, 26 and 40 days, respectively)*

The circulation of water and oxygen caused by animal activity can also influence the exchange reactions between water, the suspended sedimentary phase and the deposited sediment. Renfro [REN 73] showed that for a silty sediment contaminated by ^{65}Zn and permanently leached, the rate of desorption reached 1–3% in 18 days and 3–9% in 30 days, but that the introduction of the polychaete annelid *Nereis diversicolor* multiplied this rate three to seven times, by resuspension of the finest radioactive particles and perhaps by increasing the solubilization of zinc.

At Eniwetok, extensive bioturbation of the lagoon sediments contributed significantly to the mixing of lateral and vertical sediments, often excavating to depths greater than 2.0 m. The highest radionuclide activity levels are those associated with very fine-grained (<38 μm) sediments [SUC 86].

In the vicinity of Sellafield, *Maximuella lanskesteri* (Echinodermata) and *Callianassa subterranea* (Crustacea) are two organisms responsible for most of the bioturbation between 0.5 and 1.5 m depth in the northeast Irish Sea. The echinoderm feeds on detritus at the surface of the sediment and releases its feces into burrow at 0.5 m, transporting plutonium to depth. The second organism burrows to 1.5 m and buries the surface sediment along with the radioactive deposits [WOO 97a].

The influence of bioturbation in the remobilization of radionuclides trapped in Baltic Sea sediments was investigated in two experiments where different tracers were used in three methods, using three key macrofaunal species from this environment. The three species were the amphipod crustacean *Monoporeia affinis*, the Baltic clam *Macoma baltica* and the priapulid worm *Halicryptus spinulosus*. In both experiments, there were clear visual differences between the sediment effects of the three species. *Halicryptus spinulosus* buried deepest in the sediment, creating a network of burrows, *Monoporeia affinis* actively buried in the top few centimeters of the sediment and *Macoma baltica* was fairly stationary, but appeared to filter and deposit food on the sediment surface. Mixing depths in hypoxic sediments ranged from 4.0–3.5 cm for *M. baltica* to 7.8–2.1 cm for *H. spinulosus*. Biodiffusion rates were similar for all processes but biotransport rates were significantly different between them, primarily due to a high value of transport rate for *H. spinulosus*. Bioturbation by macrofauna both buries surface contaminants and remobilizes buried ones, but physical processes are likely to be much more important than biological processes in redistributing contaminants on a decadal time scale [BRA 06].

4.4.3.2. *Influence of organisms on the distribution of radionuclides in water bodies*

Radionuclides associated with biological material are transported by the horizontal (eel, salmon, different species of tuna, sea turtles, seabirds, marine mammals) and vertical (zooplankton diel migrations) migrations of animals.

Several authors have attempted to estimate the impact of vertical migrations of zooplankton on the biological transport of stable elements or their radioactive isotopes. According to Ketchum and Bowen (1958, *in:* [NAS 71]), in the marine environment, a CF of 340 in zooplankton is sufficient for biological transport from depth to the surface to equal physical transport. On the other hand, according to Lowman et al. (*in*: [NAS 71]), the CF should reach 6,800 to 1,000,000 depending on the oceanic area.

Radionuclides can be transferred from the surface to depth by the effect of gravity on phytoplankton, dead zooplankton, fecal pellets, exuviae and, in general, on all organic detritus. Some authors [SMA 73a, SMA 73b, BEN 74] have tried to quantify this phenomenon by developing a linear model to describe the flux of elements or radionuclides (Zn, Cd, etc.) in zooplanktonic crustaceans (*Meganyctiphanes norvegica*).

Following the Fukushima accident, Pacific bluefin tuna, *Thunnus orientalis*, carried radionuclides derived from this event throughout the North Pacific Ocean. Concentrations of ^{134}Cs (4.0 ± 1.4 Bq kg^{-1}) and elevated ^{137}Cs (6.3 ± 1.5 Bq kg^{-1}) in tuna caught in California indicate that Pacific bluefin tuna rapidly transported radionuclides from a point source in Japan to distant ecoregions, demonstrating the significance of migratory animals as vectors for radionuclide transport. These results reveal tools to trace the origin of migration (using the presence of ^{134}Cs) and potentially the timing of migration (using the ratios ^{134}Cs/^{137}Cs) in highly migratory marine species in the Pacific Ocean [MAD 12].

Young Pacific bluefin tuna (PBFT) from the Fukushima area from a recent migration had elevated concentrations in their white muscle tissue of ^{134}Cs (0.7 ± 0.2 Bq kg^{-1}) and ^{137}Cs (2.0 ± 0.5 Bq kg^{-1}). Most of the larger and older fish were not contaminated with ^{134}Cs and only background levels of ^{137}Cs, showing that one year in the California Current is sufficient for ^{134}Cs and ^{137}Cs values in tuna to reach contamination levels similar to those before the Fukushima accident [MAD 13].

4.4.3.3. *Influence of biological activity on the physico-chemical form of radionuclides*

The absorption of radionuclides by phytoplankton and their subsequent re-emission into seawater with metabolites and decomposition products has a sometimes considerable effect on their subsequent physico-chemical behavior. Thus, Gromov and Spitsyn (1974, *in:* [AMI 80]) observed that bioassimilation processes have a depressive effect on the sorption properties of ^{239}Pu, ^{106}Ru and ^{99}Tc towards clays. The sorption capacity of ^{60}Co, on the other hand, is little changed.

In animals, the consumption of contaminated food (radionuclides in “particulate” form) and the subsequent excretion result in the release of radionuclides into the environment in particulate form (feces) and in liquid form (urinary and/or gill excretion) in variable proportions. The phenomenon has been quantified for cobalt-60 [AMI 80]. For this radionuclide, the release of soluble or colloidal forms is significant. In contrast, the dietary contamination of *Meganyctiphanes norvegica* with ^{141}Ce and ^{237}Pu results in the release of almost 100% of the radionuclides with the fecal pellets. Similarly, Fowler and Heyraud (1974, *in:* [AMI 80]) showed that the biological accumulation of ^{65}Zn by the shrimp *Lysmata seticaudata* resulted in a change in the physico-chemical form of the radionuclide. The same is true for the post-mortem decomposition of radiolabeled shrimp, presumably in connection with bacterial activity.

4.5. The organotropism of radionuclides

Studies of organotropism in aquatic organisms are very numerous, and only a few examples of the distribution of the principal radionuclides introduced into the environment in organisms representative of three major zoological groups, mollusks, crustaceans and fish, will be reported.

For group I radionuclides (isotopes of stable elements with no known physiological functions), there is a striking difference depending on the physico-chemical form of the radionuclides. Easily hydrolyzable radionuclides (radioisotopes of Y, Zr, Nb, Ru, Ce, Pu, Am, Np) show a mainly external localization. The highest concentrations are found in the shells of mollusks, the exoskeleton and gills of crustaceans, the gills themselves, the digestive tract and the digestive contents of fish (*in:* [AMI 80]). For ruthenium-106, Le Gall and Ancellin [LEG 71] have shown that it was the external cuticle of the shell of *Mytilus edulis* that was highly contaminated (93% of the total radioactivity of the organism). Similarly, for plutonium-237, Fowler et al. [FOW 76] observed that the exuvia, which corresponds to the outermost layers of the exoskeleton, could represent 92–100% of the total radioactivity of the shrimp *Lysmata seticaudata*. Several experiments on the contamination of marine or freshwater fish eggs show that most of these radionuclides are bound to the membrane. Thus, before hatching, the eggs are much more contaminated (by ^{91}Y, ^{144}Ce, ^{95}Zr, ^{95}Nb and ^{106}Ru) than the larvae immediately after hatching (*in:* [AMI 80]). In eggs, this is 90% for *Pleuronectes platessa* eggs. Variations in the intensity of plutonium accumulation observed during embryogenesis of *Misgurnus fossilis* eggs could be related to changes in membrane characteristics. In eggs of the cuttlefish *Sepia officinalis* contaminated with americium (^{241}Am), this radionuclide is continuously bioaccumulated by the eggs during the development phase but does not cross the outer barrier [LAC 10].

For other group I radionuclides (Cr, Ag, Sb radioisotopes), despite significant interspecies differences, internal contamination is often predominant.

The location of group II radionuclides (*in:* [AMI 80]) is quite similar to that of the stable elements to which they are chemically related: cesium, a homologue of potassium, is preferentially concentrated in the soft tissues of mollusks, in the muscles of crustaceans and fish; strontium, chemically close to calcium, is generally more abundant in the shell than in the soft tissues of mollusks and has high concentrations in the exoskeleton of crustaceans and the calcified tissues of fish.

The distribution of group III radionuclides (*in:* [AMI 80]) in aquatic organisms generally follows that of their stable isotopes and depends on their physiological role. For Mn, Co and Zn, there is a significant accumulation in the viscera (in particular, kidneys and liver of vertebrates, similar organs of invertebrates) but the external contamination and that of the digestive tract are not negligible. The role of iron as a constituent of respiratory pigments has been demonstrated for a preferential accumulation in the blood cells and spleen – blood-forming organ – of *Pleuronectes platessa*. Eggs of the cuttlefish *Sepia officinalis* have been contaminated with various radionuclides (^{57}Co, ^{134}Cs, ^{54}Mn and ^{65}Zn). Cobalt and zinc are continuously bioaccumulated by the eggs during the development phase, while manganese bioaccumulation saturates within one month of exposure. Cesium bioaccumulates with the weight of the egg, i.e. with the influx and efflux of water into the egg [LAC 10]. Bioaccumulation of ^{60}Co, ^{54}Mn and ^{65}Zn is significantly lower in cuttlefish *Sepia officinalis* than in octopus *Octopus vulgaris* and especially squid *Loligo vulgaris*. This is related to the nature of the chorionic membrane [LAC 10].

In mussels *Mytilus edulis*, the distribution between the soft tissue and shell varies with each metal and exposure period, and ranges from 1/25 for cesium and silver to 1/1 for cobalt after nine days. After nine days of exposure, ruthenium and silver show a typical surface-type behavior, where the metal content of the mussels is proportional to their surface area. In contrast, cobalt, zinc and cesium exhibit metabolic-type behavior [NOL 91].

All mollusks, except those from Ecalgrain Bay (a site near the La Hague reprocessing plant), had higher levels of plutonium in the shell than in the soft parts. However, individuals collected in the vicinity of the plant outfall consistently had plutonium shell/soft part ratios below 1 [GUA 77a]. In addition, these authors were able to establish that there is a relationship between the rate of plutonium absorption and the calcified structures of certain marine species comparable to that which exists in the bone tissue of terrestrial mammals [GUA 77b].

In starfish, the distribution of ^{239,240}Pu and ^{238}Pu showed a strong affinity for the mucus-rich epidermal layer [GUA 82].

The organotropism of ^{60}Co in *Scrobicularia plana* contaminated via water was 54% in the shell and 22% in the digestive gland, while contamination via food (Diatoms) led to 78% in the digestive gland and 2.5% in the shell. The same radionuclide in *Carcinus maenas* contaminated via water led to 42% in the gills and 6% in the hepatopancreas. When the contamination was via food, the organotropism was different and depended on the nature of the food. While the food was *Scrobicularia*, 7% of ^{60}Co was in the gills and 39% in the hepatopancreas, while if it was *Arenicola*, 10% was in the gills and 47% in the hepatopancreas [AMI 76b].

Cobalt isotopes accumulate mainly in the hepatopancreas of crabs and in the liver and kidneys of plaice [AMI 75e].

Organotropism of polonium-210 was quantified in three different species of mollusks from the Gulf of Gdansk and Slupsk Narrow (*Mytilus trossulus*, *Mya arenaria)* and the southern coast of Spitsbergen (*Chlamys islandicus*). The highest ^{210}Po concentrations were found in the digestive gland of Baltic species, while for polar mollusks, the highest values were observed in the gills. The CF for the total soft tissue of all mollusks analyzed ranged from 3,600 to 72,000 [STE 00].

When water is the vector of contamination, the accumulation of ^{110m}Ag is mainly in the external organs, shells (34%) and mantle (16%) of the bivalve *Scrobicularia plana*, exoskeleton (40%) and gills (38%) of the crab *Carcinus maenas*. In contrast, contamination via food is responsible for a considerable accumulation of ^{110m}Ag in the digestive gland of *S. plana* (75%) and hepatopancreas of *C. maenas* (56%) [AMI 78c].

The organotropism of ^{210}Po and ^{210}Pb was quantified in muscle and hepatopancreas/digestive gland of crustaceans and cephalopod mollusks. The same nuclides were also measured in liver, pyloric cecum, stomach contents and muscle of tuna. In muscle, the CFs were an order of magnitude lower; in the digestive gland or hepatopancreas, they were an order of magnitude higher, reaching 10^6 in the hepatopancreas of shrimp of the genus *Sergestes* [HEY 79].

^{134}Cs was found in most soft tissues and was closely associated with sulfur and potassium [CRE 21].

4.6. The mechanisms for the chemical detoxification of radionuclides

In marine organisms, there are two main detoxification mechanisms: biomineralization and metallothionein induction. This results in various forms of storage of radionuclides in organisms, which will have an important influence on trophic transfers.

4.6.1. *Chemical detoxification in animals: biomineralization and induction of metallothionein*

As for stable elements, radionuclides are stored in inert forms in the cells of organisms. Two types of storage are favored by organisms, the induction of metallothioneins and biomineralization [AMI 17a]. Metallothioneins, or their plant equivalents, phytochelatins, are proteins rich in cysteine, an amino acid containing thiol groups that complex metals and their radionuclides. The induction of these proteins is carried out in vertebrates and invertebrates [AMI 06]. The role of metallothioneins in protection against ionizing radiation has been emphasized by Cai et al. [CAI 99], without a specific mechanism being described. Pre-treatment with a metallothionein inducer increases the protection of individuals to ionizing radiation. The role of metallothionein is replaced by phytochelatins in plants.

The complexation of radionuclides and their transport by metallothioneins has been demonstrated in particular by Goudard et al. [GOU 91] and by Paquet [PAQ 91] for transuranics. This preferential binding to sulfur is also observed for other elements classified as soft acids (big molecules and poorly charged acids containing weakly electropositive and highly polarizable atoms) such as cadmium, cesium or selenium.

The second major mechanism of detoxification of radionuclides in living organisms is biomineralization. Thus, the uranyl ion, which is classified as a hard acid (or Lewis acid whose electron acceptor centers are not very polarizable), will form phosphate salts in plants, invertebrates and vertebrates (Frelon et al. pp. 29–46, *in*: [MEN 09]).

4.6.2. *Storage forms and chemical speciation of radionuclides in organisms*

The various chemical detoxification mechanisms lead to storage forms of radionuclides, rendering them harmless from a chemical point of view but not from a radiological point of view. These forms of storage are very varied and according to their more or less strong bioavailability, the radionuclides will be transferred more or less abundantly from prey to predators.

With respect to uranium, most of the studies relating to the localization or speciation of this radioelement have been conducted in animals and microorganisms.

The intracellular distribution of transuranics after contamination by water or food has been studied. Thus, the subcellular localization of ^{241}Am in the digestive gland cells of periwinkles and oysters has been determined. Similar experiments have been performed on the hepatopancreas of lobsters (labeled food) with ^{241}Am, ^{238}Pu and ^{95m}Tc, and in the pyloric ceca of starfish with ^{241}Am, ^{252}Cf and ^{95}Tc. The distribution of ^{137}Cs has been studied in cytosolic fractions of eel muscle and liver and in oysters [GOU 98].

In the hepatopancreas of crabs, the localization of plutonium, americium and iron exists mainly in a dense non-mitochondrial mineral fraction composed of calcium phosphate granules. Plutonium and in particular iron are also associated with calcium phosphate microgranules sequestered in the microsomal fraction. This situation, very different from that observed in mammals, suggests that accumulated transuranics and other heavy metals may be partly extruded subsequently into the lumen of the hepatopancreas with the calcium granules [GUA 81].

Following the absorption of technetium-95m and americium-241 by lobsters (*Homarus garnmarus*) from food, the subcellular distribution and binding in the main organ of accumulation, the hepatopancreas, was analyzed. The ^{95m}Tc was mainly bound to soluble proteins, and about 40% was bound to two proteins with the same elution characteristics as Cd-thioneins. In contrast, ^{241}Am was mainly sequestered by the lysosomal and vacuolar system, indicating absorption by pinocytosis. Approximately 2–3% of americium was bound to ferritin and 20% was bound to unidentified 10- and 20-kdalton ligands [GOU 91].

The subcellular distribution of the natural radionuclide ^{210}Po was studied in the liver of Atlantic mackerel *Scomber scombrus*. The majority of ^{210}Po was found in the cytosol of liver cells. About 30% of ^{210}Po was bound to ferritin and about 28% to metallothioneins [DUR 99].

4.6.3. *Influence of chemical speciation in the biota on trophic transfers*

The cellular distributions of most metals and their radionuclides can be summarized in five compartments [WAL 03a]: metallothioneins or related proteins, heat-sensitive proteins, organelles, cellular debris (membranes, etc.) and metal-rich granules (Figure 4.3).

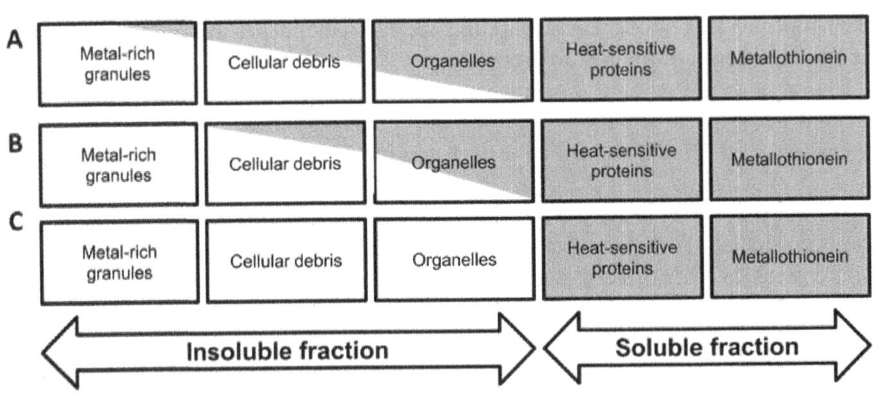

Figure 4.3. *Fractionation of a bioaccumulated metal among five compartments (from [WAL 03a]) and bioavailable fractions (shaded, with the shaded area increasing the higher the bioavailability) in the various compartments (from [RAI 11]). A. Case of a neo-gastropod mollusk; B. Case of a neo-gastropod mollusk with weak digestive powers; C. Case of a planktonic copepod feeding on phytoplankton*

Depending on such subcellular distribution, metals and their radionuclides are more or less trophically bioavailable. As a result, the absorption of trace metals and their radionuclides by predators is affected by the physico-chemical form of the metal accumulated in the prey. Based on the transfer of various elements between a unicellular alga and a copepod, Reinfelder and Fisher [REI 91] estimated that only the fraction of metal associated with proteins (metallothioneins and heat-sensitive proteins), also called the "soluble fraction", was bioavailable. Based on the transfer of Cd or Zn between a bivalve mollusk and a decapod crustacean, Wallace and Luoma [WAL 03b] considered that the trophic bioavailability of the metal (TAM, Trophically Available Metal) concerned not only the metal present in soluble form but also the metal associated with the cell's organelles. Frequently, subsequent studies involving various prey for various consumers have not validated the concept of TAM, as the bioavailable metal or its radionuclide may comprise a greater or lesser important fraction of the five subcellular compartments [NG 05, RAI 06, ZHA 06, RAI 07, RAI 10]. It is now clear that trophic bioavailability varies among foods, consumers and metals including their radionuclides [RAI 11].

4.7. Disposal of radionuclides

When a contaminated organism is transferred to an environment free of radionuclides, elimination is no longer balanced by accumulation and its level of radioactivity will then decrease more or less rapidly. This phenomenon is called decontamination or self-purification. An organism is a complex assembly of many organs, tissues and organelles, and radionuclides generally accumulate differentially in these structures and are also eliminated variably. It will follow that the general equation for the curve representing the decrease in radioactivity in the organism is in the form of a sum of exponentials [COM 55] where the compartment-specific elimination constants k_x are the quotient of Log 2 divided by the effective half-life Te_x of compartment x.

A compartment will consist of an organ, a tissue or a set of organs or tissues; sometimes, the same organ will be subdivided into two compartments such as the central plasma and the peripheral blood. The effective decontamination of an organism will depend on the physical decay of the radionuclide and the biological elimination (fecal, urinary, gill or pulmonary excretions, molting). Thus, the effective half-life (Te), which is the time required for half of the radionuclide to disappear from the body, is related to the physical half-life (Tp) and the biological half-life (Tb) [LIE 08]. Figure 4.4 represents the elimination of a radionuclide in the case of two compartments.

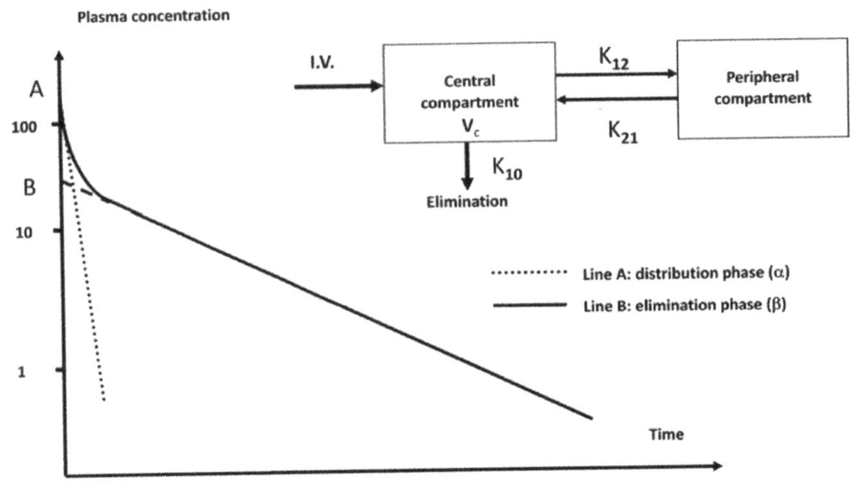

Figure 4.4. *Evolution of blood concentrations of a radionuclide as a function of time in a two-compartment model after intravenous administration (I.V.: intravenous) (semi-logarithmic coordinate representation)*

The two elimination processes (physical and biological) go in the same direction; the effective period is always shorter than the shorter period but of the same order of magnitude as the latter.

The biological and effective half-lives will be similar for long-lived radionuclides. However, for short-lived radionuclides, the effective half-life will be shorter than the biological half-life. Thus, in the case of silver-110m with a physical half-life of 250 days, the overall effective half-life (Te) of this radionuclide in the bivalve *Scrobicularia plana* is 44 days, while the effective half-lives of the two compartments (Te_1 and Te_2) are, respectively, 5 and 78 days [AMI 78a]. The elimination of ^{110m}Ag from the different organs of the animals examined depends on the mode of contamination; it is generally rapid in the external organs and slow in the digestive gland [AMI 78a].

The various elimination parameters (Te, Tb, k_n) can be estimated from measurements made in situ or during laboratory or field experiments. In the first case, it is necessary that the source of pollution stops. This is what happened (in part) after the series of aerial nuclear explosions in 1961–1962. This is how some American authors [HOD 73] were able to determine the effective half-life of some radionuclides in the liver of the tuna *Thunnus alalunga*. Thus, for cesium-137 (physical half-life 30 years), it is 12 years, 7.1 years for silver-108 (Tp = 127 years), 310 days for silver-110m (Tp = 253 days), 2.6 years for cobalt-60 (Tp = 5.26 years),

320 days for manganese-54 (Tp = 300 days), 450 days for zinc-65 (Tp = 245 days) and 3.5 years for plutonium-239 (Tp = 24,360 years). The biological elimination of silver-110m and zinc-65 through the tuna's liver is therefore zero.

However, the majority of data on the elimination of radionuclides by aquatic organisms is provided by laboratory experiments. A great number of parameters qualitatively and quantitatively influence the elimination, as well as the modes of contamination (by water, by food, etc.), the conditions under which the contamination was carried out (temperature, chemical composition of the environment, etc.) or the conditions under which the decontamination is carried out (fasting animals, season, etc.). The number of decontamination experiments is much less than the number of contamination experiments, and no bibliographic synthesis has been made of these data. Examination of some values of biological elimination periods for radionuclides in various groups of aquatic organisms [AMI 80] shows great fluctuations in this parameter depending on the radionuclide in the same species or zoological group.

Experiments show that the number of elimination phases (compartments) is most often two, sometimes one or three, very rarely more than three. It is rare that a compartment corresponds to only one organ, and one experiment of elimination is not enough to inform us about the metabolism of the radionuclide. Studies of the distribution of the radionuclide in the body at various times during elimination must be performed in parallel with elimination measurements.

The rapid elimination phase often corresponds to the unincorporated radionuclide either because it transits through the lumen of the digestive tract and is defecated rapidly, or because it is adsorbed on external surfaces with weak bonds. The long elimination phase corresponds most frequently to radionuclide incorporated into living matter in molecules with a slow turnover rate (such as bone tissue) or in storage organs (such as the liver).

The radionuclides that are eliminated rapidly will therefore be mainly those that have particulate and colloidal physico-chemical forms and have no biological role (Group I: Ce, Ru, Ag, Se, etc.). Radionuclides that have a physiological or structural function (groups II and III) may have long biological half-lives. This is the case for strontiums and cesiums which could partially replace, respectively, the stable isotopes of calcium in bone tissue and potassium in muscles. This will also be the case for transition metals, normal constituents of metalloenzymes and some vitamins, and which, moreover, are often stored in the liver or in the hepatopancreas or digestive gland in decapod crustaceans and mollusks, respectively (examples of Zn, Co, Mn, etc.). However, there are many exceptions. Thus, niobium and plutonium, present in aquatic environments mainly in particulate or colloidal forms, seem to be eliminated slowly while iodine, an element essential to life, has short biological periods. Iron, a normal constituent of some biological molecules, is eliminated rapidly in mollusks, whether it is present in the environment in soluble or insoluble forms…

The correlation between elimination kinetics and systematic ranking of the animal concerned is not as apparent as is the (negative) correlation between accumulation rate and systematic scale. This is certainly due, in part, to the smaller number of data available to us. Renfro (1972, *in:* [PRU 72]) reports that the rate of in situ elimination of zinc-65 slows down as we progress up the food chain from producers to super-carnivores with *Enteromorpha intestinalis* (Te = 8 days), *Corophium salmonis* (Te < 50 days), *Cymatogaster aggregata* (Te = 50.2 days) and *Crangon franciscorium* (Te = 68.4 days). But due to insufficient data, it is impossible to generalize.

Vives i Batlle et al. [VIV 07] note that, at least in marine organisms, there is a relationship between the half-life of radionuclides (cesium, americium, technetium and plutonium) and certain allometric characteristics. This allows the authors to extrapolate to other radionuclides. Thus, for the elements N, K, Np and Cm, the same allometric relations as those obtained for their analogs ^{99}Tc, ^{137}Cs, ^{239,240}Pu and ^{241}Am, respectively, can be applied, when no other data are available.

For cobalt-60, the self-purification capabilities of a marine mollusk and crustacean (*Scrobicularia plana* and *Carcinus maenas*) are not independent of the mode of contamination (single or chronic ingestion of radioactive food). Thus, the elimination of cobalt-60 by the bivalve and the crab is rather fast in the case of a single contamination by food and much slower when the ingestion of radioactive food is repeated many times. Specifically, in *Scrobicularia*, the effective period was 30 days for one meal and 66 days for 13 meals. This elimination had two periods, a short one of two to four days and a long one of 40 days (one meal) and 90 days (13 meals) [AMI 76c].

In tropical species, the biological half-lives are identical (37 ± 2 to 58 ± 25 d) for three bivalve species: (the oysters *Isognomon isognomum* and *Malleus regula*, and the clam *Gafrarium pectinatum)* [MET 16].

4.8. Quantification of radionuclide transfers to organisms

Transfers of radionuclides to marine organisms can occur from the two main physical compartments, seawater and sediments, or also during trophic transfers.

4.8.1. *The various methodologies for estimating radionuclide transfers*

Historically, transfers from physical media to organisms were estimated using CFs, which are the ratios of the activity in the organism to the activity in the source (generally a physical medium such as water, soil, etc.). New methodologies are currently proposed and will be detailed below [ICR 01].

Higley and Bytwerk [HIG 07] comment on the various methodologies for estimating transfer factors. The first method uses CR (concentration ratio). The CR is the ratio of the activity of the receptor compartment to the donor compartment. The CR is thus a generalization and equivalent to the CF. The CR are calculated between the concentration in an organism and a concentration in a compartment of the environment (biotic or abiotic) in relation to it. This is, for example, between a plant and soil, an aquatic organism and water, more rarely between an animal and a plant, an animal and soil or an aquatic organism and sediment. There are thousands of CF and CR values in the literature, but these ratios have significant limitations in use [MCG 96].

The second method uses kinetics. The simplest is the one that uses the time dependence to the first degree of the concentration of a radionuclide in an organism [WHI 82, GAR 00a]. The advantage is that the concentration can be predicted at any time and not only at the equilibrium between the two components (organism and medium). Many programs are proposed (FARMLAND, PATHWAY, GENII, RESCAD, etc., *in:* [HIG 07]). The major limitation of this method is that it relies on equilibrium CR data only. Is this due to lack of data or to simplification of the calculations?

The third method calculates a transfer factor as the ratio of the concentration in an organism to the daily intake $(Bq.kg^{-1}).(Bq.j^{-1})^{-1}$. The unit is $j.kg^{-1}$. The major limitation of this method is the requirement to know the daily consumption rate. The fourth method uses transfer coefficients in wild or semi-wild ecosystems. It estimates the total concentration of radionuclides in a wild species and is commonly referred to as the aggregated transfer parameter (T_{ag}) expressed in $m^2.kg^{-1}$. T_{ag} is often available for the edible portion of a wild organism [BER 04]. The latter two methods are currently only applied to transfers in the terrestrial environment.

In the literature, there is a wealth of transfer parameters; unfortunately, these data can be derived from derived, unmeasured values, and the origins of these data are not explicit. Great transparency should prevail [BER 04, SAN 04].

Garnier-Laplace et al. [GAR 00b] propose a standard procedure to calibrate dynamic models to quantitatively assess direct radionuclide accumulation and trophic transfer in fish, based on the accumulation kinetics and elimination rates of radionuclides obtained in laboratory experiments. The model takes into account the effect of the food chain, the feeding rate and the growth of organisms. This paper describes several examples concerning the transfer dynamics of ^{54}Mn, ^{60}Co and ^{137}Cs in a simple pelagic food chain (phytoplankton, zooplankton, carnivorous and super-carnivorous fish). For carnivorous and super-carnivorous fish, accumulation following direct transfer follows the following order: ^{60}Co < ^{137}Cs < ^{54}Mn. The values reached at equilibrium in L.kg^{-1} pf were for carnivorous and super-carnivorous fish, respectively: 8.7 < 27.4 < 107 and 4.14 < 6.59 < 13.4. For the trophic pathway, ^{137}Cs was the most accumulated (trophic transfer factor at equilibrium TTF (eq) was 0.485 in 291 days for carnivorous fish and 1.45 in 17 years for super-carnivorous fish). The most relevant ecological value is the aggregate CF (*Tag*), which is the ratio of the concentrations of the radionuclide in the organism and in the water at steady state and for all possible transfer pathways. The *Tag* are, at maximum and at equilibrium, respectively, for ^{137}Cs, ^{54}Mn and ^{60}Co of 550, 65 and 7.2 L.kg^{-1} pf.

4.8.2. *Transfer of radionuclides from water to organisms or bioconcentration*

The primary goal assigned to radioecologists is to define the criteria that would allow the implementation of good radiation protection for humans. To do this, it is necessary to qualitatively and quantitatively know the various ways in which radionuclides can be transferred to humans. From the environment, the main possibilities of contamination are via drinking water, inhaled air and ingestion of edible organisms.

Radioecologists have long believed that it was sufficient to know the accumulation rate of a given radionuclide in the most widely consumed edible species to estimate the radiation doses suffered by humans consuming these products, hence the use of the concentration factor concept (CF). The advantages and disadvantages of this concept and especially its limits of use have been detailed by Amiard [AMI 13], while Carvalho [CAR 18] has made a historical review. The IAEA has made several syntheses [IAE 04b].

From a biological point of view, radionuclides can be classified into three groups according to the physiological role of their stable isotopes. Group I consists of elements with no known physiological function, such as chromium, silver, antimony, ruthenium, zirconium, niobium, cerium, yttrium, transuranics, etc. Group II includes elements that are chemically close to an element that is essential for life. This category is represented by two important fission products, cesium, close to potassium, and strontium, close to calcium. Group III includes elements with an important physiological role. This is the case of a certain number of activation products, radioactive isotopes of iron, manganese, cobalt, zinc, phosphorus, etc.

At the level of the hydrosphere, we note that group II corresponds to radionuclides that are essentially in ionic form in natural waters, while most of the radionuclides of group I are characterized by complex physico-chemical forms with the predominance of particulate and colloidal forms.

4.8.2.1. *The concentration factor concept*

The need to predict the accumulation of various radionuclides in living organisms and to compare this accumulation for organisms from differently contaminated environments necessarily led radioecologists to conceive of the CF as a ratio between the amount of radioactivity present in the organism and the amount of radioactivity present in the environment.

In the aquatic environment, this CF is most frequently defined as the CR of the radionuclide in the organism to its concentration in the water, the other components of the ecosystem (sediment, food) being neglected.

The CF can be obtained by two main means: from in situ measurements or from laboratory experiments. In situ, it may be technically difficult to measure the radioactivity of organisms and/or the environment. In addition, environmental radioactivity is rarely constant due to discontinuous contamination from variations in fallout from nuclear explosions or releases from nuclear facilities (chronic, cyclic release, etc.). The ratio (CF) between the environment and the organisms will therefore not be constant over time. Moreover, some organisms move and may stay in areas of very different radioactivity. This results in large uncertainties in the determination of the CF.

When some authors tried to establish the CF from stable isotopes of the same element or of a chemically neighboring element, various difficulties arose such as isotopic discrimination, chemical discrimination or different physico-chemical forms of the elements.

Given the instability of radioactivity in the environment, the ideal solution was to reproduce the conditions of environmental contamination in the laboratory. The advantages included an ability to fix the parameters of the contamination as well as the level of contamination compatible with an easy determination of the radioactivity in the environment and the organisms. The major disadvantage of laboratory experiments is the obligation to simplify the real conditions of the environment. The evolution of certain parameters (biotic and abiotic) that are important in controlling the accumulation of radionuclides by organisms may therefore be neglected because of ignorance or technical impossibility. Here again, physico-chemical forms may play a major role.

In laboratory contamination experiments, the radioactivity of the environment frequently decreases rapidly by the fixation of part of the radionuclide on the walls of the experimental unit (aquarium, etc.) or on the sediment. Moreover, the state of equilibrium between the physical and biological environments is difficult to obtain in the short-term [WHI 82].

Whether the concentration factor is determined in situ or in the laboratory, water is considered to be the predominant vector for the contamination of aquatic organisms. This is not always the case. For benthic organisms, the possible vectors of contamination are open water, pore water or sediment. If we relate the concentration in the organisms to the concentration in each of these compartments, we will obtain very different CFs.

In the aquatic environment, the radioactivity of water can be measured with or without filtration (at 0.2 µm, at 0.45 µm, etc.). Depending on whether the physico-chemical forms of the radionuclides are mostly ionic or fine colloids, or on the contrary mostly large colloids and particles, the CFs, determined with or without filtration, will be very different.

As a result of all these aspects, CFs determined in situ or in the laboratory are rarely identical, the latter being generally lower [PAT 75]. For this reason, a critical study of the CFs reported in the literature is necessary to obtain acceptable values. The IAEA [IAE 04b] proposes a selection of CFs. Table 4.1 shows the large variability with CFs in fish.

4.8.2.2. *The concentration ratio (CR)*

As mentioned above, the CR is a generalization of the CF. Its term is more neutral. But the majority of the data concern the freshwater environment. This is the case of about 130 Russian publications that provide data on CR of radionuclides in freshwater organisms for 35 radionuclides and 11 groups of freshwater species [FES 11]. Improvements are proposed for ^{241}Am (in bivalve mollusks, gastropods and pelagic fish), ^{60}Co (in gastropods, benthic fish and insect larvae), ^{90}Sr and ^{137}Cs (in benthic fish and zooplankton).

Of the 372 CR values calculated for the marine ecosystem, 195 were identified through a literature review. Furthermore, about 70–80% of the data gaps were filled by employing "preferable" approaches such as those based on the substitution of values of taxonomically similar organisms or biogeochemically similar elements [HOS 08].

4.8.2.3. *Concentration factors of various radionuclides in organisms*

The literature providing CFs of various radionuclides is very abundant. The first radionuclides studied were cesium-137 and strontium-90, then various activation radionuclides such as cobalt-60 or silver-110m and more recently transuranics.

The oldest and most comprehensive literature review on CFs is undoubtedly that carried out in 1964–1965 by Polikarpov [POL 66]. Its most significant flaws are first of all its age and the lack of critical analysis of the reported data. Since then, many authors and various organizations such as the IAEA have proposed CFs for most radionuclides [IAE 78, IAE 94, IAE 10]. A brief analysis of this literature shows that the CFs for freshwater organisms are higher than those for marine organisms.

Elements	Minimum	Recommended	Maximum	Recommended by IAEA, 1978	Recommended by IAEA, 2004
Americium	0.5	50	200	10	100
Antimony	100	400	1,000	1,000	600
Silver	300	500	30,000	1,000	10,000
Cesium	10	20	100	50	100
Cobalt	10	50	100	100	700
Curium	100	1,000	10,000	(10)	100
Iron	300	3,000	30,000	1,000	30,000
Manganese	70	400	1,000	500	1,000
Plutonium	0.5	40	100	10	100
Ruthenium	0.1	2	20	1	2
Strontium	0.3	2	10	1	3
Technetium	20	30	200	10	80
Zinc	500	1,000	10,000	2,000	1,000
Zirconium	1	20	50	1	20

Table 4.1. *Variability in concentration factor by mode of assessment. Examples of radionuclides in marine fish (modified from [IAE 78, AMI 80, IAE 04b])*

In the smallest marine organisms (phytoplankton and zooplankton), the most concentrated elements are ruthenium, cerium, zirconium and iron. The corresponding fission products are mainly present in the environment in particulate and colloidal forms. For larger marine organisms (algae, plants, mollusks, crustaceans and fish), the most accumulated elements are transition metals (Zn, Fe,

Cu, Co, Cd, Mn, Ag) or a metalloid (phosphorus) whose radionuclides are generally activation products.

In general, CFs decrease as the complexity of organisms increases (systematic scale). Progressing from phytoplankton through zooplankton, mollusks and crustaceans to fish, biological accumulation decreases. However, there are some exceptions such as high concentrations of Mn, I and Zn in marine mollusks, P in crustacean and marine fish muscles, Co in marine crustaceans, Cs in marine fish, Zn in freshwater mollusks and fish, P, Cs, Ru in freshwater crustaceans and Fe in freshwater mollusks (Table 4.2).

Elements	Concentration factor (CF)			
	Algae	Mollusks	Crustaceans	Fish
Americium	8,000[a,b]	20,000[a]–1,000[b]	500[a]–400[b]	50[a]–100[b]
Antimony	400[a]–20[b]	200[a]–300[b]	400[a]–300[b]	400[a]–600[b]
Silver	2,000[a]–5,000[b]	10,000[a]–60,000[b]	5,000[a]–200,000[b]	500[a]–10,000[b]
Cesium	50[a,b]	30[a]–60[b]	30[a]–50[b]	20[a]–100[b]
Cobalt	10,000[a]–6,000[b]	5,000[a]–20,000[b]	5,000[a]–7,000[b]	50[a]–700[b]
Curium	8,000[a]–5,000[b]	30,000[a]–1,000[b]	500[a]–(400)[b]	1,000[a]–100[b]
Iron	30,000[a]–20,000[b]	30,000[a]–500,000[b]	5,000[a]–500,000[b]	3,000[a]–30,000[b]
Iodine	1,000[a]–10,000[b]	10[a,b]	10[a]–3[b]	10[a]–9[b]
Manganese	6,000[a,b]	5,000[a]–50,000[b]	500[a]–5,000[b]	400[a]–1,000[b]
Plutonium	2,000[a]–4,000[b]	3,000[a,b]	300[a]–200[b]	40[a]–100[b]
Ruthenium	2,000[a,b]	2,000[a]–500[b]	100[a,b]	2[a,b]
Strontium	5[a]–10[b]	1[a]–10[b]	2[a]–5[b]	2[a]–3[b]
Technetium	1,000[a]–30,000[b]	1,000[a]–500[b]	1,000[a,b]	30[a]–80[b]
Zinc	20,000[a]–2,000[b]	30,000[a]–80,000[b]	50,000[a]–300,000[b]	1,000[a,b]
Zirconium	3,000[a]	5,000[a]	200[a,b]	20[a]

Table 4.2. *Variation factors in bioconcentration as a function of the zoological group membership (modified from a: [AMI 80]; b: [IAE 04b])*

4.8.3. *Transfer of radionuclides from sediment to benthic organisms*

Sediment plays a considerable role in the life of benthic animals that come in contact with it, either on the surface or by digging their burrows in it, and that ingest more or less significant quantities of it in a usual or accidental way.

The sediment compartment is very important in the distribution of most radionuclides within aquatic environments. Therefore, it is critical to determine whether the sediment is a final repository (sink) or a potential source of contaminants to the biota. Exchanges between sediment and benthic organisms can be either direct as a result of the ingestion of contaminated particles, or indirect as a result of desorption phenomena leading to an increase in the contamination of pore water or supernatant water.

Procedures for studying exchange from sediment to organisms present numerous technical constraints. The concentrations of radionuclides in "sediment ingesters" represent the accumulation of both soluble and particulate forms. It is therefore very difficult experimentally or following field studies to distinguish between contamination routes through water or through the sedimentary phase.

For ^{95}Zr–Nb and for ^{106}Ru–Rh, both in the annelid *Nereis japonica* and in the bivalve *Gomphina melanaegis,* the aqueous pathway seems to predominate over the sedimentary pathway (Ueda et al. 1977 and 1978, *in:* [AMI 80]). The same is true for cerium-144 and lugworms (Amiard-Triquet, 1974, *in:* [AMI 80]).

According to Amiard-Triquet and Amiard [AMI 80], for all the radionuclides considered, the principal vector of contamination is water. This was verified for the transfer of cobalt-60 and cesium-137 in an annelid (*Arenicola marina* L.), an echinoderm (*Echinocardium cordatum* Pennant) and two lamellibranch bivaves (*Scrobicularia plana* da Costa, *Macoma halthica* L.). The main influence of these organisms in the sedimentary zone is the redistribution, within the sediment, of radionuclides adsorbed on the surface [AMI 74]. Similarly, sediments contaminated by plutonium do not seem to be a significant transfer vector for species living in contact with them [GUA 77b]. However, the available data do not make it possible to exclude the existence of transfer from the sediment to the organisms. However, the relative importance of direct and indirect transfer is impossible to assess. A selection of transfer factors from the sediment has been the subject of several publications [IAE 85, FOW 97].

In tropical species, Cs accumulated from sediments was absorbed in a similar manner (61.6 ± 9.7–$79.2 \pm 2.3\%$) [MET 16].

4.8.4. *Trophic transfers of radionuclides*

One possibility for radionuclide transfers in marine organisms is trophic transfer between the prey and the predator. One possibility during these transfers is the biomagnification of the radionuclide along the food chain.

The available data on the trophic transfer of radionuclides result either from in situ studies or from laboratory simulation experiments. Each of these approaches has its advantages and disadvantages, as reviewed by Amiard-Triquet and Amiard [AMI 80]. In addition, Amiard-Triquet and Amiard [AMI 80] have reviewed the data on the contamination of food chains by principal radioactive pollutants.

Lanthanides (^{141,144}Ce), actinides (^{238,239,240}Pu, ^{239}Np, ^{241}Am) and some metals (^{103,106}Ru, ^{95}Zr–Nb), which are essentially found in colloidal or particulate form in the natural environment, have a very limited capacity to pass through the wall of the digestive tract. As a result, in aquatic organisms, these products are mainly located externally and water is the main vector of contamination. Under these conditions, the observation is that of a clear decrease of concentrations for the organisms of the highest trophic levels. This appears to be the case for plutonium, but, although the experimental data are too limited to draw general conclusions, it seems that in some crustaceans and teleost fish, the accumulation of plutonium by the dietary route is not absolutely negligible (Fowler et al. 1975; Fowler and Guary, 1977; Eyman et al. 1976, *in:* [AMI 80]). In the lobster *Homarus gammarus*, the hepatopancreas is the primary site of plutonium incorporation after experimental labeling, with the cytosol being the subcellular compartment where the largest percentage of plutonium is found [TOC 95].

Organotropism also plays a decisive role in the life cycle of the alkaline earth elements, ^{90}Sr and ^{226}Ra. In aquatic environments, these radionuclides are not very concentrated in organisms of a high trophic level, with a few exceptions in mollusks, crustaceans or asteroids. The relative significance of contamination in different species seems to depend mainly on the development of calcified tissues and relatively little on feeding habits. Moreover, it should be noted that hard tissues are often neglected by the predator or are discarded with the faeces without being assimilated.

Group III radionuclides, radioactive isotopes of the transition metals essential to life (^{57,58,60}Co, ^{55,59}Fe, ^{54,56}Mn, ^{65}Zn), generally show a simple transfer in fresh water. In seawater, in situ measurements and laboratory experiments indicate a general tendency for the concentration of cobalt-60 to decrease with the trophic level, with some exceptions (Figure 4.5). Thus, some invertebrates and, in particular herbivores, could be more contaminated than primary producers. Iron accumulates in high concentrations in the liver of marine vertebrates. Its role as a constituent of respiratory pigments is evidenced by a preferential accumulation in the blood cells and spleen (blood-forming organ) of *Pleuronectes platessa*. In situ measurements of zinc-65 in the various links of the marine food chains indicate sometimes a biological amplification phenomenon, sometimes a decrease towards the top of the productivity pyramid or remains at the same level at all trophic levels. Laboratory simulations do not resolve these contradictions, as the role of the food vector varies from one species to another and from one environment to another.

Data concerning dietary transfers have since been developed. In estuarine crabs, the absorption pathway for ^{137}Cs is primarily from the diet, while that of ^{90}Sr is primarily from water. Biodynamic modeling has suggested that diet accounted for 90–97% of ^{137}Cs incorporation, while water accounted for 59–81% of ^{90}Sr [CRE 21].

Trophic transfer, accumulation, subcellular distribution and induction of metallothionein has been quantified for six metals (Ag, As, Cd, Cu, Pb and Zn) in a gastropod mollusk predator *Nassarius siquijorensis*. Snails were fed either *Ruditapes philippinarum* clams, *Perna viridis* mussels, *Crassostrea angulata* oysters or *Fistulobalanus albicostatus* barnacles, each differing greatly in their bioaccumulation of the metals. Calculated trophic transfer factors of metals for the different prey varied and were highest when clams and mussels were the prey [GUO 13].

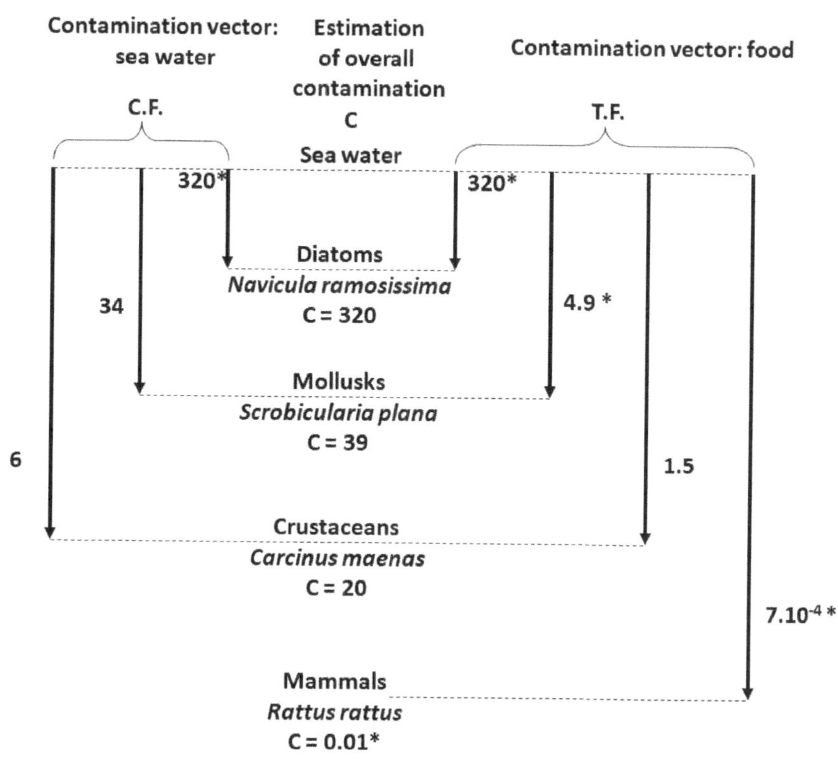

Figure 4.5. *Direct seawater contamination vector (quantified as concentration factor (CF), trophic level concentration over seawater concentration) and indirect food contamination vector (quantified as transfer factor (TF), trophic level concentration n over initial trophic level concentration) of cobalt-60 in a marine food chain leading to a mammal (modified from [AMI 77])*

From the knowledge of metal assimilation efficiency (AE), metal efflux rates and ingestion activity of the animals involved, it is possible to predict the transfer (and potential biomagnification) of metals along various marine food chains. Metals can potentially be biomagnified in benthic food chains, primarily in those for which gastropods are the primary predators, because they can effectively sequester dietary metals. The bio-diminution of metals (concentration decreases at higher trophic levels) in the classical marine planktonic food chain (from phytoplankton to copepods to fish) is largely the result of efficient metal efflux by copepods and very low metal absorption by marine fishes [WAN 02].

^{210}Po and ^{210}Pb were measured in crustaceans and cephalopod mollusks representative of a pelagic and a benthic food chain. The CFs from seawater to whole animals were approximately constant along both food chains, being on the order of 10^4 for ^{210}Po and 10^2 for ^{210}Pb [HEY 79].

Trophic transfers of polonium-210 have been the subject of a large number of studies in the euphausiids *Meganyctiphanes norvegica*; the results show higher concentrations in smaller specimens. This fact cannot be explained on the basis of surface adsorption, but is probably related to food ingestion. The distribution of ^{210}Po in euphausiids is not homogeneous, but the majority is concentrated in the internal organs, the gastrointestinal tract and hepatopancreas in particular. Food is the main source of ^{210}Po for this species, and fecal pellets are the main route of elimination [HEY 76].

The concentrations of naturally occurring radioactive radionuclides ^{210}Po in a large number of pelagic organisms in the Atlantic Ocean have been grouped into several categories. For example, decapod crustaceans were divided into four categories, two for Peneidae and two for Caridae. These groupings are explained on the basis of diet variations between the different categories [HEY 88]. Similarly, levels of ^{210}Pb in decapods consuming krill (*Euphausia superba*) showed seasonal changes that could be interpreted in terms of diet change [MIQ 93].

For plutonium, Guary and Fraizier [GUA 77b] found that there was a relationship between the concentration of plutonium in marine animals and plants and the trophic level of these organisms, and that there was a decrease in the concentration of the radioelement as the species' trophic level increased. Of the three modes of transport via water, sediment and food studied, direct contact between seawater and organisms was the primary mode of transfer to marine species. The transfer of plutonium to starfish via food also occurs, but the food chain biomagnification of plutonium by starfish does not occur in nature, as previously assumed [GUA 82].

In tropical species, trophic pathways were the main route of Cs absorption in bivalves and shrimp [MET 16].

4.8.4.1. *The concept of biomagnification*

We have seen that many pollutants are concentrated in living organisms (bioconcentration). Some of these pollutants will be transferred to successive trophic levels and to humans, the final consumer of short or long food chains. For example, in the case of food of marine origin, the short chain will include: algae – human, mollusks – human and in the case of long marine chains: crustaceans or fish – human. Two modes of contamination of food chains can be distinguished (Figure 4.6) depending on whether it is a simple transfer or "bioaccumulation", accompanied by a decrease in concentrations for the highest trophic levels, or a = biological amplification phenomenon (biomagnification), responsible for increasing concentrations for the highest trophic levels.

Biomagnification corresponds to an increase in the concentration of the pollutant at each link in the trophic chain. This phenomenon was observed for the first time for non-radioactive pollutants in Minamata Bay, where from 1953 to 1960 about 100 people died. In this case, in this bay, along the trophic chain from seawater to fish consumed by humans, the concentration of methylmercury was multiplied by 10^5.

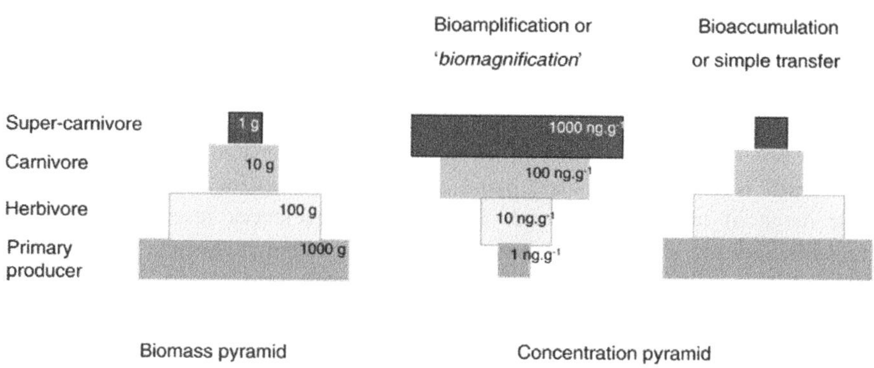

Figure 4.6. *Theoretical representation of biomass and concentration pyramids (figures are arbitrary). For a color version of this figure, see www.iste.co.uk/amiard/radioecology.zip*

The factors explaining bioconcentration, bioaccumulation and biomagnification are now well understood (Norström and Letcher, 1997, *in:* [AMI 17a]). For biomagnification, it is necessary that the pollutant has a moderate or high hydrophobicity, a low polar character and low metabolism. Few radionuclides meet these conditions except for radionuclides that can easily bind to organic matter to form stable organic molecules such as tritium. However, we will see below cases of biomagnification of radionuclides that do not follow the general rule. Indeed, in a certain number of cases, the data concerning stable elements show that the increasing concentration according to the trophic level is not directly linked to a particularly important trophic transfer but is explained by a high accumulation by a given species. The best-known examples are transition metals (Cd, Zn) in oysters, zinc in barnacles and vanadium in ascidians.

4.8.4.2. *Examples of biomagnification of radionuclides*

A certain number of biomagnification phenomena of radionuclides have been reported in the literature for tritium, cesium-137, phosphorus-32 and polonium-210.

Although released in significant quantities into the environment, tritium has been little studied. In living organisms, tritium, like hydrogen, can be associated with tissue water (about 80% of total hydrogen in fish) or organic matter (20%). Tissue-bound tritium can be subdivided into two components. Firstly, tritium bound to OH, COOH, NH, SH radicals, *ortho* and *para* positions of phenol groups that exchanges relatively quickly with tissue water; secondly, "non-exchangeable" tritium, whose turnover rate depends on the oxidation or reduction reactions of the organic fraction of the tissues, under enzymatic control. This incorporated tritium can be transferred to the food chain and used to synthesize new organic matter. While tritiated organic compounds appear to be more available to biota than tritiated water (marine sources), the importance of the dietary pathway is difficult to assess based on current knowledge.

The Severn Estuary (Wales, Great Britain) is contaminated by discharges from the Cardiff Nycomed-Amersham plant, a radiochemical plant that manufactures a wide range of labeled compounds for use in life sciences research. Very high concentrations of tritium have been detected in marine organisms living in the Severn Estuary and Bristol Channel, mainly in organic form (OBT) [PEA 99, MCC 01]. Indeed, more than 85% of tritium is found in organic form in fish, lugworms (*Arenicola marina*), mussels (*Mytilus edulis*), algae (*Fucus vesiculosis*) and in the sediment. Tritium concentrations in OBT form vary little with season: from 7 to 9 $Bq.kg^{-1}$ dw in coastal seawater (filtered at 0.45μm), from 500 to 980 $Bq.kg^{-1}$ dw for surface sediment, from 1,550 to 3,050 $Bq.kg^{-1}$ dw for algae, from 99,200 to 228,800 $Bq.kg^{-1}$ dw for mussels and from 99,700 to 240,200 $Bq.kg^{-1}$ dw for flounder. Only the activity of suspended solids fluctuated quite strongly: from 370 to 2,500 $Bq.kg^{-1}$ dw between July 1999 and May 2000. The total tritium CFs (CR in the organism to concentration in the dissolved phase of the seawater) are high: 20,000 for mussels and flounder, 300 for algae and 100 for suspended particles. These CFs are very significantly higher than those predicted by standard models (CF≈ 1) [WIL 01]. In addition, there are differences in tritium concentration in the various species studied, which appear to be related to diet [MCC 01]. Thus, benthic fish (flounder and sole) are much more contaminated than sprat, a pelagic planktivorous fish[2]. Experiments have also been carried out to evaluate the bioaccumulation of tritium in mussels. However, the water pathway seems to have been favored over the trophic pathway (consumption of phytoplankton), which makes it impossible to conclude on the question of biomagnification. Finally, at the molecular level, tritium seems to bind more frequently with lipids and carbohydrates than with nucleic acids and proteins [WIL 01].

For cesium, a very clear correlation is observed between the trophic level of organisms and their contamination rate. But the species that occupy the top of the productivity pyramid belong most often to the vertebrate group. In this group, the muscles, which are the main organ for storing cesium, represent a significant percentage of the animal's weight. However, if organotropism highlights the phenomenon of biological amplification, it is not sufficient to explain it, since Pendleton and Hanson [PEN 58] indicate that the muscles of carnivores are more contaminated than those of herbivores. The biomagnification of cesium has also been observed in aquatic environments [GAR 00c].

Using the stable isotope ^{15}N and two methods of measuring the biomagnification factor, Harmelin-Vivien et al. [HAR 12] calculated a net biomagnification of ^{137}Cs in European hake, *Merluccius merluccius*, fished in the Gulf of Lion in the Mediterranean (1.707 and 1.212).

For polonium-210, Fowler [FOW 11] reports an apparent biomagnification between level 1 (phytoplankton) and level 2 (herbivores, micro- and meso-zooplankton) in marine pelagic chains. Similarly, the transfer of ^{210}Po and ^{210}Pb in the food chain of small pelagic fish (from phytoplankton and zooplankton to the anchovy *Engraulis encrasicolus* and the sardine *Sardina pilchardus*) was studied in the Gulf of Lion by Strady et al. They demonstrated the biomagnification of ^{210}Po along the food chain of these two planktivorous pelagic fish. Under controlled laboratory conditions, the trophic transfer dynamics of the radionuclides ^{210}Po and ^{210}Pb were quantified in a food chain, with the first level of plankton feeding on brineshrimp *Artemia* sp. and the third trophic level on euphausiids *Meganyctiphanes norvegica*. The ratio $^{210}Po/^{210}Pb$ in organisms increased 5–12-fold with each trophic level (phytoplankton to grazer to carnivore), reflecting a preferential bioaccumulation of ^{210}Po over ^{210}Pb. *M. norvegica* assimilated 44% of ingested polonium-210 but only 3.5% of ingested ^{210}Pb [STE 05].

In a trophic chain of three trophic levels including at the first level inert or living particles, at the second level *Euphausia pacifica* – and at the third level either the

2 A planktivore is an animal that feeds on plankton.

lantern fish *Lampanyclus leucopsarus* or the decapod crustacean *Pasiphea pacifica*. The authors noted that the concentrations of ^{65}Zn increase at each level [SMA 97].

4.8.4.3. *Assimilation efficiency*

The assimilation efficiency (AE) is defined as the percentage of ingested metal that passes through the digestive wall [WAN 99]. It is an essential parameter for characterizing trophic inputs in the biodynamic model of metal bioaccumulation. In this framework, the trophic bioaccumulation flux I (in $\mu g.g^{-1}.d^{-1}$) is described by the equation: $I_f = IR.AE.[M]_f$, where IR is the ingestion rate ($\mu g.g^{-1}.d^{-1}$), AE is the assimilation efficiency (in %) and [M]f is the concentration of total metal in the food ($\mu g.g^{-1}$).

Assimilation by the trophic pathway is more difficult to study than the dissolved one for two reasons. On the one hand, the separate study of these two pathways is difficult and, on the other hand, the rate of ingestion is neither as permanent nor as significant as the respiratory ventilation.

The measurement of the assimilation efficiency by the trophic route can be achieved according to various methods [WAN 99]. The most commonly used is the "pulse-chase feeding" method, which consists of calculating the ratio between the quantity of the pollutant retained by the organisms after passing through a contaminated food bolus and that in the food consumed. This method frequently uses radioactive tracers [KE 01], but can also use stable isotopes [CRO 07]. In this case, the inert tracer of the digestion is frequently chromium. The advantages and disadvantages of using stable or radioactive tracers have been described by Croteau et al. [CRO 05].

This method is limited by the number of radionuclides likely to be suitable for this type of experimentation, mainly transition metals (Ag, Cd, Cr, Se, Zn).

The assimilation efficiency depends on three main parameters: the radionuclide, the species used as food (and the physico-chemical forms of storage of the radionuclide within this species, see Figure 7.9) and the consumer [LUO 08]. Thus, Ng et al. [NG 05] quantifying the efficiency of assimilation of three radionuclides (^{110m}Ag, ^{109}Cd and ^{65}Zn) for two filter-feeding bivalve mollusks (*Perna viridis* and *Ruditapes philippinarum*) and a crustacean (*Balanus amphitrite)* with seven different microalgae at their disposal (*Thalassiosira weissflogii*, *Phaeodactylum tricornutum*, *Tetraselmis* sp, *Dunaliella tertiolecta*, *Prorocentrum minimum*, *Isochrysis galbana*, and *Rhodomonas salina*) found marked differences across the three parameters (Figure 4.7).

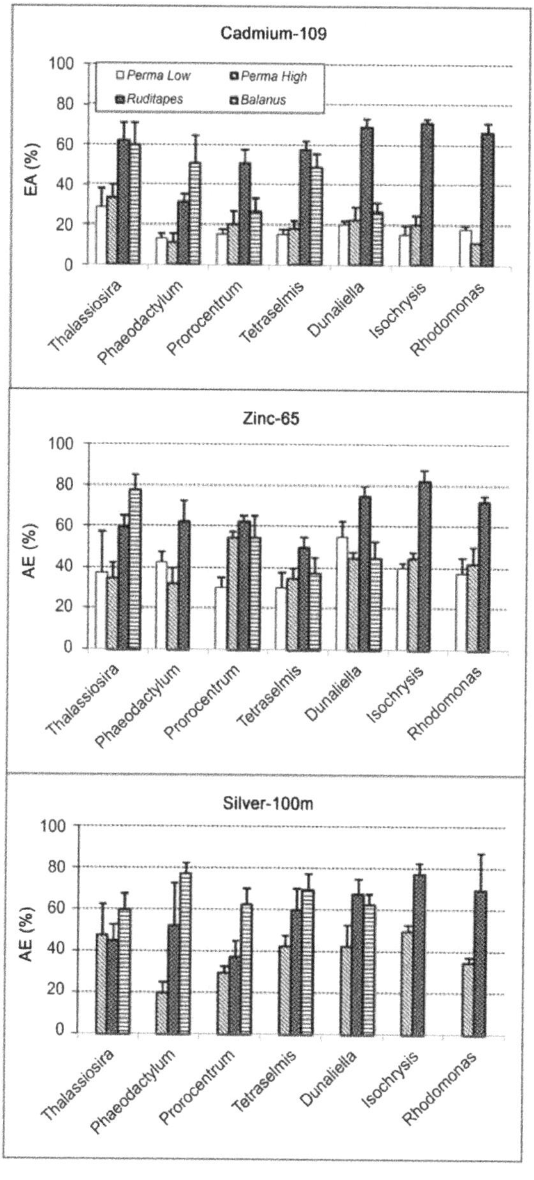

Figure 4.7. *Assimilation efficiency (AE in %) of cadmium, zinc and silver in three filter feeders (green mussel Perna viridis, clam Ruditapes philippinarum and barnacle Balanus amphitrite) fed with seven different microalgae. The concentration of phytoplankton was 0.5 mg.L^{-1} except for the batch called "Perna high" where the concentration was 3 mg.L^{-1} (modified from [NG 05])*

Rainbow et al. [RAI 06] found that for a decapod crustacean *Palaemonetes varians* feeding on an annelid *Nereis virens*, the assimilation efficiencies of silver, cadmium and zinc were about 30, 50 and 60%, respectively.

A strong accumulation of ^{60}Co was observed in the primary producer, *Navicula* sp., a diatom (CF 266–396). After a single dietary intake, primary (*Scrobicularia plana* da Costa, a lamellibranch m) and secondary (*Carcinus maenas* L., a decapod crustacean) consumers retained a significant portion of the ingested cobalt with assimilated percentages of 80–90% [AMI 75b]. In *Carcinus maenas*, the assimilation rate of ^{60}Co was high and very low in the fish *Pleuronectes platessa* [AMI 75d].

Cs absorption efficiency in bivalves and decapod crustaceans after a single meal of radiolabeled food ranged from 7.0 ± 0.4 to 40.7 ± 4.3%, with variable retention time (half-life ranging from 16 ± 3 to 89 ± 55 d) [MET 16].

Dietary assimilation efficiencies were determined to be 55 ± 1% for ^{134}Cs and 49 ± 3% for ^{85}Sr in estuarine crabs [CRE 21].

Assimilation efficiencies have been quantified in the green mussel *Perna viridis* for Ag, Cu and Zn [SHI 04a]. Several authors provide other values of assimilation efficiency factors [STE 05, FOW 05].

4.9. Conclusion

The physico-chemical forms of radionuclides in physical media appear to be one of the main parameters for the dynamics of the contamination of organisms and their level of contamination at equilibrium. The behavior of a radionuclide released into the environment will depend on the relative quantities of its different physico-chemical forms and those of its stable isotopes in the effluent and in the natural environment. The ability of these different forms to cross biological membranes will lead to their discrimination by living organisms and thus condition their association with the biota. The mechanisms become more complicated when considered in dynamic terms. Indeed, the different physicohemical forms are not stable in time but can evolve in a variable way according to the type of radionuclide and the conditions of the environment (pH, discharge of non-radioactive industrial or urban effluents, salinity, etc.). In addition, biological accumulation and metabolization of radionuclides, including by microorganisms and in particular bacteria, is a factor in the development of physico-chemical forms. This is only one example of the infinitely numerous connections that exist between the different characteristics of the environment.

In summary, the influence of natural factors on the radio-contamination of ecosystems can be summarized in two groups of phenomena. The first phenomena are physical or chemical and modify the biological availability of radioactive pollutants. The second phenomena are biological and modify the accumulation capacity of living organisms and thus, subsequently, the biological cycle of radionuclides.

The impact of many factors has been recognized and allows us to build a fairly good qualitative picture of interactions within ecosystems, but their quantitative importance is not or poorly assessed. This assessment is practically impossible from in situ measurements because many factors show concomitant fluctuations with synergistic or antagonistic effects. Although experimental radioecology allows the impact of various factors on the cycle of radionuclides in the environment to be studied individually, the extrapolation of results obtained in the laboratory to natural environments is always tricky. Only a permanent comparison of measurements made in situ and values determined in the laboratory will make it possible to better define the problem of radio-contamination of ecosystems and thus to predict the behavior of radionuclides released by a nuclear installation as a function of the ecological characteristics of the chosen site.

Estuarine areas where marine and continental waters intermingle are even more complex, with tidal variations, generally twice daily, added to seasonal variations and human activities that are generally intense in these environments. In the neritic zone, various events are likely to affect a significant fraction of the water layer: storms, terrigenous inputs, intense productivity. The oceanic zone, on the other hand, presents a high chemical and physical stability and consequently, the environmental variables are generally subject to less significant changes than in other aquatic environments.

Bioaccumulation processes are controlled by many environmental and intrinsic factors, including exposure time, physico-chemical form of the radionuclide, salinity, temperature, competing effects with other elements, organism size, physiology, life cycle and feeding habits. Once accumulated, radionuclides are transported actively by the vertical and horizontal movements of organisms and passively by the release of biogenic products, for example, soluble feces, feces, molts and eggs [FOW 97].

In marine fish, a literature review provides strong evidence that radionuclides can pass through trophic levels and in turn impact the entire food chain. In addition, there is evidence that migrating aquatic organisms can introduce trace amounts of radiation into food chains in an unaffected area as demonstrated for radiocesiums (^{134}Cs and ^{137}Cs) in tissue samples of Pacific bluefin tuna (*Thunnus orientalis*) after the Fukushima accident that were 10 times higher than measurements taken before the accident [KOL 19].

Qualitative and quantitative knowledge of the three components of biological contamination – accumulation, organotropism and elimination – is essential for predicting the fate of radionuclides in the biosphere, for assessing the harm suffered by organisms in the presence of radioactive pollutants and for estimating the risks incurred by humans via the food chain. While there is a great deal of data on accumulation (radioactivity measurements, CFs, etc.), there is much less data on the elimination of radionuclides.

Historically, bioaccumulation was estimated using the CF, which had several drawbacks. First of all, its definition is not universally recognized and some authors use divergent formulae. Moreover, the choice of the appropriate calculation method is not always obvious from both in situ and laboratory measurements. In addition, when estimating it in the laboratory, it neglects the other contamination pathways of organisms, which are notably food and sediment. In addition, a large number of biotic and abiotic parameters interfere with the accumulation of radionuclides, which, together with the differences in determination protocols, means that it is not uncommon to observe differences of a factor of a thousand (or more) between determinations of CFs for the same radionuclide, depending on the author. However, for most radioactive pollutants, a decrease in CF is observed as we progress up the systematic scale, with plants concentrating more radionuclides than animals, invertebrates more than vertebrates. Small organisms (such as phytoplankton or zooplankton) concentrate more radionuclides than large organisms (such as multi-cellular algae or larger pelagic animals).

General rules, with a few exceptions, can be identified concerning the organotropism and elimination of radionuclides. The three groups of radionuclides defined in the introduction correspond to different behaviors. The elements that have no physiological functions remain mostly external to the organisms and are rapidly eliminated. On the contrary, radionuclides whose stable isotopes have physiological functions (or are close to elements with physiological functions) are incorporated into biological molecules and are eliminated more slowly. However, there are many exceptions and it is possible that some radionuclides undesirable for the organism are stored in certain organs or organelles (liver, kidneys, etc.) and are then eliminated slowly. Moreover, the contamination pathways (water, food, etc.) will have a profound impact on organotropism and elimination.

Three main factors, the nature of the radionuclides, the vector of contamination and the mode of contamination, govern the biological accumulation of radionuclides.

While the physiological functions of the elements contribute to determining the mode of contamination, the physico-chemical form of the radionuclides nevertheless plays a considerable role. Considering accumulation in plants, organotropism, regulation phenomena or transfer in food chains, there is a relationship in the behavior of radionuclides according to their physico-chemical form. This is particularly clear for radionuclides essentially present in the environment in particulate or colloidal form. It is striking to note that an element as essential for the metabolism as iron can, under certain conditions, present a distribution in the body close to that of cerium or ruthenium. Thus, in *Pleuronectes platessa*, the polymerized forms of iron are detectable only in the digestive tract and spleen.

The ecology of living organisms depends on the nature and number of possible vectors of contamination, open water for algae; open water, interstitial water and sediment for phanerogams; open water and food for pelagic animal species. Relationships with environmental radionuclides reach their maximum complexity in burrowing animals for which sediment may play a substratum and food role. Depending on the species, the particles of sediment are more or less selected, but in all cases, the fraction of mineral matter that is ingested with organic detritus or living microorganisms is not negligible. Few data exist on the portion of the sediment that can be used by organisms and on the biological availability of a radionuclide depending on the particulate fraction to which it is bound. The absorption of radionuclides associated with organic detritus during digestion will depend on both physico-chemical factors (pH, total electrolyte content) and biological factors (consumer's enzymatic arsenal). In addition, endogenous animals are in contact with pore water which may be an additional source of radionuclides if desorption from sediment particles occurs.

In the long-term, when a source of contamination disappears (cessation or net reduction of nuclear testing, shutdown of an industrial facility), direct or indirect transfer from the lithosphere to the organisms could become the principal source of contamination for biota. It is therefore essential that this process be well elucidated and that the resulting transfers be quantified for long-lived radionuclides and in particular for transuranics. It should also be noted that in certain environments, sediment ingesters and other detritivorous animals form the basis of food chains.

Despite the qualitative and quantitative limitations of the literature concerning the transfer of radionuclides in food chains, some important points should be noted. With rare exceptions (cesium in various food chains, tritium in benthic fish in the Bristol Channel, etc.), no clear biological amplification phenomena have been demonstrated in food chains. On the other hand, even when the food vector is not very significant, the concentration of almost all the radionuclides in almost all the organisms is higher than their concentration in their physical environment. From a human radiation protection point of view, short food chains (algae–human;

mollusks–human) are likely to transfer to the final consumer short- and medium-lived radionuclides and – by ingestion of all soft tissues – radionuclides responsible for contamination limited to the external tissues and digestive contents. The radionuclides of greatest concern in this case are therefore the lanthanides, actinides and particulate metals. In the case of long food chains (crustaceans or fish–human), it is mainly the muscles that are consumed and because of its period, its organotropism and its mode of transfer in the food chains, cesium-137 probably presents the main potential danger.

5

Radioactive Contamination of the Marine Environment and Monitoring Programs

5.1. Introduction

Human activities, both military and civilian, in the nuclear field result in direct and indirect discharges of numerous radionuclides into the marine environment. These result in radioactive contamination, particularly in certain marine environments. Thus, while nuclear weapon testing has been the primary source of atmospheric contamination, resulting in low-level global contamination, sites associated with weapon production and the nuclear fuel cycle can have high levels of localized contamination, and the spread of this contamination through aquatic pathways represents a significant environmental problem [REN 11].

High concentrations may also concern natural radionuclides and certain unexpected sites. This is the case for the contamination of deep waters and in particular of the water masses of the Japan Trough and the Bonin Islands, respectively, down to 7585 m and 9750 m. This was highlighted by Nozaki et al. [NOZ 98]. It concerns the uranium/thorium series, the isotopes of thorium, protactinium, actinium and lead, and the fallout of artificial radioactive nuclides (strontium-90, cesium-137 and plutonium). Throughout the waters of the trench, the activity of $^{239,240}Pu$ is high, indicating that the plutonium is regenerated from large particles that sediment rapidly on the bottom, probably by a change in their oxidation state. The turnover time of the water in the trench has been estimated at a maximum of about five years [NOZ 98].

All states implement monitoring programs for their nuclear installations to ensure their proper functioning and to control the radioactive risk for their local populations. This is, for example, the case in the United Kingdom with CEFAS reports since 1995 [CEF 20]. In addition, various international or bilateral programs are implemented in specific areas with suspected radioactive risks.

5.2. Radioactive contamination of marine environments

The marine environment and marine organisms are susceptible to radioactive contamination from five radionuclide sources. These are marine sites of atmospheric atomic testing, sites affected by atmospheric fallout, sites contaminated by spent fuel reprocessing plants, sites affected by nuclear accidents and sites affected by submerged radioactive waste. Irradiation doses to marine organisms also depend on natural radioactive contamination by naturally occurring radionuclides. The relative comparison can be made thanks to the sites little impacted by radioactive pollution.

A major fraction of the refractory radionuclides, such as uranium (U) and plutonium (Pu) that are released during severe nuclear events, including nuclear weapon tests, reactor explosions or fires, and the use of depleted uranium ammunition, is present as particles, often ranging from sub-microns to fragments. Reprocessing facilities and civilian reactors are also sources of radioactive particles and colloids via their licensed or accidental releases [SAL 11].

5.2.1. *Radioactive contamination by natural radionuclides*

The concentrations in marine waters and sediments of various naturally occurring radionuclides are compiled in Tables 5.1 and 5.2.

Radionuclide	**Concentration in filtered seawater (Bq.m^{-3})**	
	Average	**Range**
^{3}H	50	25–100
^{14}C	6	5–7
^{40}K	18,000	9,000–56,000
^{210}Po	2	1–4
^{226}Ra	2	1.5–3
^{228}Ra	1	0.25–4
^{228}Th	0.05	0.02–0.1
^{230}Th	0.02	0.01–0.04
^{232}Th	0.001	0.005–0.002
^{238}U	40	20–80

Table 5.1. *Natural radionuclide concentrations in filtered seawater (from [IAE 88, BRO 04])*

Radionuclide	Concentration in filtered seawater ($Bq.m^{-3}$)	
	Average	Range
^{3}H	-	-
^{14}C	7	3.5–14
^{40}K	400	200–800
^{210}Po	9	4.5–18
^{226}Ra	30	15–60
^{228}Ra	-	-
^{228}Th	21	10.5–42
^{230}Th	150	75–300
^{232}Th	9	4.5–18
^{238}U	10	5–20

Table 5.2. *Concentrations of naturally occurring radionuclides in marine sediments (from [BRO 04])*

5.2.1.1. *Uranium in the marine environment*

The concentrations of uranium, normalized to a salinity of 35‰, range from 3.162 to 3.281 $ng.g^{-1}$. The average concentration of samples from the Pacific (31°4'N, 159°1'W) is about 1% higher than those from the Atlantic (7°44'N, 40°43'W and 31°49'N, 64°6'W). Concentrations of ^{232}Th in the Atlantic Ocean range from 0.092 to 0.145 $pg.g^{-1}$ [CHE 86].

Uranium concentrations and activity ratios $^{234}U/^{238}U$ were determined in 63 seawater samples (nine vertical profiles) from the Atlantic, Pacific, Arctic and Antarctic Oceans. The uranium/salinity ratio was $(9.34 \pm 0.56).10^{-8}$ $g.g^{-1}$ for the analyzed seawater with salinity ranging from 30.3 to 36.2‰. The uranium concentration of seawater at 35.35 salinity was 3.35 ± 0.2 $\mu g.L^{-1}$ and the activity ratio $^{234}U/^{238}U$ was 1.14 ± 0.03. The global average concentration of fluvial uranium was 0.3–0.6 $\mu g.L^{-1}$ and the ratio $^{234}U/^{238}U$ was 1.2–1.3) [KU 77].

In the waters of the Amazon shelf, as salinity increases to $36.5.10^{-3}$, dissolved uranium activities increase to a maximum of 4.60 $dpm.L^{-1}$. This value is much higher than the open ocean value (2.50 $dpm.L^{-1}$). Uranium activities are much lower for surface sediments on the Amazon shelf seafloor (average 0.69 ± 0.09 $dpm.g^{-1}$) than for suspended sediments in the Amazon River (1.82 $dpm.g^{-1}$). The total flux (riverine flux and the flux of desorbed-remobilized particles) of dissolved ^{238}U from the Amazon shelf (about $1.2.10^{15}$ dpm.yr^{-1}) constitutes about 15% of the uranium input to the world ocean [MCK 87].

U and Th concentrations were determined in the soft tissue and shell of various mollusks (*Mytilus edulis*, *Cardium glaucum*, *Mya arenaria*, *Macoma balthica*) and other benthic invertebrates from Gdansk Bay in the Baltic. Uranium concentrations ranged on average from 0.13 to 0.27 µg.g^{-1} dw[1] for the soft tissue and from 0.006 to 0.026 µg.g^{-1} dw for the shell [SZE 87b]. In Baltic algae, the uranium content varied from 0.09 to 0.41 µg.g^{-1} dw and thorium from 0.06 to 0.60 µg.g^{-1} dw [SZE 87a].

On the southwest coast of Sweden in 1967–1978, the algae *Fucus serratus* and *F. vesiculosus* contained the highest uranium concentrations with 13, 11 and 0.45 Bq.kg^{-1} dw, respectively, for ^{234}U, ^{238}U and ^{235}U, which correspond to dw concentration factors of about 700 for uranium isotopes. Thorium isotope activity concentrations were 100, 400 and 7,000 mBq.kg^{-1} dw of ^{232}Th, ^{230}Th and ^{228}Th, respectively. ^{239,240}Pu concentrations in these bioindicators ranged from 100 to 600 mBq.kg^{-1} dw and 10 to 50 mBq.kg^{-1} dw for ^{241}Am, depending on the year and the collection site. In unfiltered seawater, the authors found in 1978 for ^{239,240}Pu and ^{241}Am, respectively, concentrations in the range of 7–14 and 2–3 µBq.L^{-1}. The dry weight concentration factors were about 12,000 for Pu and 20,000 for Am [HOL 80].

5.2.1.2. *Polonium-210*

Polonium-210 is a natural radionuclide that is strongly present in the oceans and bioaccumulates greatly, especially in mollusks and crustaceans. This radionuclide is therefore responsible for a strong irradiation of some marine organisms, and it is necessary to control its concentrations.

Concentrations of the naturally occurring radionuclides ^{210}Pb and ^{210}Po were determined in dissolved (<0.2 mm) and particulate (>0.2 mm) fractions in surface waters of the northwestern Mediterranean Sea. Approximately 4–8% of ^{210}Pb activity was associated with particulate matter, which was determined to be primarily biogenic. The ^{210}Po content in the particulate fraction was generally higher than that of ^{210}Pb, accounting for about 20% of the total activity. The concentration of ^{226}Ra was in excess of both ^{210}Pb and ^{210}Po, with an average value of 1.72 ± 0.03 Bq.m^{-3}. The residence times of ^{210}Pb and ^{210}Po in surface waters were 0.8 and 3.0 years, respectively [MAS 02].

In the Seine estuary, the concentrations of ^{210}Po in mussels *Mytilus edulis* ranged from 90 to 700 Bq.kg^{-1} dw, while those in *Fucus vesiculosus* varied from 3 to 22 Bq.kg^{-1} dw [GER 95]. In the North Sea, the average concentration of ^{210}Po in the

1 Dry weight.

soft parts of *M. edulis* mussels was 149 $Bq.kg^{-1}$ dw [DAH 96a]. The concentration of ^{210}Po in mussels around the Irish coast ranged from 80 ± 9 Bq kg^{-1} dw at Seapark near Belfast to 468 ± 32 Bq kg^{-1} (dry weight) at Sutton and Dublin [RYA 97, RYA 99]. In mussels from Ravenglass, near Sellafield, the muscle tissue always had the lowest concentration of polonium-210 (124 Bq. kg^{-1}), while the viscera (containing the digestive gland, stomach and kidneys) contained the highest concentration (596 Bq. kg^{-1}) [MCD 93].

Polonium-210 was analyzed in the soft parts of two species of bivalves, the scallop *Chlamys varia* and the common mussel *Mytilus edulis*, from the Bay of La Rochelle and the island of Ré on the French Atlantic coast. The highest values were found in the scallop muscle with 1181 ± 29 $Bq.kg^{-1}$ dw. A size effect was seen with maxima in the smallest scallops. The highest concentrations of ^{210}Po were found in the digestive gland of *C. varia*, ranging from 3150 to 4637 $Bq.kg^{-1}$ dw. Thus, the digestive gland contained up to 60% of the soft tissue content of this radionuclide [BUS 02].

In deep-sea amphipods living in deep-sea ecosystems, the concentrations of ^{210}Po and ^{210}Pb ranged from 23 to 286 $Bq.kg^{-1}$ dw and from 12 to 67 $Bq.kg^{-1}$ dw, respectively [CHA 98]. Quantification of ^{210}Po in dry tissues (excluding intestinal contents) was performed in deep-sea lysianassid amphipods (*Eurythenes gryllus*) captured in the Arctic Ocean (Canada Basin) and in the abyssal regions of the western North Atlantic Ocean (Sohm Abyssal Plain and Bermuda Rise). Concentrations of ^{210}Po in amphipods from the Arctic Ocean were lower than the values in species from Bermuda [HAR 92].

The concentrations of ^{210}Po and ^{210}Pb were determined in over 30 species of marine decapod crustaceans. The median ^{210}Po concentration was 8.0 $pCi.g^{-1}$ dw in whole decapods and 85 $pCi.g^{-1}$ dw in the hepatopancreas. The corresponding $^{210}Po/^{210}Pb$ activity ratios were 77 and 138, respectively. Peneid decapods had median concentrations of ^{210}Po, approximately three to four times higher than those of carid decapods [CHE 81].

The concentrations of natural radionuclides ^{210}Po and ^{210}Pb were measured in krill (*Euphausia superba*), mesozooplankton, phytoplankton and seawater collected during the South African SIBEX cruise to Antarctica in the fall of 1984. The data obtained are numerous [CHE 87].

Polonium-210 was measured in the flesh of Danish fish caught in the North Sea, the Kattegat and the Baltic during 1991–1994. The mean values of 0.35, 0.65 and 0.96 $Bq.kg^{-1}$ ww were observed for cod, herring and plaice fillets, respectively [DAH 96a].

5.2.2. *The current state of radioactive contamination of the marine environment*

The IAEA led a major program called WOMARS (Worldwide Marine Radioactivity Studies) whose objective was to study the time series of concentrations of three radionuclides (^{90}Sr, ^{137}Cs and ^{239,240}Pu) in surface waters of the Pacific and Indian Oceans. For the year 2000, the estimated mean concentrations of ^{90}Sr, ^{137}Cs and ^{239,240}Pu in surface waters of the Pacific and Indian oceans ranged from 0.1 to 1.5 mBq.L^{-1}, 0.1 to 2.8 mBq.L^{-1} and 0.1 to 5.2 µBq.L^{-1}, respectively. The mean half-lives of ^{90}Sr and ^{137}Cs in surface waters were 12±1 years for the North, 20±1 years for the South and 21±2 years for the equatorial Pacific. For ^{239,240}Pu, the corresponding mean effective half-lives were 7±1 years for the North, 12±4 years for the South and 10±2 years for the equatorial Pacific. For the Indian Ocean, the mean half-lives of ^{137}Cs and ^{239,240}Pu were 21±2 years and 9±1 years, respectively [POV 05].

Tritium concentrations ranged from about 70 mBq.L^{-1} in the South Atlantic to 350 mBq.L^{-1} in the Barents Sea. The average concentration of ^{3}H in the Mediterranean Sea was about 280 mBq.L^{-1}.

Concentrations of ^{90}Sr ranged from about 0.4 mBq.L^{-1} in the South Atlantic to mBq.L^{-1} in the Barents Sea. The highest average concentrations (expressed as mBq.L^{-1}) for 2000 were estimated for the Irish Sea (~50), the Baltic Sea (~10), the English Channel (~4) and the North Sea (~4), resulting from discharges from Sellafield and Cap de la Hague.

Concentrations of ^{137}Cs (expressed in mBq.L^{-1}) ranged from about 0.5 in the South Atlantic to 5 in the Barents Sea. The highest average concentrations for 2000 were estimated for the Irish Sea (~ 60), Baltic Sea (~ 60), Black Sea (~ 25), North Sea (~ 7), English Channel (~ 4) and Mediterranean Sea (~ 3).

Concentrations of ^{239,240}Pu (expressed in µBq.L^{-1}) ranged from about 2 in the South Atlantic to about 11 in the Barents Sea. The highest average concentrations for 2000 were estimated for the Irish Sea (~500), North Sea (~15), Mediterranean Sea (~14), English Channel (~13), Black Sea (~5) and the Baltic Sea (~3) [POV 06].

Globally, the highest concentrations of ^{90}Sr and ^{137}Cs were 2 and 4 Bq.m^{-3}, respectively. The lowest concentrations were recorded off the coasts of Argentina and Chile (below 1 Bq.m^{-3} for ^{90}Sr and ^{137}Cs) and near Antarctica (about 0.1 Bq.m^{-3} for ^{137}Cs). Concentrations of ^{239,240}Pu generally varied from 0.5 to 5 mBq.m^{-3}. Occasionally in the northern hemisphere, levels up to 20 mBq.m^{-3} could be observed [BOU 97].

The concentrations of ^{210}Po and ^{210}Pb were determined in about 30 species of marine organisms collected mainly from the northeastern region of Japan to know the levels and distributions of these radionuclides and to estimate their ingestion levels in marine foods. The concentrations of ^{210}Po and ^{210}Pb showed a wide range of variation with 0.6–26 and 0.04–0.54 Bq.kg^{-1} ww in fish, 0.5–220 and 0.2–43 Bq.kg^{-1} ww in mollusks, echinoderms and chordates and 2.8–4.3 and 0.4–1.3 Bq.kg^{-1} ww in algae, respectively [YAM 94].

The radioactive contamination in ^{90}Sr, ^{137}Cs and ^{239,240}Pu of the major seas of the world was synthesized by Livington and Povinec [LIV 02]. The most radioactive seas are the Irish Sea, the Black Sea and the Baltic Sea (Figure 5.1).

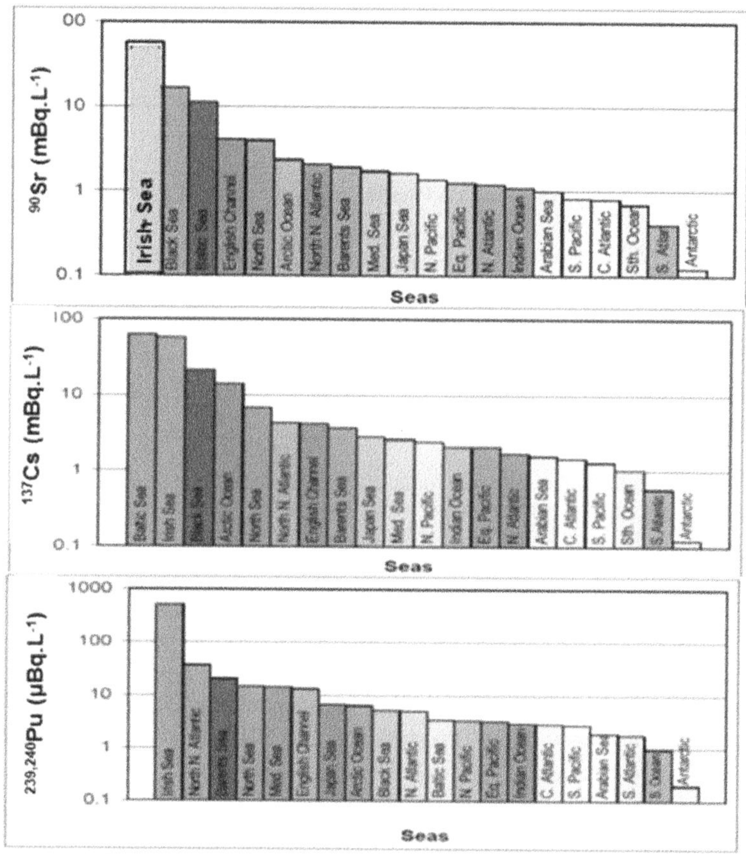

Figure 5.1. *Radioactive contamination in ^{90}Sr, ^{137}Cs and ^{239,240}Pu of the major seas of the world (from [LIV 02]). For a color version of this figure, see www.iste.co.uk/amiard/radioecology.zip*

5.2.3. *The marine sites of atmospheric atomic tests*

The marine environment will be directly impacted only by atmospheric nuclear tests and if the test site is close to the ocean, such as on an island. For underground tests, there has only been indirect contamination of the ocean during an accident during the test due to a crack in the rock or by the rupture of the closure of the atomic test well and only for tests on marine sites.

There have been nearly 270 atmospheric tests in the ocean. Nearly half were in atolls in the Pacific Ocean and the other half in the Arctic Ocean. Only four nations have conducted this type of testing (United States, Soviet Union, United Kingdom and France) (Table 5.3).

Site	Situation	Number of tests	Power (Mt)
Johnston Atoll	North Pacific	12	20.8
Malden and Christmas Islands	North Pacific	33	31
Bikini and Eniwetok Atolls	North Pacific	66 (23 + 42)	109
Novaya Zemlya	Barents and Kara Seas	91	240
Operation Argus	South Atlantic	2+2*	0.004
Mururoa and Fangataufa	South Pacific	46	10
Montebello	South Pacific	3	0.1
Hiroshima and Nagasaki	North Pacific	2	0.04

Table 5.3. *Status of aerial tests of atomic bombs with a direct link to the marine environment, with their number and power (from [AMI 13, IRS 15, WAT 15, GEC 19]). * Underwater tests*

5.2.3.1. *American sites in the Marshall Islands*

The American atmospheric atomic tests in the marine environment took place mainly in the Marshall Islands with the exception of Operation Argus, Johnston Atoll and the joint tests with the British which took place on Malden and Christmas Atolls (see section 5.2.3.4).

5.2.3.1.1. The Marshall Islands

Bikini and Eniweoak were the sites in the Northern Marshall Islands that were used by the United States as testing grounds for nuclear devices between 1946 and 1958. The chronology of the use of the various Marshall Islands atolls in atomic testing has been the subject of several detailed publications [ANO 81, DEI 91].

In November 1945, the United States launched Operation CROSSROADS, a campaign to determine how troops, ships and other military equipment would survive a nuclear attack, and to search for a site to conduct such a test. In January 1946, the US Navy announced that Bikini Atoll (Pikinni) was suitable. Following the operation, in the face of the Navy's inability to decontaminate the targeted vessels, it was decided that the 23 radioactively contaminated wrecks would be sunk in the waters of Bikini Atoll and 41 radioactively contaminated wrecks would be sunk in those of Kwajalein (Kuwajleen). Following the underwater explosion at Bikini, nearly half of the fission products from the bomb remained in the waters of the lagoon.

In April 1948, Operation SANDSTONE, conducted for weapon improvement testing, began at Eniwetok. Three more atomic bombs were exploded. This series of tests was followed by Operation GREENHOUSE, conducted to test the effects of the weapons and the improvements to be made. Three other weapons were tested using GEORGE, a device designed to test the thermo-fusion capabilities of hydrogen isotopes. In November 1952, Operation IVY began on Eniwetok, which included the testing of MIKE, the first thermonuclear bomb, with 750 times the power of the bomb dropped on Hiroshima and releasing more energy than any previous test, including those of the Soviet Union. MIKE literally pulverized the small island of Flora (Elugelab, Bokombako) and excavated a crater of 1.5 km in diameter and more than 50 m deep in the coral. Ujelang (Wu-jlañ, Ujla) was contaminated by the fallout from KING.

Operation CASTLE took place from March 1 to 15, 1954 on the atolls of Eniwetok and Bikini. It involved the detonation of six hydrogen bombs, including BRAVO, a 15-megaton experimental bomb that resulted in the most serious radiation incident in the entire US nuclear testing program.

The test conducted on March 6 on Rongelap (Roñļap) to study gamma ray doses resulted in an estimated dose rate of 37,500,000 mrem.h^{-1} (the standard is 15 mrem.$yr.^{-1}$)[2]. On Utirik, it was 4,000,000 mrem.h^{-1}, and on Bikar (Pikaar), an uninhabited small island, 16,000,000 mrem h^{-1} [RUG 60].

Very soon after the first tests, the influence of radioactivity was recognized up to about 2,000 km from Bikini Atoll [MIY 55].

From 1972 to 1978, the concentrations of $^{239,240}Pu$ in seawater varied from 30 to 43 pCi.m^{-3} at Enewetok and from 69 to 76 pCi.m^{-3} at Bikini for half-soluble and half-particulate, while the average for surface waters of the North Equatorial Pacific

2 This value was proposed by the United States Environmental Protection Agency (EPA) and adopted by the Nuclear Claims Tribunal. One rem is equal to 0.01 Sv.

was 0.38 pCi.m^{-3} [NOS 80]. The same was true for sediments from these two islands and from Utink, another of the Marshall Islands (Table 5.4).

Concentrations of ^{239,240}Pu were reported in the tissues of several species of reef and pelagic fish caught in 14 different atolls in the northern Marshall Islands. Concentrations in liver always exceeded those in bone, and concentrations were the lowest in the muscle of all fish analyzed [NOS 80].

Atoll	^{239,240}Pu	^{238}Pu	^{241}Am
Eniwetok	5,200	732	2,064
Bikini	9,500	491	7,360
Utirik	16	0.8	8

Table 5.4. *Concentrations of three transuranics in Marshall Islands surficial sediments (expressed as pCi.kg^{-1} dry) (from [NOS 80])*

Finally, Bordner [BOR 16] measured low levels of gamma radiation for the colonized island of Eniwetok (mean 0.076 mSv.yr-1), higher levels of gamma radiation for Rongelap Island (mean 0.198 mSv.yr-1) and a relatively high gamma radiation on Bikini Island (mean 1.84 mSv.yr-1). But these radiation doses do not take into account the irradiation brought about by food.

A radiological study was conducted in 1978 to assess the concentrations of persistent artificial radionuclides in 12 atolls and two islands of the Northern Marshall Islands. The concentration of ^{90}Sr in the fish muscle tissue was very low (for the most part undetectable), and there was little difference in the average concentrations of different fish from different atolls or islands. However, the concentration of ^{239,240}Pu in the muscle tissue of all reef species was higher than that of pelagic lagoon fish. In contrast, the concentrations of ^{137}Cs were the lowest in the muscle tissue of bottom-dwelling reef species, such as mullet and goatfish, and the highest in lagoon pelagic fish [ROB 81].

5.2.3.1.2. Rongelap

On March 1, 1954, the United States conducted its largest thermonuclear weapons test at Bikini Atoll in the Marshall Islands. The explosion was codenamed Castle Bravo. Following an accident, the island of Rongelap was contaminated and large values of plutonium-239–240, americium-241 and bismuth-207 were found in the top 25 cm of surface sediment [HUG 19].

Current inventories of ^{241}Am and ^{239,240}Pu, in the surface layer at Rongelap, were estimated at 3% of the respective inventories in the surface sediments of Bikini Atoll. There was a slow and continuous release of transuranics from the sediments to the water column. The concentrations of radionuclides in surface sediments collected at Rongelap in September 1978 are expressed in Bq.kg^{-1} (dry weight) for ^{241}Am 3.6–13.0 (i.e. 0.11–0.66 kBq.m^{-2}), ^{155}Eu 2.4–7.4, ^{137}Cs < 0.1–1.6 and ^{90}Sr 2.3–40.7. The ratio ^{241}Am/^{239,240}Pu is 0.67 ± 0.11. The concentrations of radionuclides in the marine surface waters of Rongelap Atoll in September 1978 are expressed in mBq.L^{-1} for ^{137}Cs 5.4 ± 0.8, for ^{90}Sr 4.1 ± 0.3 and for ^{239,240}Pu 48 ± 10 [NOS 80].

Data are available on the concentrations of ^{137}Cs, ^{60}Co and ^{207}Bi in reef and pelagic fish flesh (muscle) samples collected in Bikini and Eniwetok atolls between 1964 and 1995. The results show that the removal rates of the three radionuclides were significantly different. ^{137}Cs was eliminated from the lagoons with an effective half-life of 9–12 years. Little ^{60}Co was mobilized in the water column and its removal from the environment was mainly by radioactive decay. The properties of ^{207}Bi were different at the two atolls Eniwetok and Bikini. At Eniwetok, the radionuclide was eliminated from the environment with an effective half-life of 5.1 years. At Bikini, only radioactive decay can explain the speed at which the radionuclide disappeared from the lagoon [NOS 97].

5.2.3.1.3. Eniwetok

Plutonium has been detected in the air and in the terrestrial and aquatic environments at concentrations that varied from background levels (due to global fallout) to higher levels by several orders of magnitude. Wilson et al. [WIL 75] estimated that concentrations in marine foods were insensitive to lifestyle and exceeded those in terrestrial foods.

Damage to marine fauna seems low five decades after the beginning of the series of nuclear tests. 70% of the zooxanthellae coral assemblage of Bikini Atoll is resistant to anthropogenic disturbance [RIC 08].

From the radionuclide concentrations in fish from the lagoons of the Marshall Islands, Noshkin et al. [NOS 81] calculated the maximum dose rates to the whole body for persons consuming 200 g of fish flesh daily. For the whole body, they are 0.028 and 0.1 mrem.yr^{-1}, respectively, at Mejit and Rongelap. For the bone marrow, the dose rates are 0.024 and 0.12 mrem.yr^{-1} at Mejit and Rongelap, respectively. These annual dose rates can be compared with the guidelines of the time in the United States for the whole body and bone marrow of 500 mrem.yr^{-1} for an individual and 170 mrem.yr^{-1} for a population average.

5.2.3.1.4. Operation Argus

From August 31 to September 6, 1958, the US Navy Task Force 88, composed of nine ships and about 4,500 men, secretly conducted three nuclear tests (Argus 1, 2 and 3) at high altitude in the South Atlantic [JON 82].

5.2.3.1.5. Johnston Atoll

The intense military activity on Johnston Atoll began before World War II and intensified in the 1960s with the launch of the US nuclear testing program in the Pacific. To accommodate these operations, the Atoll's islands were expanded. The USAF turned over operational control of the islands to the Atomic Energy Commission (AEC) and their military counterpart, Joint Task Force Eight (JTF-8). The series of atomic tests codenamed DOMINIC included four successful high-altitude and five low-altitude tests. These tests were followed by three failed THOR missiles. Radioactive contamination resulting from the destruction of defective rockets occurred throughout the atoll [ANO 12, LOB 19].

The lagoon was again dredged from 1963 to 1964, and the sediments used to expand Johnston Island from 89 ha to 253 ha, again contaminating additional parts of the island [BRA 19]. Thereafter, the atoll underwent a great deal of disturbance with a lot of filling in to enlarge the small island and make four further islands. Numerous military activities were devoted to this site, such as the storage of weapons and chemicals, including Agent Orange. Decontamination efforts have been carried out periodically, including the removal and dumping of debris and surface soil onto the deep waters outside the atoll. As a result, some contaminated soil has entered the island's coastal waters and some areas of the island continue to be off limits.

The coral soil of Johnston Atoll was contaminated by three non-nuclear destruction accidents involving THOR missiles carrying nuclear warheads in 1962.

Characterization of the actinide content of a contaminated coral soil sample revealed that over 99% of the total actinide content is associated with discrete bomb fragments. Environmental exposure of these fragments caused the conversion of relatively insoluble actinide oxides to more soluble actinide oxides. This process led to the dissolution of actinides from discrete particles and migration to soil surfaces [WOL 97].

Johnston Atoll hosts turtles for their egg-laying. The flashes of the explosions must have blinded the turtles and their radioactive contamination is poorly estimated [BAL 85].

5.2.3.2. *Soviet sites in Novaya Zemlya*

The Soviet atomic test site has been the subject of several IAEA reports [IAE 04a]. Complete information on the tests from 1955 to 1992 has been compiled by Andrianov et al. [AND 04b].

The IAEA Monaco Marine Environment Laboratory is collaborating in the assessment of the radiological situation in the Barents and Kara Seas. In Stepovovo Bay (Novaya Zemlya), the sediments contain at the surface 1.2 $Bq.kg^{-1}$ (dry weight and granulometry < 63 µm) of ^{137}Cs and the ratio $^{238}Pu/^{240}Pu$ is between 0.8 and 1.0 [POV 95].

Oughton et al. [OUG 04] found ratios of $^{240}Pu/^{239}Pu$ in Arctic sediments that differ from the ratios due to fallout from civilian activities with high burning rates. This is the case at the entrance of the Kara Sea (Kara Gate south of Novaya Zemlya), in the Yenisei estuary and in the Abrossimov Fjord. On the other hand, the report is close to the civil reports in the Stepovogo Fjord.

In 2016, it was established that the content of radioactive cesium in the seawater of the bays of the Novaya Zemlya archipelago and in the open sea was at a level corresponding to global fallout (1 $Bq.m^{-3}$). The plutonium content in the water was slightly higher. The activity of neptunium in water was an order of magnitude higher than that of plutonium (0.76–1.89 $Bq.m^{-3}$), although its content in global fallout was almost two orders of magnitude lower than the plutonium content. The plutonium content of the bottom sediments of the Kara Sea was 0.2–3.8 $Bq.kg^{-1}$. The radioactive cesium content was highest at the mouth of the Yenisei River (up to 21 $Bq.kg^{-1}$). Neptunium content in benthos samples was above the detection limit and amounted to 1–80 $Bq.kg^{-1}$ wet weight. The neptunium content of benthos has been shown to be one of the indicators of radioactive contamination of the marine environment [GOR 19].

The Severny ice dome serves as a secondary source of radionuclides on the surface of the ice cap: this source originates from previous nuclear weapon tests in the atmosphere above the Severnaya Zemlya test site. The first data on the glacier, not yet examined, indicate that the radioactively contaminated layer is located at a depth of 15–20 m at the limit of the feeding zone of the glacier [MIR 17].

Chernaya Bay, on the southern coast of Novaya Zemlya, was the site of two underwater nuclear tests conducted in the 1950s. Levels of $^{239,240}Pu$ in sediments in the central region of Chernaya Bay exceeded concentrations of 15,000 $Bq.kg^{-1}$, and were among the highest ever reported for the marine environment. Smith et al. [SMI 00] estimated that approximately 11 TBq of $^{239,240}Pu$ from the tests were retained in the Chernaya Bay sediments. The atomic ratios of $^{240}Pu/^{239}Pu$ of 0.03 were well below the ratios of 0.18 typical of global fallout. High levels of ^{137}Cs (> 300 $Bq.kg^{-1}$)

and ^{60}Co (> 100 Bq.kg^{-1}) were also measured in surface sediments in the central regions of Chernaya Bay near the presumed epicenter of the explosions. High levels of $^{239,240}Pu$ were measured in mollusks *Macoma* (104 Bq.kg^{-1}), macroalgae *Fucus* (15 Bq.kg^{-1}) and polychaetes (1,292 Bq.kg^{-1}) living in Chernaya Bay. These authors estimated that 2 TBq of $^{239,240}Pu$ present in the sediments of the Barents Sea originally came from Chernaya Bay [SMI 00].

5.2.3.3. *The British sites in the Montebello Islands*

The archipelago of the Montebello Islands is made up of a large number of islands and small islands (Trimouille, Alpha, Hermite, North West, South East, North Delta, South Delta, etc.) and is located on the west coast of Australia. Three atomic tests have taken place on this archipelago, each with a power of about one kiloton. The first two (Hurricane, October 3, 1952 and Mosaic G1, May 16, 1956) were carried out on Trimouille Island, the third (Mosaic G2, June 19, 1956) on Alpha Island. The first test took place on an old ship, the HMS Plym (K271). It was a River class frigate that served in the Royal Navy between 1943 and 1952. With the nuclear bomb inside the ship, it was destroyed in this first nuclear weapon test. The next two tests were carried out from the top of a 30 m tower.

The principal radionuclides that have had an environmental impact on the Montebello Islands are activation products (^{60}Co, ^{152}Eu and ^{154}Eu, with half-lives of 5.3, 12.4 and 7.8 years, respectively), radionuclides from atmospheric fallout (^{90}Sr–^{90}Y, ^{137}Cs, ^{155}Eu, $^{239,240}Pu$ and ^{241}Am, with half-lives of 27.7, 30, 4.6, 24,390, 6,580 and 432.2 years, respectively) [COO 79].

In 1952 and 1956, the local soil levels still exceeded the maximum dose limits recommended for continuous occupancy [COO 79]. The dose rate in μSv.h^{-1} varied at site G1 at Trimouille (epicenter of the second test) from 35 in May 1962 to 18 in October 1978. The epicenter of G2 reached 160 μSv.h^{-1} in May 1962. Several sites on the various islands still had dose rates above 30 μSv.h^{-1} in 1982 [MOR 82].

The role of particles at Montebello is significant because in this marine site the particles are calcareous and ferruginous sands and therefore rich in Ca, Sr and Br. The plutonium at this site persists for a long time in organisms [SAL 18].

The activity ratios $^{239}Pu/^{241}Am$ were measured for selected samples. The average values were 40 for the HURRICANE (HMS Plym) test, 20 for MOSAIC Gl and 15 for MOSAIC G2 [COO 90].

Samples of oysters and clams (*Tridacna* sp., Giant Clams) were collected and analyzed for ^{155}Eu, ^{152}Eu, ^{137} Cs and ^{60}Co [COO 79]. There was no significant radioactive contamination of oyster soft tissues or shells. For example, at Cocoa Beach near HMS Trimouille and at the G1 site of Trimouille, the concentration in the flesh of oysters was 0.4–1 $mBq.g^{-1}$ and in the shells 0.3 $mBq.g^{-1}$ fresh in May 1983 [COO 83].

Johansen et al. [JOH 19] estimate that on the Montebello archipelago, the dominant plutonium in all marine and terrestrial wildlife was of local origin and that this radionuclide remains a long-term exposure problem for human visitors and wildlife. They found that activity levels of fission and neutron activation products have decreased about 100-fold near the Ground Zero locations. However, Pu concentrations remain elevated, some of which are high compared to most other Australian and international sites (up to 25,050 $Bq.kg^{-1}$ for the three plutoniums-239, 240 and 241. Moreover, the three sites are distinguished by their atomic ratios of $^{240}Pu/^{239}Pu$ and $^{241}Pu/^{239}Pu$ [JOH 19].

5.2.3.4. *The British and American joint sites of Malden and Christmas*

Between 1957 and 1962, the United Kingdom and the United States conducted 33 atmospheric nuclear weapon test detonations on or near the Malden and Kiritimati (Christmas) Islands (total yield 31 megatons) [ALE 21].

5.2.3.5. *The French sites of Mururoa and Fangataufa atolls*

At the request of the French government, the IAEA has undertaken a vast study of the monitoring of the marine environment and the assessment of the potential effects of radiation on marine fauna [WOO 98, WOO 99].

This study was the subject of an extensive publication of a six-volume IAEA technical report and a general synthesis [IAE 98b].

The radiological inventory in the lagoons of Mururoa and Fangataufa were in 1996, respectively, of 1,800, 1,000, 30,000 and 1,200 GBq for ^{3}H, ^{137}Cs, $^{238,239,240}Pu$ and ^{241}Am. The annual flux of tritium varied from 7 to 36 TBq between 1987 and 1996. The export of plutonium associated with the Mururoa sediment to the ocean was 8 GBq annually for a wind overflow and 0.69 TBq in the case of a storm. The quantities of radionuclides expelled from the lagoons of Mururoa and Fangataufa were estimated in 1996 at 5,700, 27, 11 and 11 GBq, respectively, for ^{3}H, ^{90}Sr, ^{137}Cs and $^{238,239,240}Pu$ [IAE 98b].

While transuranics have been found in their places of origin, radionuclides with low Kd values such as ^{3}H, ^{90}Sr and ^{137}Cs have already been transported to the sediments (their reservoirs), as well as to the lagoons of the atolls and the open sea [MUL 99].

The dominant radionuclide in both lagoons is plutonium in the sediment, with a total inventory of about 30 TBq. A decrease in radionuclide concentrations has been observed in recent years in the lagoon waters, with the exception of ^{3}H and ^{90}Sr, for which a contribution from underground sources is expected [POV 99].

5.2.4. *Sites affected by atmospheric deposition*

The majority of atmospheric atomic testing has been conducted in the Northern Hemisphere, and as a result, atmospheric fallout on the marine environment is also in this hemisphere on the Arctic Ocean and its adjacent seas, the Baltic Sea, the Mediterranean and its adjacent seas, and the North Pacific.

5.2.4.1. *The Mediterranean Sea and adjacent seas*

A synthesis of the Ligurian Sea has been carried out by Cossa et al. [COS 20]. The first complete vertical profiles of ^{237}Np (0.12–0.27 μBq.L^{-1}) showed a similar behavior of Cs and Np, confirming that Np behaves conservatively [BRE 17].

Concentration levels of $^{239,240}Pu$ in the surface waters of the northwestern Mediterranean have decreased over the past four decades by a factor of 5.

Flux measurements of $^{239,240}Pu$ suggested that sinking particles accounted for 60–90% of the upper layer (0–200 m) plutonium inventory loss over the period 1989–2013. The residence time of Pu in the upper layer is estimated to be about 28 years [BRE 17].

Deep marine sediment cores were collected from the abyssal plains of the Mediterranean Sea (depth > 2000 m) and analyzed for $^{239,240}Pu$ and ^{137}Cs to elucidate the concentrations, inventories and sources of these radionuclides in the deepest zones of the Mediterranean (Table 5.5). The excess ^{210}Pb inventory was used to normalize the $^{239,240}Pu$ and ^{137}Cs inventories for variable sediment fluxes. The $^{239,240}Pu/^{210}Pbxs$ ratio was uniform across the sea, with an average value of $1.24.10^{-3}$, indicating a homogeneous deposition of $^{239,240}Pu$. The ratio $^{137}Cs/^{210}Pbxs$ showed differences between the eastern (0.049) and western (0.030) basins, due to the clearly significant impact of the 1986 Chernobyl accident on deep-sea sediments [GAR 09a].

The Black Sea is a highly contaminated body of water, whose main sources of radionuclide contamination of human origin were the atmospheric nuclear tests in the 1950–1960s and the accident at the Chernobyl nuclear power plant (1986) [ALI 06, LAP 06, YES 06].

In the Sea of Marmara, the sediment concentrations of $^{239+240}Pu$ varied between 0.45 ± 0.04 and 0.10 ± 0.01 Bq.kg^{-1}, those of ^{40}K between 296 ± 47 and

437 ± 60 Bq.kg^{-1} and those of ^{137}Cs were between 12.6 ± 2.6 and 44.1 ± 6.2 Bq.kg^{-1} [VAR 06].

Pathway	**^{137}Cs (PBq)**	**^{239,240}Pu (TBq)**
Atmospheric deposition before 1986	10.9	190
Chernobyl accident 1986	2.3	0.02
Atmospheric deposition from 1986 to 1996	0.3	7.5
Atmospheric deposition from 1997 to 2000		0.4
Discharge from the Rhône before 2000	>0.05	0.44
Black Sea from 1986 to 2000	0.26	0.4
Exchanges with the Atlantic Ocean from 1980 to 2000	-	-6.8
Total	13.8	192

Table 5.5. *Activities in ^{137}Cs (corrected decay in 2000) and in ^{239,240}Pu introduced into the Mediterranean Sea until 2000 (from [GAR 09a])*

5.2.4.2. *The Arctic Ocean and adjacent seas*

The Nordic seas are heavily contaminated with various radionuclides from three main sources: atmospheric fallout, discharges into rivers and possible leaks of Soviet radioactive waste.

In the Arctic basin and the Russian Arctic seas, the maximum concentrations of marine waters in ^{137}Cs are the Barents and Kara seas. In contrast, the least polluted waters are located in the Laptev and East Siberian seas [MAT 19].

As far as Russian rivers are concerned, two carry quantities of radioactivity, the Ob and the Yenisai, with annual flows of 530 and 603 km^3.yr^{-1}, respectively.

The highest concentrations of ^{137}Cs (expressed in Bq.m^{-3}) are in the sediments of the Barents (160), Kara (270), White (300) and Pechora (170) seas.

The concentrations of plutonium (in Bq.kg^{-1}) in the sediments of Chernaya Bay (Novaya Zemlya), which was an atomic bomb test site, reached 5,000 in Yenisey Bay (3) and in the Norwegian (0.7–1.6), Kara (0.3–1.7) and Baltic (1.5–6.8) Seas. The bays could be classified into seven classes based on sediment contamination (Table 5.6) [MAT 04].

Class	^{137}Cs	^{60}Co	^{239,240}Pu	Bays
1	1–15	0	0	Yarnyshnaya, Teriberka, Titovka
2	20–70	0		Dvina, Onega, Pechora
3	80–400	0	0.2–3.0	Taganrog, Finnish, Bothnia
4	40–100	0.5–6.0	1–6	Yenisey, Ob
5	30–120	4–80	2–9	Kola (naval bases), Motovsky
6	50–1 440	92–618	1,000–5,000	Chermaya
7	0.6–100,000	30–3,000		Abrosimov, Stepovoy, Tsivolka, Kola

Table 5.6. *Sediment concentrations (expressed in Bq.kg^{-1} dw) of various radionuclides in northern seas (from [MAT 04])*

In the Arctic Ocean in 2015, radionuclide activity concentrations were low compared to levels measured between the 1970s and 1990s, and continue to decline [AMA 16].

The highest contamination of ^{137}Cs, ^{60}Co, ^{90}Sr and ^{239,240}Pu was observed in sediments collected near submerged containers in the Abrosimov and Stepovogo fjords. Indications of leakage from submarine No. 601 in the Stepovogo Fjord observed in 1993 could not be verified in 1994 because only an increased level of ^{60}Co could be observed. In the Tsivolky fjord, traces of ^{60}Co could indicate the presence of leaks [JOI 96].

In the Barents Sea and along Norwegian coasts, activity concentrations of ^{99}Tc in seawater in 2007–2009 were low but remain high compared to levels prior to the increased release of this radionuclide from Sellafield in the 1990s. Activity concentrations of the natural radionuclide ^{226}Ra in Barents Sea seawater were comparable to expected background values. Activity concentrations of ^{137} Cs in surface sediments in 2007–2009 were low, with higher values observed in sediments from coastal areas along the Norwegian mainland than from open ocean sites. Activity concentrations of ^{137}Cs and ^{99}Tc in marine biota were low and up to an order of magnitude lower than in previous decades [GWY 12].

In the Svalbard marine environment in 2001, seawater concentrations of ^{99}Tc activity were 0.13–0.36 Bq.m^{-3}, five times higher than in 1994, reflecting the increase in discharges of ^{99}Tc from Sellafield in the mid-1990s. Activity concentrations of ^{137}Cs in seawater were 2.23–2.43 Bq.m^{-3}, 10 times lower than in the 1980s, reflecting reduced discharges of this radionuclide. Activity concentrations of ^{238}Pu, ^{239,240}Pu and ^{241}Am in seawater were <0.3–0.7 mBq.m^{-3}, 5.6–8.9 mBq.m^{-3}

and 0.6–2.4 $mBq.m^{-3}$, respectively, with activity ratios suggesting that global fallout was the dominant source. Activity concentrations of ^{99}Tc in brown algae were up to 18 times higher than in the 1980s with the highest concentrations in *Fucus distichus* (25.7–58.7 $Bq.kg^{-1}$ dw) [GWY 04].

5.2.4.3. *The Baltic Sea*

In the southern Baltic Sea ecosystem, the study of ^{210}Po and $^{239,240}Pu$ in Baltic biota revealed that these radionuclides are highly accumulated by some species. Their average bioaccumulation factor values ranged from 9.10^2 to $3.5.10^4$. Baltic Sea algae, benthic animals and fish concentrate uranium isotopes only weakly, and mean values of the bioaccumulation factor for this element ranged from 1 to 55 [SKW 97].

In the Baltic Sea, the concentrations of ^{239,24} Pu (expressed in $mBq.kg^{-1}$ ww) for various fish species were for plaice 0.94, herring 2.22, cod 2.35 and sprat 0.33. Based on the activity ratio $^{238}Pu/^{239,240}Pu$ in organs and tissues, the proportion of Chernobyl-derived plutonium in the Baltic Sea was calculated to be 60% in the muscle, skeleton and stomach of herring and about 100% in the gills, skeleton, liver and skin of cod [SKW 01].

In the Baltic, technetium-99 concentrations in *Fucus vesiculosus* were 1.3 ± 0.1 $Bq.kg^{-1}$ dw except on the south coast where the concentration was higher 4.5 $Bq.kg^{-1}$ dw [HOL 86b].

In 2000, the concentration of ^{129}I in seawater from Borholm and Moen in the Baltic Sea reached $6.0.10^{-13}$ and 16.10^{-13} $g.L^{-1}$, respectively. These values were more than two orders of magnitude higher than the level of global atmospheric deposition. The highest value of 270.10^{-13} $g.L^{-1}$ was found in the seawater of the Kattegat [HOU 02].

5.2.4.4. *The North Pacific Ocean*

The Northwest Pacific Ocean has been affected by both global (stratospheric) and tropospheric (from nuclear weapons testing at Bikini and Enewetak atolls) fallouts [LIV 00], as well as from Soviet testing. Following the 1961 Russian nuclear weapon tests, euphausiids, salps and sergestid decapod crustaceans sometimes showed strong peaks due to zirconium-95 and niobium-95, ruthenium 103 and 106[3] and cerium-141[4]. The maximum activity measured was 618 $pCi.g^{-1}$ dry weight ^{95}Zr–^{95}Nb in euphausiids taken 45 miles off Astoria, Oregon, USA in November 1961 [OST 63].

---

3 Ruthenium-103 and 106 are in equilibrium with the respective rhodiums.

4 These radionuclides have short physical half-lives, 32.5 days for ^{14} Ce and 39 days for ^{95}Zr-Nb.

The Pacific Ocean is a significant reservoir of plutonium released from atmospheric nuclear weapon testing. The maximum depth of the $^{239,240}Pu$ concentration in the central Northwest Pacific Ocean increased from about 500 m in 1973 to 800 m in 1997. Similarly, the inventory of $^{239,240}Pu$ in water increased from about 130 to 100 $Bq.m^{-2}$ [LIV 00].

5.2.5. *Sites contaminated by spent fuel reprocessing plants*

Spent fuel reprocessing plants are the most polluting stage of the nuclear fuel cycle. Few states have developed this stage. It is mainly the states that have acquired nuclear weapons and at the beginning these plants were military. Then, other countries, such as Germany, Japan and those of the European Union, started reprocessing. Most of these plants are within continents. However, there are four exceptions: the US military plant at Hanford, which is relatively close to the Columbia River estuary and has an impact on the Pacific Ocean; the British plant at Windscale, now Sellafield, in the Irish Sea; the French plant at La Hague on the English Channel; and the Japanese plant at Rokkasho.

5.2.5.1. *Monitoring of the Hanford plant*

Wildlife concentrations of ^{65}Zn were highest at the mouth of the Columbia River in 1963–1964. However, pelagic organisms further south showed high concentrations during the summer. ^{65}Zn bioaccumulated throughout the food chain, while the concentration of ^{95}Zr–^{95}Nb decreased rapidly after the first two trophic levels [CAR 66].

Zinc-65 and chromium-51[5] (mainly from the nuclear reactors at Hanford, Washington) and zirconium-95–niobium-95, ruthenium-103 and cerium-141 (primarily from Russian fallout) were measured in marine organisms and particulate matter caught off the Oregon coast. The crustacean *Euphausia pacifica* constituted the largest biomass; euphausiids near the mouth of the river contained up to 150 $pCi.g^{-1}$ dw of zinc-65, while even those taken 165 miles off Coos Bay (#250 miles from the mouth of the river) contained about 10 $pCi.g^{-1}$, depending on the season [OST 63].

In a three-tiered food chain with phytoplankton as the first trophic level, euphausiids represented a second trophic level, while lanternfish (whose stomachs contained primarily euphausiids) represented a third trophic level. Zinc-65 was found in the upper trophic levels, but only appeared in samples near the mouth of the Columbia River. Fission products (Zr95–Nb95, Ru103–106 and Ce-141) appeared in both phytoplankton and euphausiids, but were strongly discriminated against at

5 These two radionuclides are activation products.

the third trophic level. Chromium-51, formed at Hanford by neutron activation, was the most prominent gamma emitter in phytoplankton samples collected near the river mouth, but was almost absent at higher trophic levels [OST 63].

The Columbia River flows as a plume off Oregon only in the summer. This does not prevent marine animals on this part of the coast from being contaminated by ^{65}Zn throughout the year [OST 64].

5.2.5.2. *Monitoring of the Sellafield plant*

The first British spent fuel reprocessing plant was at Windscale, renamed Sellafield after a serious fire in 1957 in the fuel of a plutonium reactor at the site. The site is located in Cumbria (UK), and liquid discharges are made into the Irish Sea.

Data on temporal and spatial variations in ^{99}Tc and ^{137}Cs in UK coastal waters are derived from surveys conducted over the period 1993–1998 before and after the release of elevated amounts of ^{99}Tc from the Sellafield nuclear fuel reprocessing plant. Releases of ^{137}Cs were relatively small (about 0.3% of their peak values in the 1970s). The annual transfer factor was estimated for the Sellafield releases in this stream. It was 52 Bq.m^{-3} per PBq with transit times from Sellafield to the Pentland Firth and Lowestoft (on the east coast) of about 9 and 24 months, respectively [MCC 02].

The changes of plutonium isotope ratios in *Fucus vesiculosus* from the west coast of Ireland were used to estimate transport times from the Sellafield nuclear fuel reprocessing plant to various locations on the Irish coast. For conservative radionuclides, transit times of five to six months to the northeast coast of Ireland, one to three years to the south coast of Ireland and three to eight years to the west coast were calculated. In contrast, for plutonium, the Sellafield signal was only observed on the west coast in the late 1980s and early 1990s [MCM 05].

Mitchell et al. [MIT 99] provided chronological data on the concentrations of plutonium, americium and ^{137}Cs in surface sediments and seawater in the Irish Sea.

Radiocesium released into the Irish Sea from the nuclear fuel reprocessing plant at Sellafield, Cumbria, on the English coast, can be traced through the North Sea and further via the Norwegian Coastal Current and the West Spitsbergen Current to the Arctic Ocean, and even into the cold East Greenland Current, which flows southward. The transport time from the North Sea to the East Greenland Current is estimated to be four to five years [DAH 96b].

The concentration of ^{99}Tc in *Fucus vesiculosus* in Scottish waters ranged from 1,000 to 10,000 Bq.kg^{-1} dw [HOL 86a].

The peak in seawater and *Fucus vesiculosus* Tc activity concentrations occurred in 1997, about two years after the peak of the ^{99}Tc releases, with a 29-fold increase over the average concentration for the period 1988–1993. Technetium-99 activity concentrations were measured in fish, lobsters, shrimp, mussels and oysters landed at major fishing ports on the east and northeast coasts of Ireland between 1996 and 1998 [SMI 01].

In Cumbria, Ravenglass mussels (*Mytilus edulis*) had ^{239,240}Pu concentrations ranging from 43 Bq.kg^{-1} dry in muscle tissue to 1658 Bq. kg^{-1} dry in byssal tissue, with a corresponding ^{137}Cs range of 131–1340 Bq. kg^{-1} [MCD 93].

5.2.5.3. *Monitoring of the La Hague plant*

The reprocessing plant at La Hague (now ORANO, formerly COGEMA, then AREVA) is located in the Normandy-Brittany Gulf near the Channel Islands and directly discharges various radionuclides. The monitoring of the site of the French reprocessing plant at La Hague has been the subject of numerous publications, and this operation has been analyzed in a thesis [CAL 86].

A large number of radionuclides (^{144}Ce–Pr, ^{125}Sb, ^{7}Be, ^{103}Ru, ^{106}Ru–Rh, ^{137}Cs, ^{95}Zr–Nb, ^{110m}Ag, ^{134}Cs, ^{54}Mn, ^{65}Zn, ^{60}Co, ^{140}La, ^{131}I, ^{147}Nd and ^{40}K) were detected. ^{106}Ru–Rh and ^{144}Ce–Pr are quantitatively the dominant elements [GER 79].

By North Cotentin, in the early 1980s, radioactivity levels for ^{239,240}Pu, ^{238}Pu and ^{241}Am were about 1–10, 0.5–7 and 1–19 pCi.kg^{-1} fresh weight in biological samples, 24–90, 11–28 and 24–31 pCi. kg^{-1} dry weight in sediment, 1–7, 5–40 and 2–15 fCi.L^{-1} in seawater. For sites far from the La Hague outfall (Seine and Mont Saint Michel bays), the levels of ^{239,240}Pu, ^{238}Pu and ^{241}Am were approximately 0.3–5, 0.1–2 and 0.2–3 pCi. kg^{-1} fresh weight in biological samples, 30–80, 5–26 and 14–40 pCi. kg^{-1} dry weight in sediment, and 1–3, 3–4 and 3–8 fCi.L^{-1} in seawater [GER 84].

In 1983, at Goury, very close to the La Hague outlet, technetium-99 concentrations in seawater varied from 15 to 19 mBq.L^{-1}, those of the algae *Fucus* sp. from 2,600 to 3,100 Bq.kg^{-1} dry (CF 49,000) and those of the mollusk *Patella* sp. averaged 210 Bq.kg^{-1} (CF 3,400) [PAT 86].

At La Hague, cyclic seasonal variations in the levels of radioactivity of three radionuclides (^{127}I, ^{40}K and ^{99}Tc) were followed. The maximum values were in winter and the minimum values in summer (differences between 2- and 3-fold). At Roscoff, a site with little exposure to discharges, the levels varied with the same annual cycle as the natural and stable isotopes. At Herquemoulin, a site close to the point of discharge, the trend was mainly related to the activities released with a delay of about one year. At Wimereux, a remote site more strongly affected by marine currents, the cycle of peaks appeared in 1983 and 1984 but was masked in 1985 and 1986 by the arrival of more strongly contaminated waters because of the larger releases in 1985 (26 TBq), instead of 12 TBq in 1983 and 1984 [PAT 90].

Time series of ^{99}Tc as well as ^{125}Sb concentrations in seawater and algae clearly confirm a considerable decrease in release rates at La Hague during the years 1986–1993. The transit times and transfer factors derived from the series are generally in good agreement with the results of modeling efforts [MAS 95b].

The spatial distribution and temporal evolution of plutonium isotopes (^{238}Pu, ^{239,240}Pu) in indicator species (*Fucus serratus*, *Patella* sp., *Mytilus edulis*) were studied along the Channel coast. The concentrations of the different isotopes were lower than 1 Bq.kg^{-1} dw for the most recent years. A decrease in these concentrations for several years was observed and associated with the decrease in the quantities of plutonium released by the La Hague plant [GER 00].

In the 2000s, iodine-129 releases from the two European spent fuel reprocessing sites, Sellafield and La Hague, had an observable radioecological impact from Cap de La Hague to the Norwegian coast. Thus, the concentrations of ^{129}I in *Fucus* sp. from Cap de La Hague were 60 Bq.kg^{-1} dw, and at Wimereux, 300 km from the outlet, 10 Bq.kg^{-1} dw [MAR 99]. A large retrospective study of the releases from La Hague has been carried out by a pluralistic group of experts who took into account more than 500,000 land and marine samples [GRN 99a]. Currently the traces of radionuclide releases from the La Hague plant are clear for ^{3}H, ^{14}C and ^{129}I. Modeling tools have been used to understand how radioactive releases propagate from the input source [FIÉ 20].

5.2.5.4. *Impacts of the two European reprocessing plants on the Nordic seas*

In the surface waters of the northern European seas, the mean concentrations of ^{137}Cs were estimated for the year 2000 to be 60 ± 50 Bq.m^{-3} for the Irish Sea, 3 ± 1 for the Celtic Sea, 2.1 ± 1.2 for the English Channel, 4 ± 2 for the North Sea, 50 ± 20 for the Baltic Sea, 2 ± 1 for the Norwegian Sea, 2.4 ± 0.5 for the Barents Sea and 4 ± 1Bq. m^{-3} for the Kara Sea. Based on the time series data, the corresponding effective half-lives of ^{137}Cs in surface waters were estimated to be 4 ± 1 years for the Irish Sea, 2 ± 1 for the Celtic Sea, 1.2 ± 0.5 for the English Channel, 3 ± 1 for the North Sea, 13 ± 2 for the Baltic Sea, 4 ± 1 for the Norwegian Sea, 5 ± 1 for the Barents Sea and 13 ± 4 years for the Kara Sea. The mean transit times of the water masses were estimated to be 0.5 ± 0.1 years from the Irish Sea to the North Channel, 2 ± 0.2 years to the north of Scotland, 2.5 ± 0.2 to the east coast of England, 3 ± 0.5 south of the North Sea, 4 ± 0.5 to the central and eastern North Sea, 1.5 ± 0.2 to the Celtic Sea, 270.2 to the English Channel, 5 ± 1 to the Baltic Sea, 4.5 ± 0.5 to the Norwegian Sea, 5 ± 1 to the Barents Sea and 6 ± 1 years to the Kara Sea [POV 03].

Mean concentrations of the anthropogenic radionuclides ^{99}Tc, ^{90}Sr and ^{137}Cs were measured at 0.005, 1.6 and 2.5 Bq.m^{-3} in oceanic surface waters of the Northeast Atlantic, east and northeast of the Azores in 1992, respectively. Comparison with older data indicates an observed half-life for ^{90}Sr and ^{137}Cs in the surface waters of the Northeast Atlantic of 20 years, which corresponds to an average residence time of 80–100 years for the stable elements [DAH 95b].

Iodine ratios ^{129}I/^{127}I were determined using accelerator mass spectrometry on 34 Arctic seaweeds collected between 1930 and 1993. This ratio increased by three orders of magnitude, from an average of <1.10^{-11} (atom/atom) in the pre-nuclear era (before 1945) to nearly 1000.10^{-11} in 1993 for seaweeds collected in the Novaya, Zemlya Archipelago separating the Barents and Kara Seas. The predominant basis for the higher ratios in the Novaya Zemlya kelps appears to be the increase in current sources of ^{129}I from the nuclear fuel reprocessing facilities at Sellafield, UK, and La Hague, France[6]. Relatively high ^{241}Pu/^{239}Pu ratios (compared to bomb fallout observed in boreal latitudes) also corroborated the influence of industrial sources. In contrast, seaweeds collected between 1969 and 1993 in the Bering, Beaufort and East Siberian Seas had much lower ratios ^{129}I/^{127}I, on average 14.04.10^{-11} ± 3.15 SD than those observed in the European Arctic [COO 98]. In 2000 at Utsira, Norway, the concentration of iodine-129 was 0.1 Bq.kg^{-1} dw, 17.5 times the contamination observed in 1982 (0.006 Bq.kg-1 dry) [YIO 02].

Algae *Fucus vesiculosus* collected at three sites along the Norwegian coast since 1980 show the changes of coastal anthropogenic radionuclide concentrations. The authors estimated that an annual discharge of 1 ton of a conservative substance into the English Channel results in concentrations of about 55 ng.m^{-3} off the southwest coast of Norway within a time span of 15–18 months [DAH 97].

5.2.5.5. *The Japanese plant of Rokkasho*

The Rokkasho spent fuel reprocessing plant is located in the Kamikita District of the Aomori Prefecture, in the Tōhoku region, north of the island of Honshū, Japan's main island.

The concentrations of ^{129}I in seawater samples from two sites (off Sekine and Toyama Bay) in the Sea of Japan were determined by accelerator mass spectrometry. The fraction of natural origin and global fallout being 2% and 8.9–13.8%, respectively, the residue of more than 80% of the concentration must have originated mainly from nuclear fuel reprocessing plants [SUZ 08].

---

6 The French operator decided to release all the iodine into the sea because the dilution obtained is not comparable, from the point of view of dosimetric impact on humans, to that of an atmospheric release and the consequences of fallout.

Concentrations of ^{129}I in coastal surface sediment samples from the east coast of Aomori, Japan, between 2 and 65 km from the wastewater outlet of the Rokkasho nuclear fuel reprocessing plant in Japan were measured from 2006 to 2010. No clear effect was observed on the atomic ratios $^{129}I/^{127}I$ in marine sediments. However, the ratio in sediments collected from a fishing harbor at the mouth of a brackish lake increased in 2007, showing that ^{129}I migrated from the lake to the harbor [SAT 19a]. Fishery products (flatfish and brown algae) were collected from localities around Japan's first commercial spent nuclear fuel reprocessing plant at Rokkasho and various radionuclides were measured from 2006 to 2016. Statistically significant increases in the concentration of ^{129}I and the ratio of $^{129}I/^{127}I$ were observed during this period in farmed products and flatfish [SAT 19b].

5.2.6. *Sites affected by nuclear accidents*

Four nuclear accidents have impacted the marine environment.

5.2.6.1. *Palomares and Thule accidents*

A plane crash occurred in 1966 in Palomarès (Almería, Spain) in which four nuclear bombs were involved. As a result of the chemical explosion of two of them, there was a dispersion on the ground of plutonium (Pu) and uranium (U) over an area of about 2.3 km^2. Fifty years of study and monitoring have been carried out by the Spanish CIEMAT (*Centro de Investigaciones Energéticas Medioambientales y Tecnológicas,* Center for Environmental and Technological Energy Research). The main results concluded that most of the Pu contamination in the area affected by the Palomarès accident was in the form of particles and that the bioavailability of Pu in the area was quite low. A three-dimensional map of Pu has been constructed to update the estimate of the plutonium inventory [SAN 19]. Despite the proximity of the marine environment, the latter does not seem to be impacted.

Following the 1968 Thule (Greenland) accident involving four US atomic bombs, most of the plutonium at Thule was in the dissolved phase as Pu (V, VI) (mean 68 ± 6%, n = 6), with little or no apparent distinction between surface and bottom waters. The ratio $^{238}Pu/^{239,240}Pu$ was 0.06 ± 0.02 close to that of atmospheric deposition. Thus, there is little evidence of weapon-grade plutonium in the water column at Thule at this time [MCM 00].

5.2.6.2. *Chernobyl accident*

In general, it can be concluded that the Chernobyl accident has had a measurable impact on the marine environment. The radionuclide levels (mainly cesium-137) reached two to three orders of magnitude higher than the pre-Chernobyl levels [POV 96].

A sample of mussels, *Mytilus galloprovincialis,* taken six days after the accident in Monaco (in the Mediterranean) showed near maximum levels of most radionuclides. Maximum concentrations in the flesh were transiently 480 $Bq.kg^{-1}$ ww for ^{103}Ru. Other radionuclides detected included ^{132}Te, ^{129m}Te, ^{131}I, ^{106}Ru, ^{134}Cs, ^{137}Cs, ^{110m}Ag, ^{140}Ba, ^{125}Sb, ^{95}Nb and ^{141}Ce. Biological half-lives for elimination in this environment were generally about 10 days or more. These results suggest that radionuclides were absorbed very rapidly from fallout particles, rather than from solubilized radionuclides. *Patella lusitanica* specimens had activities about 20–50 times higher than those of the mussels [WHI 88].

The impact of Chernobyl was detectable from mid-May 1986 in the samples of particles collected with the sediment traps set in the Mediterranean. Many radionuclides were detected (^{95}Zr–Nb, ^{103}Ru, ^{106}Ru, ^{134}Cs, ^{137}Cs, ^{144}Ce, ^{239,240}Pu and ^{241}Am) [FOW 97].

Following the Chernobyl accident, the concentrations of ^{137}Cs in 2005 in the deep basins of the North and South Aegean Sea (3.7–5.5 Bq m^{-3}) are significantly higher than in the deep waters of other basins of the Mediterranean Sea [DEL 06].

5.2.6.3. *The Fukushima accident*

Since the accident at the Fukushima Dai-ichi nuclear power plant on March 11, 2011, the number of scientific publications has been significant. For example, in 2016, Kaeryama [KAE 17] counted more than 70 papers published in the previous four years.

Interpretation and analysis of the behavior of radiocesium released into the marine system after the FDNPP accident was greatly facilitated by the availability of long-term data from the systematic monitoring that was conducted in coastal seawaters near nuclear power plants in Japan. Immediately after the accident, the monitoring of seawater, sediments and aquatic organisms in the areas affected by the accident was intensified [IAE 20].

Only the Fukushima accident has had an impact on the bioaccumulation of radionuclides in the marine environment. For example, for cesium-137, the CRs (concentration ratio in fish to the concentration in seawater) were obtained on the Japanese coast from 1984 to 2016. After the Fukushima reactor accident in 2011, the CRs were increasing sharply. The impact of the accident was prolonged in the waters of eastern Japan, and the effective half-lives of apparent CRs (CRa) ranged from 100 to 1,100 days [TAK 19].

In isopod crustaceans, *Ligia* sp. in October 2011 and June 2012, relatively high concentrations (expressed as $Bq.kg^{-1ww}$ ww) of ^{110m}Ag (2.1–127), ^{134}Cs (2.6–61) and ^{137}Cs (3.5–92) were detected in specimens from eastern Honshu regions. Significantly lower concentrations of ^{137}Cs (0.7–1.6) were detected in specimens from western and northern Honshu. The decay-corrected concentration of ^{137}Cs was significantly inversely correlated with distance from the Fukushima Dai-ichi nuclear power plant [QIU 17].

During an accident, typically a non-equilibrium situation, numerical simulations are difficult. According to Vives i Battle et al. [VIV 18], biokinetic models can better represent the transfer of radionuclides within biota, bringing more realism to the predictions, especially when combining physical, chemical and biological interactions that occur in an open and dynamic environment such as the ocean.

For its part, the IAEA [IAE 20] developed a compartmental (boxed) model including the deposition of ^{137}Cs in seawater and bottom sediments. This organization found that the pre-accident geometric mean of K_d values around Japan increased from 600 in 1984 to 1,000 in 2010. The CR values in the affected areas were disturbed due to the imbalance between ^{137}Cs in marine organisms and the water surrounding them. Furthermore, the CRs became apparent concentration ratios (CR_a) that increased sharply after the accident in the affected areas. Post-accident, the effective half-lives of ^{137}Cs (T_a) ranged from 200 to 1,200 days. The imbalance persisted for the coastal areas of eastern Japan even five years after the accident. Marine K_d values in the affected areas increased sharply to a maximum and thereafter did not decrease significantly, but remained relatively constant, indicating the realization of (near) equilibrium conditions.

The mobility of plutonium released during the Fukushima accident was low. This radionuclide could not be detected in marine sediments collected from 1.5 to 57 km from the Fukushima power plant. However, the atomic ratios ^{240}Pu/^{239}Pu in the sediments (0.21–0.28) were significantly lower than those at the power plant (0.330 ± 0.032) and higher than that of global fallout (0.182 ± 0.005) [WEN 17].

Madigan et al. [MAD 12] provided unequivocal evidence that Pacific bluefin tuna, *Thunnus orientalis*, carried Fukushima-derived radionuclides throughout the North Pacific Ocean. In 15 tuna caught in California in August 2011, levels of two cesium radionuclides were significantly higher in these migratory fish compared to their resident congeners (Table 5.7). These results indicate that Pacific bluefin tuna can rapidly transport radionuclides from a point source in Japan to distant ecoregions and demonstrate the significance of migratory animals as vectors for radionuclide transport.

Radionuclide	^{134}Cs	^{137}Cs	$^{134+137}$Cs	^{134}Cs/^{137}Cs
PBFT 2008	0	1.5	1.11	0
PBFT 2011	4.0	6.3	10.3	0.62
YFT 2011	0	1.2	1.10	0

Table 5.7. *Average cesium concentrations (Bq.kg^{-1}) of migrant (PBFT, Pacific bluefin tuna) and resident (YFT, yellowfin tuna) tuna caught before (2008) and after (2011) the Fukushima accident (from [MAD 12])*

Vertical transport of radionuclides by pelagic and benthic fish was evident after the Fukushima accident, as well as trapping by phytoplankton and their migration for fish [BUE 17].

Johansen et al. [JOH 15] described the distribution of ^{134,137}Cs and ^{90}Sr in the tissues of a fish two years after the accident. 73% of cesium-134 and 137 was in the muscle mass, while for strontium-90, 38% was in bone tissue, 37% in the digestive tract and 18% in the skin and scales.

The impact of the Fukushima accident has been felt far from the site. For example, the concentration of ^{137}Cs in fish in northeastern Finland averages 14 Bq.kg^{-1} ww (5–46) and shows an increase of 1.6% following the Fukushima accident [AMA 16].

In Korea, while in late March and early April 2011, the first contaminated aerosols were observed for ^{131}I, ^{134}Cs and ^{137}Cs, on the other hand, the significant increase in concentrations of these three radionuclides was not observed in seawater and marine biota samples [KIM 12].

^{137}Cs can enter Indonesian waters from the Pacific Ocean through Indonesian through flow (ITF). The average concentration of ^{40}K, ^{226}Ra, ^{212}Pb, ^{214}Bi, ^{228}Ac, ^{214}Pb and ^{137}Cs in the sediments of Bali Strait waters are, respectively, expressed in Bq.kg^{-1} 165.16, 25.11, 8.36, 7.51, 7.05, 6.68 and 0.15. In Lombok waters, the concentrations of radionuclides ^{40}K, ^{226}Ra, ^{212}Pb, ^{214}Bi, ^{228}Ac, ^{214}Pb and ^{137}Cs in sediments were expressed in Bq.kg^{-1} 172.00, 25.88, 10.10, 6.50, 6.39, 5.70 and 0.162, respectively. The concentrations of ^{137}Cs in seawater from Bali and Lombok were 0.113 and 0.644 Bq.m^{-3}, respectively [PUT 17]. The two radionuclides of cesium-134 and 137 were also found on the coast of Kalimantan in Borneo from the western part of Aceh to Papua in the east [PRI 20].

5.2.7. *Sites affected by submerged radioactive waste*

Several nuclear reactors have been submerged, sometimes with their spent nuclear fuel still inside. Among the many radionuclides present, tritium is extremely mobile. A tritium anomaly has been recognized in the surface layer of the Kara Sea in the Novaya Zemlya basin (228 ± 9.1 TU[7], while the average in the Arctic Ocean is 10–15). This anomaly may be caused by spent nuclear fuel from the reactors of the nuclear submarine K-140 or the icebreaker *Lenin* which is located 100 miles to the south [SOY 11].

Releases to the environment from the Russian civilian nuclear breaker base (RTP "ATOMFLOT"), located in Kola Bay in northwest Russia, were not negligible. Mean (dw) activity concentrations for surface sediments of 63 5.8 and 0.45 $Bq.kg^{-1}$ for ^{137}Cs, ^{90}Sr and $^{239\ 240}Pu$, respectively, were measured. Activity levels in filtered seawater were in the range of 3–6.9, 2.0–11.2 and 16–40 $mBq.m^{-3}$ for ^{137}Cs, ^{90}Sr and $^{239\ 240}Pu$, respectively [BRO 02].

Measurements of sediments taken at the Novaya Zemlya disposal sites showed a deviation from the global deposition ratios $^{240}Pu/^{239}Pu$ [OUG 04].

The study sites are (a) former Soviet dumping sites in the Kara Sea, (b) former European dumping sites in the Northeast Atlantic, (c) the location in the Norwegian Sea where the Soviet nuclear submarine Komsomolets accidentally sank in April 1989 and (d) former Soviet and Russian dumpsites in the Far Eastern seas. Some dumping sites (Stepovovo and Tsivolki Bays, Northeast Atlantic) had higher concentrations of ^{137}Cs, ^{60}Co or $^{239,240}Pu$ in bottom waters and sediments, indicating a possible contribution from the disposed waste [BAX 95].

Increased levels of ^{137}Cs, ^{60}Co, ^{90}Sr and $^{239.240}Pu$ in sediments in the vicinity of submerged containers in the Abrosimov and Stepovogo fjords demonstrated leaching from submerged materials. The radionuclide ^{137}Cs was found to be strongly associated with the sediments, while ^{90}Sr was more mobile [SAL 97a].

The increase in the level of radionuclides observed in the sediments in the vicinity of the Stepovogo Fjord submarine in 1993 could not be confirmed in 1994. Otherwise, traces of ^{60}Co in the sediments were observed in the immediate vicinity of all the objects located [SAL 97a].

Surface sediments (1–2 cm) in Stepovoy and Abrosimov Bays (Nova Zemla) were contaminated with ^{137}Cs and ^{60}Co. Thus, in Stepovoy, the values for the period 1992–1994 for cesium and cobalt were, respectively, 260–5,450 and

7 The tritium unit (TU) *is* defined as a dilution of one tritium atom per 10^{18} hydrogen atoms.

31–3,150 Bq.kg^{-1} and for the period 2002–2004 were 20–1,800 and 1–26 Bq.kg^{-1} (Stepanets et al. 2007) [STE 07].

Surface, intermediate and near-bottom water samples were collected from two of the Northeast Atlantic radioactive waste disposal sites (47° 14'N, 05° 33'W and 45° 27'N, 06° 16'W) in the course of a sampling campaign in June–July 1998. Analysis of ^{239,240}Pu concentrations shows no significant differences [SMI 11b].

In order to assess the level of radioactive contamination in the Kara Sea and to evaluate the short- and long-term consequences of the dumped radioactive wastes, joint Russian–Norwegian expeditions have been carried out each year since 1992. The 1993 expeditions to the dumping sites confirmed the local contamination of the Stepovogo and Tsivolki fjords due to leaks of dumped radioactive waste [STR 95].

In Andreeva Bay, on the Murmansk Peninsula, there is a radioactive waste storage facility, which is considered to be the main source of radioactive contamination of the Barents Sea coastline and thus the cause of the strongest impact on the radioecological status of the marine environment [III 19].

5.2.8. *Sites with little impact from radioactive pollution*

Anthropogenic ^{129}I emissions from nuclear activities are now detected in the surface waters of the Antarctic seas. This has been verified in surface seawater samples from the Drake Passage, Bellingshausen, Amundsen and Ross Seas [XIN 17].

Some sites with little impact from radioactive contamination, such as Brazilian coastal environments, have been radioecologically monitored to quantify groundwater inflow (GWI) using radioactive tracers [SAN 08].

The Turkish marine environment, with little radioactive contamination, was the subject of radioecological monitoring from 2001 to 2009 [TOP 10].

In Kuwait in 2012–2013, the seawater concentrations of ^{3}H, ^{210}Po, ^{210}Pb, ^{137}Cs and ^{90}Sr ranged from 130 to 146, 0.48 to 0.68, 0.75 to 0.89, 1.25 to 1.38 and 0.57 to 0.78 mBq.L^{-1}, respectively, The seawater concentration of ^{40}K ranged from 8.9 to 9.3 Bq L^{-1}. The sediment concentration of ^{40}K, ^{210}Pb in total, ^{137}Cs, ^{90}Sr, ^{226}Ra, ^{228}Ra, ^{238}U, ^{235} U,^{234}U, ^{239,240}Pu and ^{238}Pu ranged from 353 to 445, 23.6 to 44.3, 1.0 to 3.1, 4.8 to 5.29, 17.3 to 20.5, 15 to 16.4, 28.7 to 31.4, 1.26 to 1.30, 29.7 to 30.0, 0.045 to 0.21 and 0.028 to 0.03 Bq.kg.$^{-1}$dw, respectively [UDD 18].

The concentrations of ^{40}K, ^{210}Pb, ^{137}Cs, ^{90}Sr, ^{226}Ra, ^{228}Ra, ^{238}U, ^{235}U, ^{234}U, ^{239,240}Pu and ^{238}Pu in marine sediments from the Gulf of Kuwait were 386–489,

32.3–48.8, 1.5–2.9, 4.53–5.42, 18.3–23.1, 18.8–23.0, 22.3–30.5, 0.99–1.33, 25.6–34.8, 0.30–0.93 and 0.0008–0.00018 $Bq.kg^{-1}$ dw, respectively [UDD 18].

The concentration in whole fish of ^{40}K, ^{226}Ra, ^{224}Ra, ^{228}Ra, ^{137}Cs, ^{210}Po and ^{90}Sr ranged from 230 to 447, 0.7 to 7.3, <0.5 to 6.6, <0.5 to 15.80, <0.17, 0.88 to 4.26 and 1.86 to 5.34 $Bq.kg^{-1}$dw, respectively. The highest ^{210}Po concentration was found in the bivalve mollusk *Marica marmorata* (193.5–215.6 $Bq.kg^{-1}$dw) [UDD 15].

In Kuwait, in 2014–2015, a large number of samples of various marine organisms covering several trophic levels, from microalgae to sharks, were analyzed. The highest polonium-210 concentrations in muscle tissues were observed in the following order: tunicate> bryozoan> mollusk> crustacean> algae> fish. Thus, the ^{210}Po concentration in soft tissues of tunicate and bryozoan samples was 872–1012 and 402–527 $Bq. kg^{-1}$ww[8], respectively, much higher than that of fish muscles (0.04–44.9 $Bq.kg^{-1}$ww). Among all biota analyzed, the highest overall concentration of ^{210}Po was noted in yellowfin tuna (*Thunnus albacores*), indicating potential biomagnification of ^{210}Po in this top predator species. The highest concentrations of ^{210}Po were found in yellowfin tuna tissues, 37.3–44.9, 451–548 and 1511–1693 $Bq.kg^{-1}$ ww in muscle, digestive tract and liver, respectively [UDD 17].

In Malaysian mangroves, the surface concentrations of sediment cores were for ^{210}Po and ^{137}Cs, respectively, from 40 to 70 and from 1 to 1.5 $Bq.kg^{-1}$ and for ^{228}Th from 95 to 110 $Bq.kg^{-1}$ [ALO 04].

Tritium concentrations in Southern Ocean waters were about 100 mTU t in deep waters with varying concentrations that ranged from the detection limit of about 5 mTU to about 80 mTU in bottom waters near areas of new Antarctic bottom water formation [SCH 06].

In the Adriatic, radioactivity measurements began in 1963 as part of an extensive program of monitoring radioactivity in the Croatian environment. ^{90}Sr data obtained from long-term monitoring were used to estimate the upper limit of the turnover time of the Adriatic Sea, which was found to be 3.4 ± 0.4 years [FRA 06].

5.3. Environmental radiological monitoring networks

5.3.1. *General information*

The protection of the environment and of humans are the reason for environmental monitoring. As early as the 19th century, it was realized that the

8 Wet weight.

quality of the environment had repercussions on human health in the short-term. Any damage to the quality of the environment is likely to have deleterious effects on humans in the short-term, as well as in the medium- and long-terms. Moreover, a quality environment is a reserve of important resources for humans (food, medicine, well-being, etc.).

For a long time, the central tenet of the nuclear industry's monitoring philosophy was that if environmental levels of radioactivity were kept low enough to protect humans, then other species would also be protected[9]. Already, Thompson in 1988 [THO 88] showed that, for species such as large marine predators, there were no supporting data.

In the case of radionuclides, the quality of the environment can be threatened by the introduction of so-called artificial or anthropogenic isotopes that are not naturally present or by an increase in the concentration of natural radionuclides that are normally present but whose quantity is increased by human activities[10].

To maintain biodiversity and human health at an optimal level, it is necessary to ensure the quality of all ecosystems. Since it is impossible to identify and quantify all radionuclides, a selection of those to be retained must be made. The criteria for selecting the radionuclides of greatest concern are their persistence in the environment, their high capacity to bioaccumulate and their significant radiotoxicity.

To monitor and evaluate the quality of the environment, there are four main methods. Firstly, the detection and quantification of radionuclides in physical media (water and sediments). Then, the detection and quantification of radionuclides in organisms (bioaccumulators), followed by the determination of the floristic and faunistic composition of ecosystems (quality indices or bioindicators). Finally, the identification and quantification of changes in biological parameters (biomarkers). In marine radioecology, only the first two methods are commonly used.

---

9 This was based on the positions of the ICRP which wrote in its publication ICPR 27 (1977), paragraph 14 "*The Commission therefore believes that if man is adequately protected then other living things are also likely to be adequately protected.*" This ICRP 26 inspired the Euratom Directives of July 1980 and September 1984, transcribed in France in a decree on October 2, 1986. The ICRP 60 (1991) has a more subtle approach by saying that humans are the most fragile living species, so by protecting them, we protect all other species: "*The Commission believes that the standards of environmental control needed to protect man to the degree currently thought desirable will ensure that other species are not put at risk.*"

10 This is particularly true of uranium mining, which, by fracturing the rock, releases radon and increases the leachable surface by releasing radium-226 (the father of 13 radioactive sons).

The last three methodologies are based on living things at various levels of biological organization. Studies at the ecosystem level have high ecological relevance, but respond after a lag time of several years. Responses at the sub-individual level (biomarkers) are very rapid and are therefore an effective warning system but with low ecological relevance. Responses at the organism level (bioindicators and bioaccumulators) are moderate both temporally and in terms of ecological relevance [AMI 17a].

The objectives, purposes, design and development of marine monitoring programs have been reviewed by Preston and Wood [PRE 71]. They pointed out that monitoring programs can be carried out prior to start-up of the nuclear facility or during normal operation. In the nuclear field, most monitoring programs are imposed by the legislation of the state where the nuclear installation is located. It is then a series of checks to verify the good condition of the installation and its proper functioning. The majority of the programs simply quantify the radioactivity in some components within the environment of the installation such as, for the marine environment, seawater, sediments and some organisms considered as indicator species of radioactive contamination such as algae, mollusks or fish. There are also sometimes obligations to carry out an annual ecological report of the nuclear site[11].

The information from these monitoring programs can then be used for other purposes, for example, to confirm data on carbon-14 [LEG 09, GUI 12], or to evaluate the radioactive risk for children as at La Hague and to validate numerical simulation models [GRN 99c].

5.3.2. *Environmental monitoring using bioaccumulators*

Some living organisms have the ability to accumulate, usually with a lag time of several days, the radionuclides to which they are exposed. These organisms are therefore good indicators of the quality of their environment. In addition, the question arises as to the possible radiotoxicity of this bioaccumulation for these organisms as well as for the consumer.

Living organisms can be indicators of exposure and these species are short-, medium- or long-term integrators. Moreover, the analytical problem is reasonably difficult. The bioaccumulated fraction of the radionuclide is that which is most

11 Since BNIs make discharges into the environment, when applying for authorization to discharge effluents and withdraw water from a BNI, the facility operator must, among other things, provide a study of the radiological impact of its radioactive effluent discharges on the environment and the health of populations. The first request thus leads to a "zero" point of the environment.

likely to have an adverse effect on the organism or its consumer. The bioaccumulation mechanisms are known for many radionuclides. The routes of entry and elimination are varied as well as known.

The choice of bioaccumulators must meet certain criteria. The species chosen must have the capacity to bioaccumulate, a sufficient lifespan to allow the integration of the radionuclide, be abundant and of a reasonable size for easy sampling, and have as wide a distribution as possible. Bioaccumulative species will make it possible to compare the degree of contamination of different sites and to highlight long-term trends, thus making it possible to provide warning about the possible worsening of pollution and to evaluate the effect of regulations aimed at improving the quality of the environment. In marine waters, the organisms selected are benthic algae (notably *Fucus* sp., [ILU 06]) and especially bivalve mollusks (notably mussels, [BOU 06]).

However, there are difficulties in interpreting the results obtained. Indeed, two orders of factors intervene in the determination of the concentrations resulting from bioaccumulation: the physical and biological factors related to the organism in interaction with the environment, and the factors related to the experimental protocols. It is necessary to take them into account in the estimation of pollutant concentrations.

Many biological factors are involved in bioaccumulation phenomena: species, gender, tissues or parts of the organism, sexual cycle, age, sex, life stage, in connection with the characteristics of the environment: radionuclides present in the environment and their physico-chemical form, radionuclide resuspension phenomena, temperature and seasonal variations. Thus, for example, under certain conditions, the temperature of the environment is strongly suspected of promoting the transfer of radionuclides into living organisms.

The difficulty of interpretation is particularly true for cumulative radionuclides, i.e. those that accumulate throughout an individual's life (such as Cd or the Hg/Se couple in dolphins), antagonistic pollutants such as chemical and biochemical analogs (such as potassium and cesium), and certain chemically distant elements (such as cadmium competitively inhibited by calcium and non-competitively by magnesium). Another difficulty concerns possible interactions between radionuclides or between a radionuclide and another substance present in the environment.

Similarly, the difficulty of interpretation is high for radioactive isotopes of essential metals that interfere with metabolism. For example, copper in various crustaceans and mollusks is bound to the respiratory pigment (hemocyanin) and replaces iron in species with hemoglobin; this results in very high concentrations of this element. Similarly, zinc is the activator of many metalloenzymes and in oysters the detoxification mechanisms result in the storage of the metal in cellular organelles without elimination over time and, therefore, in cumulative bioaccumulation throughout the individual's life.

In addition, the sampling and sample preparation protocols, as well as the quality of the measuring equipment (technical characteristics, calibration, maintenance) are also likely to induce variations in the measured concentrations. Hence, the need to take precautions during sampling to control as many factors as possible and to perform international intercalibration exercises. Radioactive measurements have been the subject of a very comprehensive review [VAL 00].

5.3.2.1. *Improvement of controls with transplantation from control populations*

Increasingly, monitoring networks are using the transplantation methodology. It consists of selecting a population of a bioaccumulative species as homogeneous as possible from a specific site and transplanting these individuals in various sites to quantify their quality. The species used with this methodology are quite numerous (mosses, mollusks, fishes).

Active biomonitoring has several advantages over conventional monitoring [AND 04a]. The transplanted animals have a known history, the exposure time is controlled, the choice of plant location is independent of bathymetry and the positioning of individuals in the water column is controlled. Moreover, the measurements are optimized because the samples are more homogeneous thanks to a selection of the individuals used in the experiments (parental origin, size, age, originating from a healthy site). Various constraints exist such as heavy logistics, interpretation of the data depending on the physico-chemical and trophic variability of the stabling sites, and the need to determine complementary biometric parameters. The abundant literature in this field allows for transplantation protocols that take into account the time required for equilibrium with the transplantation environment, the initial stress and the trophic factors of the implantation site.

5.3.2.2. *Data management*

With the increasing use of databases for environmental radiological monitoring, there is a demand for integration and availability of this type of information. The data management tool must allow for the greater ease of research on a wide variety of topics. Efforts are underway to consolidate marine radionuclide data from monitoring programs, published literature and special studies into a large information database (GLOMARD) managed by the IAEA [IAE 00]. In Brazil, ENVIRAD is a new database proposing to integrate a set of previous databases [GON 20].

Friedlander et al. [FRI 05] provided a synthesis of the various national, regional and international monitoring programs designed to periodically assess marine waters, sediments and biota (fish, shellfish, algae, etc.) in the oceans and several northern hemisphere seas and bays. They found a reduction in cesium-137 concentrations in the marine environment over the past several decades, although the mobilization of sedimentary deposits continues to contribute somewhat to some environmental burdens, and concentrations of technetium-99 vary over time.

5.4. International and regional marine environmental monitoring networks

Several international and regional marine environmental monitoring networks exist, such as those of OSPAR, or have existed, such as those led by the IAEA.

5.4.1. *The work of the IAEA in the field of the marine environment*

The IAEA does not coordinate any continuous monitoring program of the radiological quality of the marine environment. On the other hand, this international organization, at the request of a state or on its own initiative, organizes studies of the radioactive contamination of various sites and the resulting radiation doses for humans, rarely for other organisms. Among the studies carried out by the IAEA in the marine environment are the study of the Bikini site [IAE 98a] and the Mururoa and Fangataufa site [IAE 98b] or the Kara Sea [IAE 98c], the site of much radioactive waste.

The International Arctic Seas Assessment Project (IASAP) was established in 1993 by the IAEA to address concerns about the potential health and environmental impacts of radioactive waste discharged into the shallow waters of the Arctic seas [SJO 95, SJO 98, SJO 99].

5.4.2. *OSPAR radiological monitoring*

The Convention for the Protection of the Marine Environment of the North-East Atlantic (Oslo and Paris Convention or OSPAR Convention) was opened for signature at the Ministerial Meeting of the Oslo and Paris Commissions on September 22, 1992 in Paris. It was adopted with a Final Declaration and an Action Plan. It entered into force on March 25, 1998 and replaced the Oslo and Paris Conventions. It was installed by France in 1992 [GUI 13].

The International Convention has 16 members (Belgium, Denmark, Finland, France, Germany, Iceland, Ireland, Luxembourg, Netherlands, Norway, Portugal, Spain, Sweden, Switzerland, United Kingdom and the European Union).

The OSPAR Convention has several major principles such as the application of the polluter pays principle and the adoption of the best available techniques and the best environmental technique (limitation of discharges and their health and environmental impacts). For radioactive substances, the progressive and substantial reduction of marine discharges to achieve concentrations in the environment close to ambient values for natural substances and close to zero for man-made substances.

The OSPAR Convention is very involved in the protection of the waters of the Northeast Atlantic. These have been divided into five regions, which are in turn subdivided into several monitoring areas. The regions are: I Arctic waters, II the wider North Sea, III the Celtic Seas, IV the Bay of Biscay and the Iberian coast and V the open Atlantic Ocean [OSP 18].

A number of indicator radionuclides have been selected by OSPAR. For each sector, indicator radionuclides for discharges are selected. These indicators allow the assessment of progress towards the objectives of the OSPAR "Radioactive Substances Strategy". The selection has been made by establishing the composition of discharges from the various sectors and the magnitude of the radiation dose from the radionuclides concerned. OSPAR has collected data on annual discharges of indicator radionuclides from the nuclear sector since 1990 and from non-nuclear sources since 2005.

The main radionuclides monitored in the OSPAR program are selected to trace the impact of three sectors, nuclear, non-nuclear and medical. The radionuclides used as tracers for nuclear industries (^{99}Tc, ^{137}Cs, ^{239}Pu, ^{240}Pu, ^{3}H), the radionuclides released by human offshore oil and gas activities (^{210}Pb, ^{226}Ra, ^{228}Ra, ^{228}Th) and those with medical uses (^{99}Tc, ^{131}I). For the nuclear sector, it was agreed to consider the period between 1995 and 2001 as a reference period against which to assess progress towards the objectives of the "Radioactive Substances Strategy". The OSPAR Strategy on Radioactive Substances aims to achieve "concentrations in the environment close to background values for naturally occurring radioactive substances" [HOS 12].

The Radioactive Substances Committee (RSC) is chaired by Justin Gwynn (Norway). Each state is required to provide a report on the implementation of PARCOM Recommendation 91/4 on radioactive discharges to the OSPAR Commission on the application of the best available technology in nuclear power plants. This recommendation has been revised with the OSPAR Recommendation 2018/01 on radioactive discharges. This has led, for example, to a 12-fold reduction in discharges from La Hague (Orano) between 1991 and 2018 for alpha emitters and 12.7-fold for activation and fission products (excluding tritium)[12].

For France, the controls are carried out in 12 plants distributed among three regions 1 (Roscoff to Arcachon), 2 (Carteret to Barfleur) and 3 (Honfleur to Wimereux) in three compartments (seawater, mollusks and algae) with a variable periodicity (annual, semi-annual or quarterly) [OSP 19].

For the marine environment, reprocessing plants and nuclear fuel fabrication and enrichment plants are responsible for 98% of radionuclide releases from the nuclear sector. Inputs of radionuclides to the sea are related to liquid discharges and, to a lesser extent, to solid waste and atmospheric emissions [AMI 13].

For its fourth assessment period (2007–2013), OSPAR noted a 2.5-fold reduction in direct discharges to the marine environment for alpha emitters, a 12-fold reduction for beta emitters (excluding tritium) and a 38-fold reduction for ^{99}Tc emissions compared to its baseline period (1995–2001) [OSP 17].

OSPAR's annual collection of data on discharges from the non-nuclear sector has only taken place since 2006 and has only been published since 2009. For 2018, the oil sector discharges 0.80, 0.60 and 0.10 TBq of ^{226}Ra, ^{228}Ra and ^{210}Pb, respectively, i.e. a total of 6.7 TBq for alpha emitters and 4.8 TBq for beta emitters. Although incomplete, it is possible to evaluate the relative contribution of the medical sub-sector for which the total reported iodine-131 releases for 2018 were 14.6 TBq. The non-nuclear sector released in 2018 6.5 TBq of alpha emitters, 19.5 TBq of beta emitters and 0.4 TBq of tritium [OSP 20].

5.4.3. *The northern sea networks*

Three networks (AMAPJRNC and HELCOM) have been implemented to monitor the radioactive contamination of the Nordic seas.

The Arctic Monitoring & Assessment Programme (AMAP) is one of six working groups of the Arctic Council: (intergovernmental forum) established in 1991 comprising eight countries ranging from the High Arctic to the sub-Arctic regions of Canada, the Kingdom of Denmark (Greenland and Faroe Islands), Finland, Iceland, Norway, the Russian Federation, Sweden and the United States, including associated

12 This has been done by means of evaporators that treat the radioactive effluents and allow vitrification of the dry residue. This method, which has replaced the treatment by co-precipitation (which produced radioactive sludge), produces standard medium-level waste packages in containers identical to those for high-level waste.

marine areas. Some of the activities of this program concerns radioactivity in the Arctic [AMA 16].

The collaboration between the Russians and the Norwegians (JRNC, Joint Russian–Norwegian Collaboration) has carried out several maritime expeditions to monitor the impact of Soviet radioactive waste on the marine environment [NIK 97, SAL 97a, GWY 16].

Following the Helsinki Convention of March 22, 1974 and its various extensions, the Helsinki Commission (HELCOM) was established for the protection of the marine environment of the Baltic Sea Area. Ten States are involved (Germany, Denmark, Estonia, Finland, Latvia, Lithuania, Poland, Russia, Sweden and the European Union). One part of the research focuses on the import of radionuclides into the Baltic Sea. The assessment of the quality of water and biota (herring, flounder) in the Baltic Sea is not good for several regions with regard to the concentration of cesium-137. The threshold values retained by HELCOM for ^{137}Cs are seawater and 2.5 and 2.9 $Bq.kg^{-1}$ ww for herring and flounder, respectively [HEL 18a, HEL 18b, HEL 18c, NIE 18b].

5.5. Radiological monitoring in France

Because of their particular nature, radioactive pollutants are subject to specific monitoring. In France, the control of nuclear safety and radiation protection is essentially the responsibility of three actors: parliament, the government and the Nuclear Safety Authority (*Autorité de sûreté nucléaire*, ASN).

The ASN is responsible for monitoring nuclear safety and radiation protection in France: it carries out this monitoring on behalf of the State to protect workers, patients, the public and the environment from risks related to nuclear activities. The ASN monitors basic nuclear installations (BNIs), from design to dismantling, pressure equipment specially designed for these facilities, radioactive waste management and the transport of radioactive substances. The ASN also monitors all industrial and research installations as well as hospitals where ionizing radiation is used.

The field of nuclear safety and radiation protection is constantly evolving, with new generations of nuclear installations, new medical applications of ionizing radiation, new uses of radioactivity, etc. In this changing environment, the ASN is careful to prioritize its actions and allocate its resources in proportion to the issues at stake.

The ASN relies on approved organizations to carry out systematic and standardized inspections when, in particular, the size of the facilities concerned, the standardization of equipment or the existence of standards to which they must conform allow it.

5.5.1. *Principles of control*

Nuclear activities are governed by a certain number of principles enshrined in legislation and regulations and applicable to a specific field of activity. These are the principles of responsibility, polluter pays, optimization, justification and limitation.

The principle of responsibility states that the responsibility for risky activities lies primarily with those who undertake or carry them out. It is the responsibility of operators for the safety of basic nuclear installations; the responsibility of users for the radiation protection of the public; the responsibility of suppliers for the return of radioactive sources; the responsibility of employers for the radiation protection of workers; the responsibility of the doctor performing the procedure for the radiation protection of patients; the responsibility of polluters for environmental damage and the responsibility of producers for waste disposal.

The Environmental Code introduces the polluter-pays principle, according to which the costs resulting from pollution prevention, control and abatement measures must be borne by the polluter. The polluter is therefore responsible for the environmental damage caused by their activity.

The French Public Health Code sets out three principles for radiation protection. The principle of optimization states that the exposure of persons to ionizing radiation resulting from nuclear activities must be kept as low as reasonably achievable, taking into account the state of the art, economic and social factors, and, where appropriate, the medical objective pursued. The principle of justification enshrined in the Public Health Code states that a nuclear activity may only be undertaken if it is justified by the benefits it provides, particularly in terms of health, social, economic or scientific benefits, in relation to the risks inherent in exposure to ionizing radiation to which it may subject individuals. The principle of limitation provides that the exposure of a person to ionizing radiation resulting from a nuclear activity may not exceed the sum of the doses received, except when the person is exposed for medical or biomedical research purposes.

5.5.2. *The IRSN monitoring network*

Radiological monitoring of the French environment is one of IRSN's public service missions. Implemented by a combination of telemetry and laboratory analyses of environmental samples, it is carried out throughout the country using plants located near or away from facilities likely to release radionuclides. Environmental samples are analyzed in IRSN accredited laboratories [IRS 18].

Telemetry networks provide real-time environmental monitoring using autonomous devices that continuously measure the radioactivity of the environment in which they are installed. As soon as an increase in radioactivity is detected, an alarm is sent to IRSN and an investigation is carried out. The Téléray network monitors the air by continuously measuring the ambient γ dose rate in the air. Téléray is made up of nearly 400 probes distributed in metropolitan France and in the French overseas departments and territories. They transmit data in real-time to the IRSN remote monitoring room. The Hydrotéléray network is used to monitor the water in rivers that receive effluents from nuclear power plants. The data acquired by these networks is automatically transmitted to a centralized IT infrastructure, the remote monitoring room at Le Vésinet, which is dedicated to the storage, exchange and consultation of data as well as to the management of networks during radiological emergencies. Each year, several hundred thousand measurements are acquired and archived.

The sampling networks complement the remote monitoring and make it possible to follow the changes of the levels of radioactivity in the environments that have an impact on the population, through samples from the air, water, soil, flora and fauna. Each year, approximately 30,000 samples are analyzed.

In addition to these networks, the IRSN has 40 Opera-Air plants (*Observatoire permanent de la radioactivité,* Permanent Radioactivity Observatory) that enable it to carry out measurements of radioactivity down to trace levels in samples collected in the environment. Its objective is to understand the processes governing the distribution in time and space of radionuclides of natural and artificial origin in the various environmental media, including the food chain. The knowledge of radioactivity levels is necessary to model the transfer processes of radionuclides in the environment. This network also enables the constitution of measurement databases that are essential for scientific studies on the variability of the observed levels.

A large number of other nuclear actors are statutorily required to have radiological monitoring networks for their immediate environment. These include operators (Andra, Orano, and EDF), research centers and the French Navy. Thus, EDF [EDF 20] is developing monitoring programs for all its nuclear power plants[13].

5.5.3. *RNM (Réseau national de mesures de la radioactivité de l'environnement)*

The *Réseau national de mesures* (RNM, French National Network of Environmental Radioactivity Measurements) was developed under the aegis of the *Autorité de sûreté nucléaire*, (ASN, French Nuclear Safety Authority), in coordination with the *Institut de radioprotection et de sûreté nucléaire* (IRSN, French Institute for Radiation Protection and Nuclear Safety). The other actors in the network include representatives of the main ministries concerned, health agencies, the nuclear industry, and environmental and consumer protection associations.

The IRSN is responsible for the technical management of the *Réseau national de mesures de la radioactivité de l'environnement* and is responsible for the secretariat of the Steering Committee, the centralization, processing and archiving of analysis results, and the provision and dissemination of these data.

The mission of the *Réseau national de mesures de la radioactivité de l'environnement* is to contribute to the estimation of doses resulting from ionizing radiation to which the population is exposed. The network gathers and makes available to the public the results of environmental radioactivity measurements and summary documents on the radiological situation of the territory.

The data on environmental radioactivity include the results of measurements carried out: 1) within the framework of legislative or regulatory provisions aimed at assessing the doses to which the population is subjected, particularly those resulting from nuclear activities; these measurements must be carried out by approved laboratories or by the IRSN; 2) at the request of local authorities, state services and its public establishments, if these measurements have been carried out by an approved laboratory or by the IRSN; and 3) at the request of any public, private or associative body, if the measurements have been entrusted to an approved laboratory or to the IRSN and the body holding the results asks to publish them on the National Network.

The *Réseau national de mesures de la radioactivité de l'environnement* includes an accreditation commission. The mission of this body is to ensure that the measurement laboratories have the organizational and technical skills to provide the

13 It is a decree of December 22, 2016 that homologates a decision of the ASN on the controls to be carried out in the environment of all BNIs in France.

network with quality measurement results. Indeed, the measurement laboratories are the privileged partners of this network. Since 2007, in application of the "TSN" law, they must be approved by the ASN so that their measurements of radioactivity in the environment can be published on the portal of the *Réseau national*. The commission is responsible for proposing approvals to the ASN. It makes its decision on the basis of an application file submitted by the applicant laboratory and its results in the interlaboratory tests organized by IRSN.

5.6. Conclusion

Quantifying environmental quality using measurements in the physical environment (air, water, soil, sediment) only tells us about the exposure of living organisms to pollutants in their environment, not about the capacity of these pollutants to penetrate organisms and contaminate them. Similarly, the quantification of environmental quality using measurements in bioaccumulators reveals the fraction of contaminants that are available for bioaccumulation in organisms, but does not inform us about the possible damage that this contamination can cause. The use of biotic indices has a strong ecotoxicological relevance since we can observe imbalances in communities and ecosystems, but this observation is only visible after a latency period of several years and it is then too late to intervene. The use of biomarkers to estimate the quality of the environment allows for early and sensitive warning systems, but the ecotoxicological relevance is low because the biomarker response is not necessarily followed by the disappearance of a population or a modification of the ecosystem. The selection of damage biomarkers, especially those with an impact on behavior, reproduction, energy metabolism and genetics, should significantly improve these warning systems.

Environmental quality monitoring networks in France are well structured for atmospheric and aquatic environments (freshwater and marine), even if we must regret a weakening of the marine network, the ROCCH replacing the RNO. Similarly, the radiological monitoring networks are efficient. On the other hand, soil monitoring is not sufficiently developed in France and an effort must be made, particularly for polluted soils.

6

Radiation Doses Received by Marine Organisms

6.1. Introduction

The principal radiations affecting the environment are electromagnetic radiations (photons) without charge or mass (X-rays and gamma rays), or particles of low mass and mostly negative charge (electrons, beta radiation), or particles of fairly high mass and positive charge (helium nuclei, alpha particles); protons and neutrons are less frequently encountered in the environment [BLA 73]. The pathways of the various radiations through living matter will vary according to the energy (eV), mass and charge of the particle. Alpha emissions travel only a few fractions of a millimeter, whereas gamma rays can travel several meters. Estimating the exposure dose and the dose received by an organism are therefore not always the same.

6.2. Methodologies for estimating the dose given to marine organisms

The relevant estimation of the radiation dose to an organism begins with an objective analysis of the exposures and thus a good evaluation of the radioactivity of the various physical environments and of the contamination of the organism by all radionuclides, whether they are of natural or artificial origin. In some cases, especially in the absence of radioactive contamination data for certain compartments, the values are estimated by modeling the flow of radionuclides between the various compartments (Adam-Guillermin et al. pp. 47–62, *in*: [MEN 09]). This will allow the calculation of an irradiation dose.

Strictly speaking, there is a similarity between external exposure and delivered radiation dose only for photons (X- and γ-rays) and in the air. Depending on the energy of the photons and the nature of the tissues, the physical dose (joule/kg) delivered and the equivalent dose can be quite close or very different. To take into account the biological effect, a quality factor called relative biological effectiveness (RBE) should be applied.

The absorbed dose (amount of energy/kg) is much more difficult to measure than the exposure. Dose measurement can be direct by calorimetry or indirect using the number of ionizations. Ionization can be estimated by chemical dosimetry (ferrous sulfate, ceric sulfate, electron paramagnetic resonance spectrometry of alanine) or physical dosimetry, using RPL (radio-photoluminescence), OSL (optically stimulated luminescence) or TLD (thermoluminescence) dosimeters. More and more often, the dose estimation is carried out by calculation using models.

For this calculation in the case of electrons, it will be necessary to take into account the radioactivity of the organism (or organ) and the surrounding environment expressed in mass, volume or surface activity (e.g. $Bq.g^{-1}$, $Bq.cm^{-3}$, $Bq.cm^{-2}$, etc.), the energy of the electrons (MeV), as well as the geometry of the organism (to estimate the number of electrons that have effectively reacted with the living matter compared to those that escape to the outside).

In the case of photons, the approach is quite similar with generally a calculation of the energy flow rate for a given energy, i.e. the sum of the energies of all the particles that penetrate a given volume and are absorbed there. With this last method, the geometrical forms (volume or surface) retained can be more diversified (infinite medium, semi-infinite, infinite plate, sphere, cylinder, truncated cone, disk, etc.) and the coefficients of absorption and attenuation of photons in the material are applied with more rigor and precision. Conversion factors enable passing from flow rates to absorbed doses.

The calculation of the dose or dose rate to an organism in the environment is not obvious. A practical approach to calculating dose rates from radon-222 to terrestrial flora and fauna is presented by Vives i Batle et al. [VIV 12]. Beresford et al. [BER 10] compare seven approaches to predicting whole body concentrations of various radionuclides (^{90}Sr, ^{137}Cs, ^{241}Am and the isotope ^{239}Pu) and absorbed dose rates for various terrestrial species living in the Chernobyl exclusion zone.

In the case of humans, the dose calculation will involve various radiation and tissue weighting factors, but this is not yet the case for non-human organisms. However, more and more scientists are recommending the use of corrective factors for non-human organisms as well as for humans, particularly in the case of internal contamination by low linear energy transfer alpha-emitting radionuclides. UNSCEAR [UNS 08] lists weighting factors ranging from 1 to 50, and Chambers et al. [CHA 06] report weighting factors as high as 377. Gariel et al. [GAR 06] propose average weighting factors following a literature review.

The physical half-life of the radionuclide has a great influence on the radiation dose received, as it is inversely proportional to the rate of decay (quantified by the measurement of radioactivity). The time the radionuclide remains in the body is dependent on the rates of bioaccumulation and elimination, as well as on the changes of radioactive contamination in the environment.

6.2.1. *The units of radiation doses*

Some terms need to be defined [IAE 92, UNS 96, UNS 08]. An acute exposure is delivered in a short time compared to the long time period of the biological response of the damage. A high exposure corresponds to an acute response, most often with a severe pathology that can be fatal. On the contrary, a chronic exposure is undergone by the organism during a long period of its life. A low exposure will correspond to a marginal effect and with a very delayed effect on mortality but with an observable effect on normal biological processes without necessarily producing a serious harmful effect. UNSCEAR [UNS 10] considers radiation doses of 200 mGy or less and dose rates of 0.1 mGy per minute (spread over an hour or less) for radiation such as external X-rays and gamma rays as low. Since its 2010 report, UNSCEAR [UNS 18] has reported an excess relative risk in humans for thyroid cancers from 100 mSv.

For non-human organisms, only two dosimetric quantities are currently used: absorbed dose and dose rate. Absorbed dose is the amount of energy deposited in biological tissues at a point per unit mass of living matter. This fundamental quantity is expressed in Gray (Gy, equivalent to one joule per kilogram). The dose rate corresponds to the increase in absorbed dose per unit of time. The unit of time is the second, the minute, the hour or the year ($Gy.s^{-1}$, $Gy.min^{-1}$, $Gy.h^{-1}$, $Gy.yr^{-1}$). The old unit was the rad (Radiation Absorbed Dose) (1 rad = 10^{-2} gray)[1].

The roentgen (R) is the unit of radiation exposure. One roentgen corresponds to $2.58.10^{-4}$ coulomb per kg of ions generated in the air and an exposure of one coulomb per kilogram is equivalent to 3,876 roentgens. The dose depends on the radiation exposure (expressed as R) and the nature of the radiation. Calculation of the radiation dose (in Gy) from a radiation exposure of 1 R depends on the energy of the X- or γ-rays and the composition of the material being irradiated. For example, if soft tissue is exposed to 1 R of γ radiation, then the radiation dose will be approximately 9.3 milligray (mGy).

1 The old units were converted according to the following relationships: 1 Ci = $3.7.10^{10}$ Bq, 1 rad = 0.01 Gy.

6.2.2. *The difficulties of estimating the radiation dose*

While absorbed doses due to external exposures are relatively calculable for high-energy emitters, absorbed doses due to internal exposures, particularly for low-energy emitters, are very difficult to estimate. In fact, in the case of low-energy beta or X and gamma emissions, the assumption is made that all the energy has been absorbed, given the short path of these radiations. For example, an electron of about 40 keV has a path of less than 0.2 mm. If it is emitted into the body by a radionuclide, the assumption that it is totally absorbed can be made without risk of error.

Some authors have measured the radiation doses to aquatic organisms indirectly using thermo-luminescent dosimeters containing lithium fluoride powder. For example, Lappenburst et al. [LAP 71] measured the radiation dose to periphyton in the Columbia River. Similarly, Pentreath et al. [PEN 73] found that adult plaice living near Windscale receive an average overall radiation dose (on the skin) of 0.35 $\mu Gy.h^{-1}$ with a few cases at 25 $\mu Gy.h^{-1}$, corresponding to a gonadal radiation dose between 1.16 and 2.55 $\mu Gy.h^{-1}$. Similarly, Watson and Templeton [WAT 73] measured radiation doses (1968–1969) in the benthos of the Columbia River ranging from 2,200 $\mu Gy.d^{-1}$ to 340 $\mu Gy.d^{-1}$, respectively, 7 and 22 km downstream of the release; water contamination was responsible for 30 to 55% of the total dose. Fish received a radiation dose of 110 to 170 $\mu Gy.d^{-1}$. However, the disadvantage of these dosimeters is that they do not detect low-energy emitters, nor internal emitters that can sometimes be of great importance for microorganisms or for a critical organ.

6.3. Examples of natural radiation doses to marine organisms

As with radioactive contamination, we can distinguish between natural irradiation (cosmic rays and radionuclides naturally present in the environment) and anthropogenic irradiation (radionuclides coming from the fallout of atomic explosions or from the peaceful use of nuclear energy). It is common to distinguish between external and internal irradiation depending on whether the radionuclide generating the radiation is located outside or inside the body.

The first estimates of natural radiation doses reported by Folsom and Harley (1957 in: [AMI 80]) showed that large organisms are more irradiated than microorganisms because of internal β emissions of potassium 40[2] (Table 6.1.).

2 This is due to the high energy of the beta radiation of potassium-40: 1.996 MeV and 3.5211 MeV. The 2 MeV beta has a path length of about 8 mm. It will not be able to give up much of its energy in an organism that is measured in hundredths or tenths of a millimeter. It will do so more easily in an organism that is measured in centimeters.

Cosmic rays are rapidly attenuated by the thickness of the water, which explains the low radiation doses experienced by aquatic organisms living at depth. These estimates neglected the irradiations due to internal alpha emitters and in particular, polonium-210 intensely concentrated in certain tissues of aquatic organisms, which can multiply by five the irradiation dose (Table 6.1). The natural irradiation of microorganisms has been estimated at 0.39 mGy.yr^{-1} at the ocean surface and 0.05 mGy.yr^{-1} at 100 m depth [POL 66]. According to Woodhead [WOO 73], the natural irradiation is 0.028 to 0.082 μGy.h^{-1} and the irradiation due to atmospheric fallout corresponds to 0.0026 to 0.245 μGy.h^{-1}, while the Windscale (Sellafield) effluents cause an additional irradiation of 2 to 21 μGy.h^{-1}.

	Cosmic rays	External sources	Internal ^{40}K	Internal transmitters	Internal ^{210}Po
		Large organisms			
On the surface	350	10	280		1,600–3 200
At 100 m	5	10	280		1,100
		Small organisms (< 0.01 mm)			
On the surface	350		40	230–2,800	200–500
At 100 m	5		40	230–2,800	
In the sediment		400–6,200			8,000

Table 6.1. *Estimates of natural irradiation of marine organisms as a function of their spatial location in ecosystems (irradiation doses in μGy.yr^{-1} (modified from [AMI 80])*

The majority of the calculated absorbed dose to marine and freshwater organisms is from internally incorporated alpha emitters, with ^{210}Po and ^{226}Ra being the major contributors (Table 6.2).

	Macroalgae	Crustaceans	Fish
Total absorbed dose (μGy.h^{-1})	0.26	1.7	0.14
Interval	0.16–0.95	0.27–27	0.08–0.71

Table 6.2. *Calculations of absorbed dose rates to marine organisms (μGy.h^{-1}) for naturally occurring radionuclides in the environment (from [BRO 04, IAE 15]).*

The highest doses are in bacteria (0.67 μGy.h^{-1}), crustaceans (0.21 μGy.h^{-1}) and mollusks (0.19 μGy.h^{-1}) (Jones et al. *in*: [IAE 03b]).

Fish from the Baltic Sea (cod, sprat, herring, plaice), mollusks and crustaceans living mainly near the bottom of two sub-zones (Gdansk Basin and Bornholm Basin) contaminated mainly with ^{137}Cs, ^{226}Ra and ^{239}Pu had total maximum annual doses that did not exceed one percent of the recommended dose limits of 3.6 $Gy.yr^{-1}$ [KRA 03].

6.3.1. *Pelagic marine organisms*

In 11 pelagic species (cephalopod mollusks and crustaceans) analyzed, alpha radiation doses have been estimated to be of the order of 0.1 Gy per year in the digestive gland or hepatopancreas, which contains about 50 to 90% of the entire animal's ^{210}Po [HEY 79].

Polonium-210 bioaccumulates preferentially in the hepatopancreas of decapod crustaceans and the natural radiation dose received by this organ in these individuals will be of the order of 1 Gy per year [CHE 81]. This value is impressive.

In cod, halibut, redfish and shark, the main contributors to the radiation dose were ^{210}Po and ^{40}K, with anthropogenic radionuclides making only a small contribution. For their part, deep-sea fauna do not live in an environment protected from ionizing radiation [CAR 10].

The naturally occurring radionuclides potassium-40, radium-226, polonium-210 and lead-210 were measured in commercial fish species such as cod, halibut, rockfish and shark from several North Atlantic fishing areas, as well as the anthropogenic radionuclides ^{137}Cs and plutonium isotopes (^{238}Pu and $^{239,240}Pu$). The major contributors to the radiation dose were ^{210}Po and ^{40}K. The radiation doses from absorbed natural radionuclides still exceeded the radiation doses caused by anthropogenic radionuclides introduced into the Northeast Atlantic [CAR 11].

On the Chinese coast of Qingdao, the doses were dominated by natural ^{40}K and the doses due to ^{137}Cs were negligible compared to the doses derived from ^{40}K. The total doses to marine biota ranged from 16.55 to 62.41 $nGy.h^{-1}$ [YAN 15].

The unweighted natural absorbed dose rates for reference marine organisms range from 0.02 to 1.2 $\mu Gy.h^{-1}$, while the weighted dose rates range from 0.08 $Gy.h^{-1}$ to 8.8 $\mu Gy.h^{-1}$. With the exception of the special case of sea anemones and corals, the most significant radionuclides for total dose rates are ^{210}Po and ^{226}Ra for heterotrophic organisms (i.e. all vertebrate and invertebrate animals), while ^{226}Ra and ^{228}Th contribute the most for autotrophic organisms (i.e. macroalgae, phytoplankton and aquatic plants). In addition, ^{40}K has a significant impact on the unweighted dose rates for all organisms [THØ 09].

The highest natural radiation doses from ^{210}Po were calculated for mussels and commercially valuable pelagic fish from the Black Sea such as sprat (0.87 mGy.yr^{-1}) [LAZ 03].

6.3.2. *Benthic marine organisms*

Natural radiation doses to fish differ between habitats, and the majority of the dose is due to sediment contamination (Table 6.3).

Radiation source	**Dose rate (μGy.h^{-1})**
Internal radionuclides	0.094–0.10
Radionuclides in seawater	0.0005–0.013
Radionuclides in sediment (only)	0.37–33.1
Total	***0.47–33.2***

Table 6.3. *Exposure of flounder in the Irish Sea (from [WOO 97b])*

6.3.3. *Deep-sea marine organisms and hydrothermal sources*

Concentrations of the naturally occurring radioactive nuclide polonium-210 have been determined in crustaceans and fish at depths up to 1,500 meters. Unusually high levels have been found in some categories of organisms, indicating that these organisms are exposed to a particularly high dose of natural radiation [CHE 82].

Natural radionuclides resulting from the convective circulation of seawater within the oceanic crust and its interaction with basaltic or ultramafic host rocks are in high concentrations at hydrothermal vents. The tissue distributions of ^{210}Po and ^{210}Pb, two naturally occurring radionuclides in the ^{238}U decay chain, in *Bathymodiolus azoricus* specimens from the Mid-Atlantic Ridge (Menez Gwen field) explained that the ^{210}Po levels encountered in *B. azoricus* were not exceptionally high, leading to a weighted internal dose rate between 3 and 4 μGy.h^{-1}. These levels are slightly higher than levels characterizing coastal mussels (about 1 μGy.h^{-1}) [CHA 11].

The minimum and maximum dose rates have been estimated at 0.1 nGy.h^{-1} in the cold seep of Sagami Bay, central Japan, and at about 2,000 nGy.h^{-1} near the Okinawa Trough hydrothermal spring, southwestern Japan [FUR 06]. The IAEA [IAE 88] provides the doses absorbed by marine organisms living at depth (Table 6.4).

	Fish	Mollusks	Large crustaceans	Small crustaceans
Bathypelagic	0.028–0.46	-	0.93–3.6	0.25–2.9
Benthic	0.085–1.4	0.93–3.4	0.99–4.6	0.33–4.2

Table 6.4. *Estimated equivalent dose rates (in µSv.h-1) from deep-sea organisms under natural exposure conditions (from [IAE 88])*

6.4. Examples of anthropogenic radiation doses to organisms

Many human activities using nuclear energy for peaceful or military purposes have led to the radioactive contamination of the marine environment resulting in the irradiation of organisms. The Arctic area is an example. The estimated radiation doses (in $mGy.d^{-1}$) are, respectively, $3.6 \pm 1.0.10^{-6}$, $1.3 \pm 0.6.10^{-6}$, $4.0 \pm 1.1.10^{-6}$ and $3.4 \pm 1.1.10^{-6}$ for benthic fishes, pelagic fishes, aquatic plants and bivalve mollusks living in the Arctic region [AMA 16].

According to the IAEA [IAE 76], the absorbed radiation doses to marine organisms are significantly greater for individuals living in the vicinity of the Windscale (Sellafield) fuel reprocessing plant, while the natural radiation doses and those due to atmospheric fallout are roughly similar (Table 6.5).

Site	Phyto-plankton	Zoo-plankton	Mollusk	Crustacean	Pelagic fish	Benthic fish
Marine environment						
A	0.65–2	0.8–4.1	2.4–2.8	2.5–9.5	0.8–1.2	0.2–5.2
B	0.7–6.5	0.4–3.6	0.03–2	0.09–0.12	0.04–0.5	0.04–0.05
C	50–525	133–1,725	13–850	10.8–113	0.15–1	9.3–835

Table 6.5. *Estimated natural and artificial radiation doses for various groups of aquatic organisms (µGy.d-1) (modified from [AMI 80]). A: natural radioactivity; B: fallout from atomic testing; C: Windscale (Sellafield) reprocessing plant (Irish Sea); The irradiation for marine organisms is calculated for a depth of 20 m and away from the sediment except for benthic fish at 20 m but in contact with the sediment*

6.4.1. *Irradiation doses resulting from the testing of atomic bombs*

At the Bikini site, the natural radiation dose due to ^{210}Po and ^{210}Pb accumulated over 50 years is equal to 0.29 mSv. The doses due to anthropogenic radionuclides $^{239,240}Pu$ and ^{137}Cs from the bombs are, respectively, 0.000045 and 0.005 mSv [ROB 98].

At the Mururoa site studied by IAEA experts [IAE 98b], the highest absorbed dose rates from internal sources were about 200 nGy.h^{-1} from alpha radiation to the lagoon zooplankton and benthic crustaceans. Although higher than the absorbed dose rate (12 nGy.h^{-1}) from the internal source of ^{40}K, these values are lower than the 4,000 nGyh^{-1} estimated at the genitalia of a small oceanic decapod crustaceans in the middle of the radionuclide contaminated waters, exposed to the naturally accumulated alpha emitter ^{210}Po [UNS 96]. The local absorbed dose rate to benthic organisms due to contamination of the coral sand may be as high as about 900 nGy.h^{-1} from the presence of ^{60}Co and ^{137}Cs in an area of about 1 km^2 at the Dindon area of Mururoa, and perhaps 200 nGy.h^{-1} from gamma radiation from ^{241}Am in another small area, on the Colette sandbar, since organisms there were potentially exposed to alpha radiation from active particles. It was also possible to estimate absorbed dose rates up to 300 Gy.h^{-1} limited to small volumes of tissue [IAE 98b].

Radionuclides released during weapons testing in the 1950s on the Montebello Islands, Western Australia, may have had an impact on sea turtle embryos that incubate in eggs laid in contaminated sand. In addition, these radionuclides were absorbed into the body tissues of adults. These contaminated sands caused most of the dose to the eggs (external dose of ^{137}Cs, ^{152}Eu), while most of the dose to the adults came from internalized radionuclides (98%). While current dose rates are relatively low, local dose rates were high for a decade after the 1950s detonations and may have resulted in death or health effects in a generation of turtles that likely carry biomarkers today [JOH 20]. Internal dose accounts for 37% in embryos and 98% in adults of turtles undergoing irradiation from British atomic test fallout. The doses from ^{90}Sr, ^{137}Cs, ^{238}Pu, ^{240}Pu and ^{241}Am are for each radionuclide about 0.001 μGy.h^{-1}, about 0.01 μGy.h^{-1} for ^{239}Pu and 0.1 μGy.h^{-1} for ^{210}Po [JOH 20].

The risk of radiobiological effects from plutonium in marine biota at the main entrance to the Indonesian through flow (ITF) can be considered negligible, as the estimated total dose level is below the screening dose of 10 μGy.h^{-1} [MAK 20].

6.4.2. *Radiation doses resulting from nuclear accidents*

It is mainly the Fukushima nuclear accident that has had significant impacts on marine organisms. An international study conducted within the framework of the United Nations Scientific Committee on the Effects of Atomic Radiation (UNSCEAR) was carried out to assess the radiological impact of the nuclear accident at the Fukushima-Daiichi nuclear power plant. In this study, a large number of samples were collected including 500 sediment and 6,000 seawater samples, providing 5,000 data points during the period from May 2011 to August 2012. The main findings of this study were that doses to marine organisms were generally below population effect levels. The only exception was the presence of high concentrations of iodine-131 in macroalgae near the plant early after the accident [VIV 14]. In this accident, the internal dose dominated the external dose from seawater contamination by two or three orders of magnitude [VIV 13].

Dose rates to the most affected fish species near the Fukushima plant (median 1.1 mGy. d^{-1}, 2012–2014 data) remained above baseline levels for potential dose effects at least three years longer than indicated by previous data-limited assessments. Dose rates due to ^{134}Cs and ^{137}Cs were highest in demersal species[3]. In addition to ^{134}Cs and ^{137}Cs, the radionuclide ^{90}Sr contributed up to about half of the total 2013 dose rate to fish in the vicinity of the accident reactor. Mesopelagic fish 100–200 km east of the accident, coastal fish in the Aleutian Islands (3,300 km) and transpacific migratory species all experienced increased dose rates as a result of this accident, but their total dose rates remained dominated by the background from naturally occurring radionuclides [JOH 15].

A hypothetical human consumer of 50 kg of fish, collected 3 km from the FDNPP in 2013, would have received a total committed effective dose of about 0.95 $mSv.yr^{-1}$ due to combined exposure to the accident and ambient radionuclides, of which 0.13 $mSv.yr^{-1}$ (14%) was from reactor radionuclides alone and was below the 1 $mSv.yr^{-1}$ reference for public exposure [JOH 15].

The equilibrium concentration ratio is generally the parameter used to estimate organism activity concentrations in wildlife dose assessment tools. Although this is assumed to be fit for purpose, there are scenarios such as accidental or irregular, fluctuating releases from permitted facilities where this may not be the case. In such circumstances, the concentration ratio approach may underestimate or overestimate radiation exposure as a function of time since release. To perform assessments of such releases, a dynamic approach is needed. The simplest and most practical option is to represent the uptake and turnover processes by first-order kinetics, for which organism- and element-specific biological half-life data are required. Biological half-life values are reported for 52 elements across a range of wildlife groups, of which about nine consist of marine organisms [BER 10].

During the Fukushima accident, the dose rates during the period from May 10, 2011, to August 12, 2012, at their highest, were between 0.10 and 0.25 $Gy.h^{-1}$. These levels were proportional to the background dose rates in the marine environment. For fish, the maximum estimated dose rate occurred in the first month (approximately 140 $\mu Gy.h^{-1}$) and the accumulated dose over one year was approximately 0.32 Gy. The maximum calculated exposures for macroalgae (exceeding 20 $mGy.h^{-1}$) at the same location occurred 23 days after the accident, but

3 Demersal species live in contact with the sediment of the ocean floor.

dropped rapidly, with iodine-131 being the dominant component. The cumulative dose to macroalgae over one year was about 7 Gy [UNS 14].

Dose calculations for the 30 days post-accident and for the three most abundant radionuclides present (^{131}I, ^{134}Cs and ^{137}Cs) provided very high irradiation doses for the coastal marine ecosystem near Fukushima. For example, the maximum dose rates ranged from 210 to 4,600 mGy.d^{-1} for seabirds and macroalgae, respectively, with intermediate values of 2,600 mGy.d^{-1} for benthic fauna (fish, shellfish) (Figure 6.1). At such high dose rates, marked reproductive effects and even mortality for the most radiosensitive taxa are expected for all groups of marine fauna whose life history characteristics confine them to the vicinity of the contaminant release area [GAR 11].

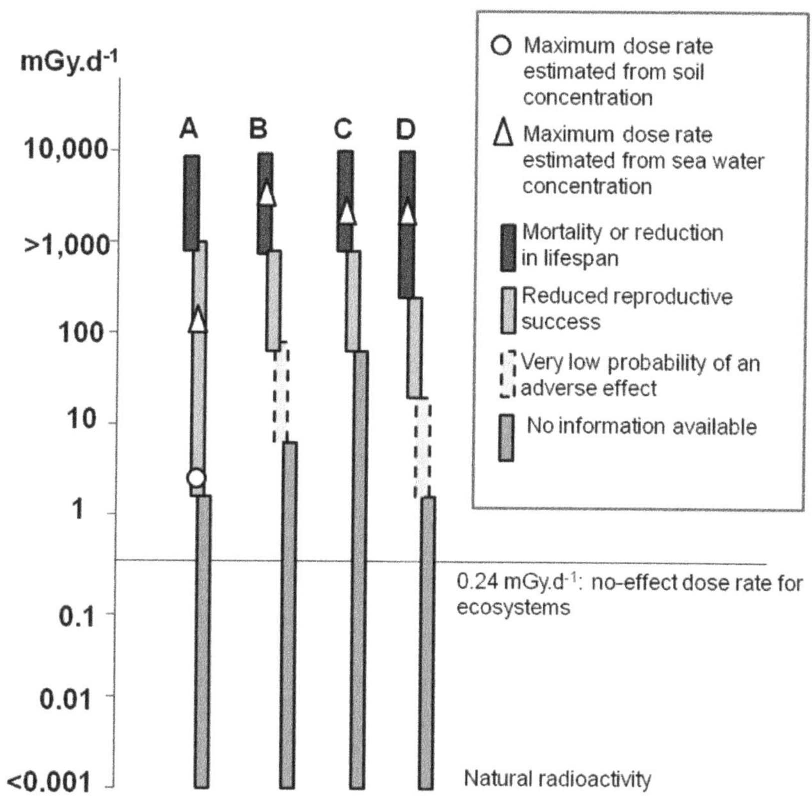

Figure 6.1. *Estimates of the dose rates received by various marine ecosystems in areas impacted by the Fukushima accident and assumed consequences on the health of organisms. A: birds; B: brown algae; C: marine benthic invertebrates; D: marine benthic flatfish (modified from [GAR 11]). For a color version of this figure, see www.iste.co.uk/amiard/radioecology.zip*

6.4.3. *Irradiation dose from spent fuel reprocessing plants*

The estimated artificial radiation doses around the Bradwell (Eastern England), Gundremmingen (Germany) and Trawsfynydd (Wales) nuclear power plants are of the same order of magnitude or lower than the natural irradiation for all groups of organisms. In contrast, marine organisms in the Irish Sea near Windscale-Sellafield and freshwater organisms in the Columbia River receive (or rather have received) radiation doses up to 4,000 times greater than the natural background dose (Table 6.6).

BNI	Aquatic plants	Invertebrates	Fish
Mines	630	1,000	12
Ore purification plant	12,000	3,500	220
Conversion plant	14	1.4	0.14
Enrichment plant	17	1.7	0.18
Manufacturing plant	0.13	1.6	0.26
PWR reactor (marine)	69	14	0.39
BWR reactor (marine)	31	290	5.9
Reprocessing plant	1.5	0.27	0.12

Table 6.6. *Contribution of the different stages of the nuclear fuel cycle to the irradiation of various groups of aquatic organisms (mGy.yr-1) according to Blaylock and Witherspoon, 1975, in: [AMI 80])*

It is the benthic organisms that are the most irradiated because of the presence of large quantities of radionuclides (natural or/and anthropogenic) in the sediments. The effect of size that was previously reported also occurs, and it is the large organisms that absorb, on average, the highest doses of external radiation. Small mollusks such as *Nucula turgida* and *Vultellus pellucidus*, present in the Sellafield discharge area, receive 0.20 mGy.h^{-1} (of which 0.18 mGy.h^{-1} is due to the sediment). A flounder (*Pleuronectes platessa*) living 4 km north of the Windscale release receives a natural radiation dose of 0.084 µGy.h^{-1} and receives a radiation dose from fission products alone of 58.9 µGy.h^{-1} (of which 40 µGy.h^{-1} is due to emitters in the sediment) [WOO 71]. However, the high radiation doses to zooplankton around Sellafield and especially around the Hanford reactors should be noted.

In marine fish (flounder) or freshwater fish (trout), irradiation is due almost exclusively to the presence of incorporated natural or anthropogenic radionuclides, except in the case of flounder living near Sellafield and in contact with highly radioactive sediments [WOO 74].

Blaylock and Witherspoon [BLA 75] estimated the internal radiation doses to various groups of aquatic organisms according to the stage of the fuel cycle. To do this, they took into account theoretical releases, estimates of concentration factors for each radionuclide and a single geometry for all organisms (a sphere with a radius of 30 cm) in order to model each parameter. It appears that organisms living near mines and ore purification plants receive the highest radiation doses. BWRs will cause higher internal radiation doses to aquatic organisms on average than PWRs, and irradiation will be more intense in marine environments than in fresh waters for BWRs. Kaye's estimates [KAY 73] are based on actual releases from 19 PWRs and 10 BWRs in the United States, and on average concentration factors: while there is generally a decrease in the dose received as one moves up the systematic scale, several exceptions appear, in particular for birds and aquatic mammals, which could receive up to 620 $mGy.yr^{-1}$. These internal irradiations are mainly due to ^{134}Cs then ^{54}Mn, ^{58}Co, ^{60}Co, ^{131}I, ^{133}I and ^{137}Cs.

In the Esk estuary adjacent to the Sellafield nuclear spent fuel reprocessing planta, a comprehensive survey of 576 samples of gamma dose rates was undertaken in 2007 and the data were compared to a similar survey carried out in 1989. There was a highly statistically significant decrease in external gamma dose rates in the Esk estuary between 1989 and 2007. The total gamma dose rate range measured in 2007 was 0.064 to 0.235 $\mu Gy.h^{-1}$ with an average of 0.123 $\mu Gy.h^{-1}$. This equates to a maximum exposure of 0.200 $\mu Sv.h^{-1}$ [WOO 11].

In the case of phosphorus-32, accumulation and elimination were studied in two bivalve species. Bioaccumulation of ^{32}P is significantly higher in the digestive gland. Moreover, tissue-specific dose rates can be in some cases significantly higher than whole body averages [VER 18].

In estuarine crabs, ^{85}Sr is more concentrated in the gonads than in other organs, resulting in a proportionally higher radiation dose to the reproductive organs and requiring further investigation [CRE 21].

At La Hague, the major contribution to the dose to marine organisms is due to ^{106}Ru-Rh [CHA 05].

6.4.4. *Radiation dose from oil and gas operations*

The exploitation of oil and gas fields leads to radioactive contamination of the marine environment (due mainly to natural radionuclides, ^{226}Ra, ^{228}Ra, ^{210}Po, ^{210}Pbn, i.e. enhanced natural contamination) and an increase in the internal radiation dose rate that is negligible compared to the natural dose, which is greater than 0.2 $\mu Gy.h^{-1}$

for benthic crustaceans and mollusks (Figure 6.2). The majority of the additional dose from natural radionuclides is internal, and the external dose is negligible.

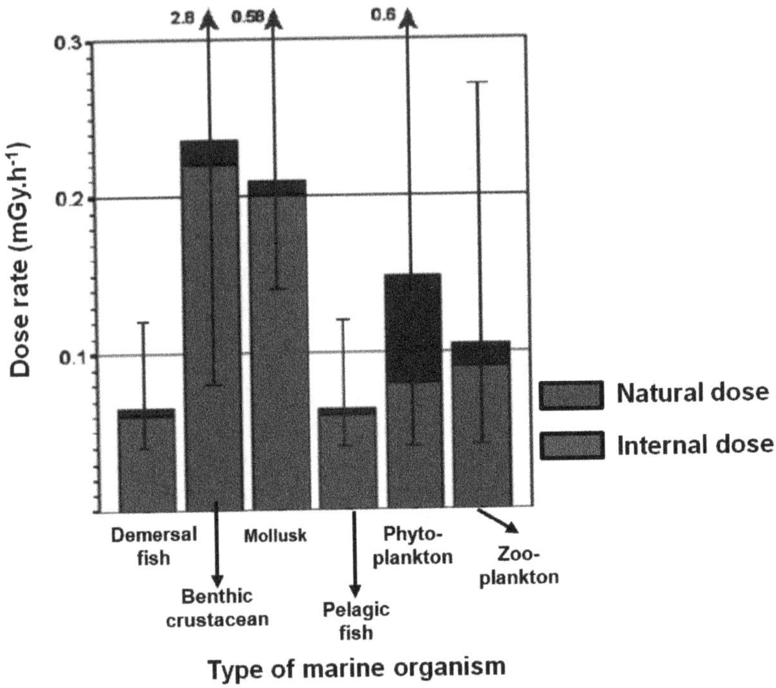

Figure 6.2. *Unweighted absorbed dose rate (μGy.h-1) for various marine organisms due to natural exposure (red bars) and due to internal exposure to naturally occurring radionuclides whose concentrations are increased by oil and gas operations (blue bars) (from [BRO 05]). The bars define the range of dose rates derived for natural radionuclides from [BRO 04]). For a color version of this figure, see www.iste.co.uk/amiard/radioecology.zip*

6.4.5. *Irradiation dose due to experimental contamination*

Experimental contamination of fish eggs and embryos by some fission products (^{144}Ce-Pr, ^{137}Cs, ^{106}Ru-Rh, ^{90}Sr-Y, ^{95}Zr-N) results, according to Woodhead [WOO 71], in an irradiation dose of only 0.0009 μGy.h^{-1}, whereas potassium-40 alone causes a natural irradiation dose of 0.007 μGy.h^{-1}. On the contrary, Fedorov (1964 *in*: [POL 66]) estimates that, depending on the size of the fish eggs and the radionuclide (^{90}Sr-Y or ^{137}Cs), the increase in the radiation dose level will be a factor of 30 to 300. For young plaice living near Sellafield, the absorbed dose to the gonads is estimated to be between 11.3 and 14.6 μGy.h^{-1} by Pentreath et al. [PEN 73].

Contamination of algal cultures *Dunalieela bioculata* at 370 Bq.L^{-1} and 37 kBq.L^{-1}, respectively, for ^{51}Cr and ^{109}Cd results in an irradiation dose to the cells of 0.32 μGy.h^{-1} for ^{51}Cr and 5.50 mGy.h^{-1} for ^{109}Cd [AMI 78b]. According to the author of this experiment [SAR 75], irradiation with cadmium-109 causes ultrastructural damage (see Chapter 7).

Radiation doses for shellfish were calculated following experimental contamination by silver-110m through their food (diatoms subjected to 5 μCi.L^{-1} or 185 kBq.L^{-1}). In *Scrobicularia* and Ccrabs, the digestive gland or hepatopancreas is, respectively, 47 and 72 times more irradiated than the whole body, i.e. 3 088 and 1 936 μGy.h^{-1} [AMI 78c]. Thus, while the dose to the whole body is tolerable, it becomes very high in the digestive gland or hepatopancreas.

6.5. The ICRP approach

Following its new policy of radiological protection of the environment, the ICRP has created a fifth commission in charge of this problem. This commission of the ICRP has decided to adopt an approach similar to that of humans to estimate the radioactive risk to non-human organisms [PEN 09].

6.5.1. *Reference organisms (RAP, Reference Animal or Plant)*

The ICRP considers that it is impossible to quantify transfer and doses for all species, so it advocates an approach using representative groups such as "reference organisms" (RAP, Reference Animal or Plant) [ICR 08, HOW 11]. Of the reference animals and plants, seven organisms are aquatic and only three are marine.

The dosimetric models, adopted in Publication 108, provide dose coefficients (DCs) for a group of reference organism entities [ICR 08].

In ICRP Publication 136 [ICR 17], conversions from concentrations (media and reference organisms) to doses are calculated. These dosimetric models pragmatically assume simple body forms with uniform composition and density, homogeneous internal contamination, limited sets of idealized sources of external exposure to ionizing radiation for aquatic and terrestrial animals and plants, and truncated radioactive decay chains.

The dominant radionuclides contributing to exposure in RAPs are ^{40}K, ^{210}Po and ^{226}Ra. The unweighted and weighted average dose rates for aquatic PARs are 0.07–0.39 μGy.h^{-1} and 0.37–1.9 μGy.h^{-1}, respectively [HOS 10]. The internal dose accounts for 52, 52 and 61% of the total dose to flatfish, crabs and algae, respectively [HOS 10]. The weighted and unweighted absorbed doses from naturally occurring radionuclides have been estimated for the reference organisms (Tables 6.7 and 6.8).

	Flatfish	Crab	Algae
Total dose	$1.5.10^{-1}$	$1.4.10^{-1}$	$2.7.10^{-1}$
5th percentile	$8.7.10^{-2}$	$6,.7.10^{-2}$	$7.2.10^{-2}$
95th percentile	$2.4.10^{-1}$	$2.8.10^{-1}$	$2.1.10^{-1}$

Table 6.7. *Absorbed dose rate (expressed in µGy.h-1) from natural radionuclides, calculated for marine flora and fauna (from [HOS 10])*

	Flatfish	Crab	Algae
Total dose	$5.8.10^{-1}$	$5,.9.10^{-1}$	$8,.7.10^{-1}$
5th percentile	$2.4.10^{-1}$	$1.2.10^{-1}$	$5.2.10^{-1}$
95th percentile	$1.2.10^{0}$	$1,.4.10^{0}$	$1.1.10^{0}$

Table 6.8. *Total weighted absorbed dose rate (expressed in µGy.h-1) from natural radionuclides, calculated for marine flora and fauna (from [HOS 10])*

6.5.2. *Limitations of the methodology*

The dose to humans from radiation exposure can be quantified using a well-defined system: its environmental equivalent, however, is still in a state of development.

These calculations are far from perfect because of some simplifying assumptions that are not always appropriate. Assuming uniformly contaminated organisms or assuming that a cylinder or disk will represent an entire class of organisms, or even all organisms, can result in significant errors. The validity of the estimates may be limited by the concentration factors used and by some of the theoretical effluent discharge assumptions made. In in situ contamination, some radionuclides may be neglected. Thus, at Sellafield, the radionuclides ^{238}Pu, ^{239,240}Pu and ^{241}Am determine non-negligible internal irradiation doses for various marine animals, 0.66 µGy.h^{-1} for the visceral mass of *Mytilus edulis*; 1.3 µGy.h^{-1} for the gills and 0.20 µGy.h^{-1} for the hepatopancreas gland of *Cancer pagurus*; 0.0071 µGy.h^{-1} for the liver, 0.0061 µGy.h^{-1} for the kidneys of *Pleuronectes platessa* [HET 75]. Many authors [WOO 73, WOO 74, IAE 76, etc.] have completely neglected this contamination due to transuranics and the resulting irradiation.

In addition, the results of several papers published during the years 2005–2015 were criticized for poor dosimetry. As a result, a workshop on environmental dosimetry was organized by the STAR (Strategy for Allied Radioecology) Network of Excellence. For example, in screening model calculations that estimate total body dose to biota, organism shapes are simply represented as ellipsoids, whereas recently developed complex voxel phantom models allow for organ-specific dose estimates. Thus, while Stark et al. [STA 17] believe that although major progress has recently been made in environmental radiation protection, substantial improvements are needed to reduce uncertainties and increase the reliability of environmental dosimetry.

For their part, Howard and Beresford [HOW 11] emphasize that protection is focused on populations rather than on individuals, so some of the approaches used for the assessment of chemical pollutants are also adopted for radionuclides.

6.5.3. *Towards essential improvements*

Since these historical estimates of absorbed dose, ecosystems have been described in a simplified manner with, for example, three physical components for aquatic systems: the bottom sediment, the water column and the air. Reference organisms have been selected with simple geometric shapes and standard sizes allowing approximate modeling of the true absorbed dose. This is the case of the EDEN tool [BEA 06a]. This results in the production of dose conversion coefficients (DCC), expressed as absorbed dose in Gy by the reference organism as a function of time and mass activity ($Bq.kg^{-1}$) or volume activity ($Bq.L^{-1}$) delivered by the source, which is either the physical medium or the organism itself. For the entire life cycle of fish, the concentrations of 70 radionuclides in water, sediment and fish were estimated from a previous study [GRN 99b]. The dosimetric evaluation was performed with EDEN (Elementary Dose Evaluation for Natural Environment), a tool developed at the IRSN to estimate the dose rate to biota per unit concentration of radionuclides. A risk of 10^{-3} was estimated for a release of 10^{15} Bq. The predicted no-effect dose rate (PNEDR) was estimated for flounder to be 1 $mGy.d^{-1}$ [BEA 06a].

UNSCEAR [UNS 08] provides DCCs for 16 radionuclides and 12 reference organisms (terrestrial species, soil-living species and aquatic species).

During the last decade, a number of models and approaches have been developed for the assessment of exposure of non-human biota to ionizing radiation. In some countries, these are now used in regulatory assessments. However, to date there has been no attempt to compare the outputs of the different models used. Beresford et al.

[BER 08b] present the work of the IAEA EMRAS BiotaWorking Group that compares the predictions of a number of these models.

In the Environmental Protection from Ionizing Contaminants in the Arctic (EPIC) project for impact assessment, the basic elements of information needed to calculate dose rates to organisms are: 1) activity concentrations of radionuclides in reference organisms and their habitat; 2) dose conversion factors (DCFs); and 3) occupancy factors defining the time spent by biota at different locations representative of their habitat (Figure 6.3) [THØ 06].

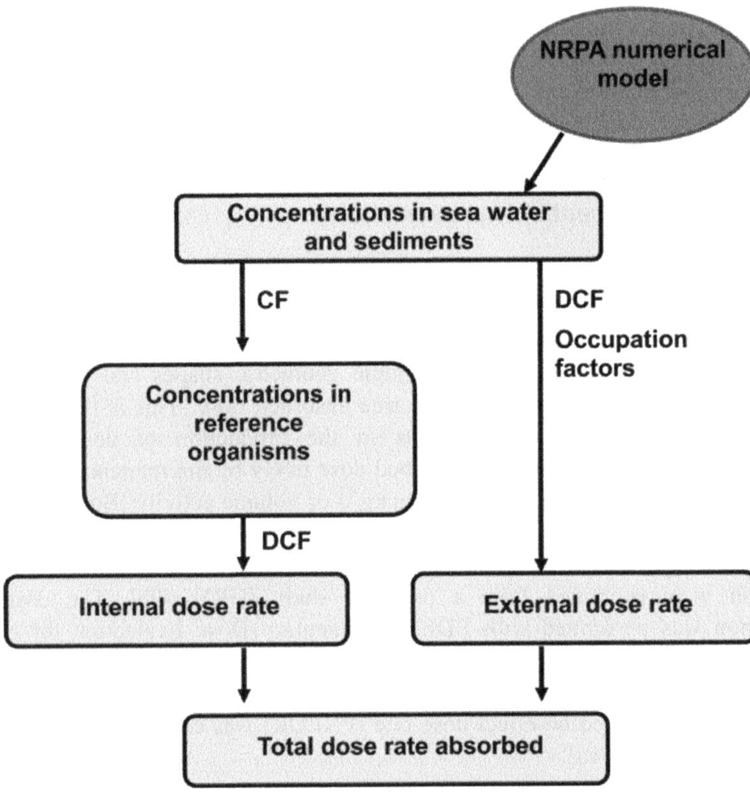

Figure 6.3. *Various steps in the EPIC approach for exposure assessment of non-human organisms (from [THØ 06]). For a color version of this figure, see www.iste.co.uk/amiard/radioecology.zip*

According to Delistraty [DEL 08], radiation protection has historically focused on humans with the assumption that human protection confers protection to all non-human biota[4]. However, this needs to be scientifically and independently demonstrated for effective protection of non-human biota. A good approach to the impacts of irradiation on non-human biota involves an ecological risk assessment, the definition of dose limits, the definition of reference organisms and this assessment will be limited to a geographical area. The author recommends harmonization of a single radiation protection framework for both humans and non-human biota, a single and consistent methodology for assessing both radionuclides and non-radioactive pollutants, a graded assessment approach, development of dosimetric models for reference organisms, compilation of a database of radiological effects and periodic review by methodological experts.

In addition, the dose rates to non-human organisms for a certain radioactive contamination of the environment will be significantly different from that to humans. Table 6.9 lists the dose rates to organisms for each individual radionuclide in an environment where humans would receive an annual dose of 1 mSv. Thus, tritium delivers a dose rate nearly six times greater to plants and animals than to humans. Strontium-90 irradiates fish 67 times more than humans. On the other hand, plutonium-239 is more dangerous for humans.

Radionuclide	Dose rate ($\mu Gy.h^{-1}$)		
	Plants	Animals	Fish
^{3}H	5.8	5.8	0.59
^{32}P	32	28	4.8
^{90}Sr	2.0	0.042	67
^{137}Cs	5.4	3.1	0.72
^{239}Pu	0.023	0.00055	0.49

Table 6.9. *Estimated dose rate to various organisms equivalent to an annual human dose of 1 mSv (modified from [UNS 08])*

6.6. Systematic underestimation of the dose received by organisms

The estimate of the dose received by organisms is systematically underestimated, and this is also true for marine organisms. This bias is due to several causes, first of all to oversimplifying assumptions.

Firstly, the hypothesis of a homogenization of radioactive contamination of organisms is unrealistic. In reality, there is a strong heterogeneity in the

4 This statement is due to the ICRP (publication 26 of 1977), see Chapter 1, section 1.1.1. A caution appears only with publication 60 of 1990.

contamination of organisms at various levels of biological organization, with preferential bioaccumulation in certain tissues or organs[5], as well as at the sub-cellular level with accumulation in certain cellular organelles.

The calculations of absorbed doses are carried out for organisms and organs with simple but unrealistic geometrical shapes. Voxelization seems to be the solution to remedy this drawback.

This is also due to the fact that the characteristics of certain very short-lived emitters, such as alpha emitters and low-energy beta emitters (commonly called soft betas), are not taken into account. In addition, a majority of radionuclides with one or two primary emissions have secondary emissions that are not taken into account. Efforts are underway to remedy this shortcoming.

For all these reasons, it is very difficult to estimate internal radiation doses with good accuracy.

6.7. Conclusion

The radiation doses absorbed by marine organisms are mainly due to natural radionuclides present in the marine environment and bioaccumulated by the organisms. Some populations of marine organisms are however impacted by human nuclear activities. This is particularly true around the Windscale-Sellafield spent nuclear fuel reprocessing plant or following nuclear accidents affecting the marine environment such as Fukushima.

However, there is a systematic underestimation of absorbed doses due to oversimplifying assumptions and neglected radiation sources.

5 For example, iodine goes to the thyroid, cesium-137 to the muscles, plutonium to the bone surfaces and liver, etc.

7

Effects of Irradiation on Marine Organisms

7.1. General information on the effects of irradiation

Historical approaches to radiation protection have been based on the assumption that human safeguards are adequate to protect non-human organisms. This view differs from that of other toxic substances where well-developed frameworks exist to minimize exposure to biota. In the late 1990s, this view evolved slightly. However, this explains the severe lack of studies on the effect of ionizing radiation in non-human organisms. This aspect has become a priority area for future research within the scientific and regulatory communities.

Most regulatory agencies use four endpoints to estimate the adverse effect of ionizing radiation: mortality, morbidity, reproduction and mutation. Based on the limited data, mutation was found to be the most sensitive endpoint of radiation exposure, with mortality being the least sensitive.

The marine environment is the richest in faunal taxa and one of the least studied from the point of view of the effect of ionizing radiation. For example, crustaceans include more than 66,000 ecologically important species and there is a lack of data on the effects of relevant radiation doses on the biology of this zoological group [FUL 15].

7.1.1. *Modes of action of ionizing radiation*

When the radiation emitted by a radioactive source comes into contact with matter, it can remove electrons from atoms and thus cause the loss of their electrical neutrality. As a result, the atoms are transformed into ions. This phenomenon of ionization is the main mechanism by which radioactivity interacts with matter. Hence the term "ionizing radiation" used to describe radiation caused by radioactivity. When molecules are ionized and excited, they form free radicals (ROO° peroxide radical, among others) which are very reactive due to the presence of unpaired electrons. They will then react with each other and with neighboring molecules, which will lead to a disturbance of the chemical state of the cellular constituents. The cellular constituents will be affected either directly by the radiation or indirectly, via the free radicals resulting from the direct ionization of the cellular constituents. However, the effect of an attack on macromolecules by the free radicals resulting from the radiolysis of water has a dominating role.

The various radiations lose their energy all along their path in the tissues, and this released energy is at the origin of the biological effects. The tissue's degree of ionization is inversely proportional to the path taken. Thus, a 5 MeV alpha emitter generates 4,000 ion pairs per μm and a 1 MeV electron only six ion pairs per μm (*in:* [POL 66]).

The most numerous data concern responses at the individual level on increased mortality or decreased lifespan. Other responses such as changes in reproductive organs, abnormalities in gamete development or juvenile stages have been used to demonstrate the potential effect of radiation on reproductive success. Also of great interest are the subcellular and cellular responses of the major molecules, including genetic and physiological effects. The doses required to elicit a response at the individual level are generally much higher than those that elicit responses at lower levels of biological organization (cellular and molecular). At the ecosystem level, the most relevant responses are those that affect the ability of organisms to reproduce, such as fertility and sterility. Direct responses will be the number of gametes produced, gamete survival, increased mortality or juvenile stage aberrations. Indirect responses will be revealed through alterations in genetic material that can change reproductive parameters and the general condition of the individual.

7.1.2. *Deterministic and stochastic effects*

Exposure to ionizing radiation leads to two types of effects: deterministic effects and stochastic effects.

Deterministic effects occur with certainty. They usually occur rapidly and their severity increases with the dose. They are threshold effects. Deterministic effects usually occur for a single dose. They often occur in accidental situations. In addition to mortality, these effects concern, for example, fertility, morbidity, behavior, etc.

Stochastic effects (or random effects) are those for which the probability of occurrence increases with the dose delivered, while the severity remains constant whatever the dose. They include both cancer induction and transmissible genetic effects. The initial damage results from the effect at the molecular level; it takes the form of structural modifications of the DNA molecules, the effect being dependent on the type of radiation and the tissue exposed. The evolution of the DNA damage leads to various situations. A damage repair process is initiated and the cell survives and resumes normal function; a damage repair process is initiated but does not proceed normally. Latent damage is expressed in the affected cell and its progeny; damage leads to cell death. There is sufficient evidence to suggest that the induction of cancer by low doses is plausible.

7.2. Effects of ionizing radiation at the molecular level

Ionizing radiation can alter all biochemical molecules, starting with the most precious of them, DNA, which guarantees the survival of the species.

7.2.1. *Effects on DNA*

Deoxyribonucleic acid (DNA) is the essential molecule of life since it contains all the genetic information and ensures its transmission from generation to generation at each cell division. DNA is mainly found in the nucleus of cells, in direct interaction with proteins (histones) with which it forms a structure called chromatin, a constituent element of chromosomes. Hence, the attack on DNA by ionizing radiation can have severe mutagenic, carcinogenic and lethal effects on living organisms. Organisms that have been subjected to high levels of radiation since the beginning of life have developed systems for repairing their precious DNA.

DNA is a macromolecule in the form of a double helix interacting with water molecules. The action of radiation can therefore be direct by absorption of energy by components of the double helix, or indirect by their reaction with radiolytic species formed in the surrounding water [DOU 08] (Figure 7.1). The alterations are either single-strand breaks (SSBs), considered non-lethal because they can be repaired by ligases and polymerases, or double-strand breaks (DSBs), which are more or less repairable depending on the position and number of these breaks. There may also be base alterations, which are not lethal but may be the cause of a mutation in the case of non-repair or faulty repair (Figure 7.2). These lesions are repaired by excision-resynthesis mechanisms involving various enzymes (endonucleases, glycosylases, polymerases and ligases). Other lesions involve inter-strand, intra-strand or protein bridges. Some products may add up a radical binding onto a base ("adducts") or the intercalation of a substance in the DNA helix. The distribution of such damage along the DNA molecule strongly depends on the accessibility of the reactive sites of sugars and bases for the OH radical. This accessibility depends on the three-dimensional structure of the DNA molecule and thus on the sequence of the DNA as well as on environmental factors such as the concentration of metal ions.

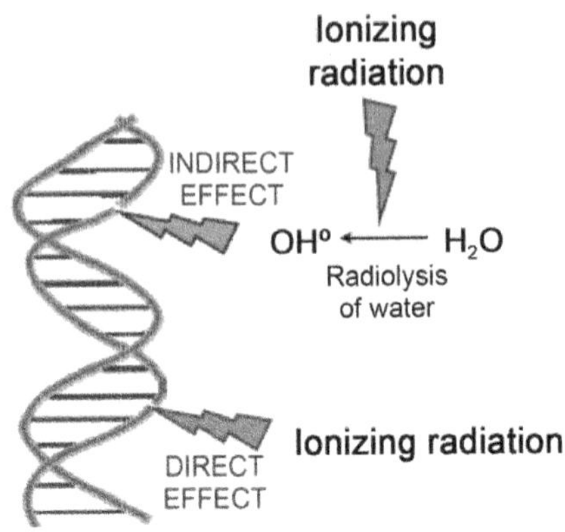

Figure 7.1. *Direct and indirect effects of ionizing radiation on DNA. For a color version of this figure, see www.iste.co.uk/amiard/radioecology.zip*

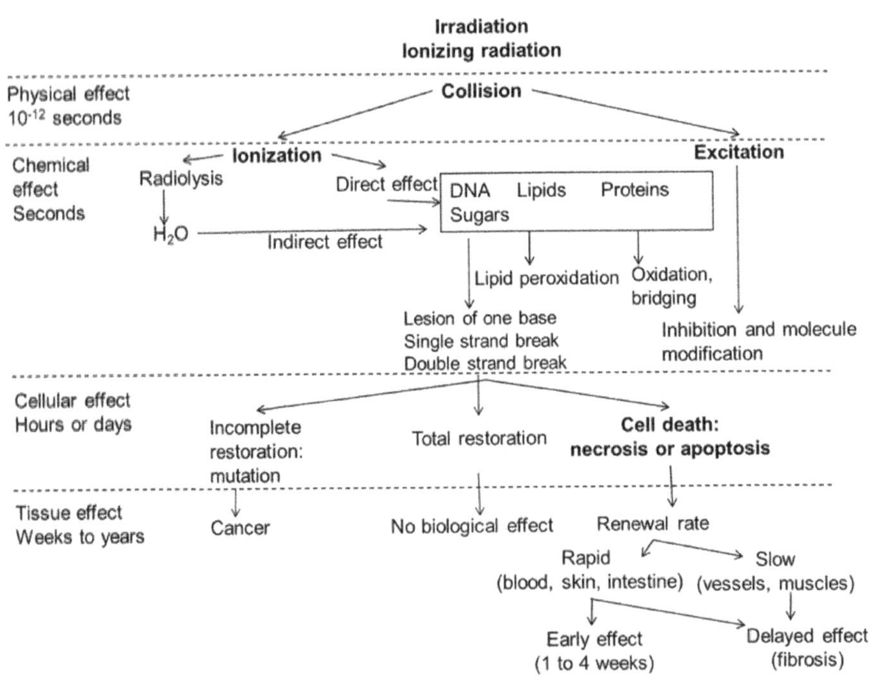

Figure 7.2. *Biological effects of ionizing radiation. Possible changes at the cellular level as a function of time (modified from [HAT 05])*

DNA repair occurs in a multi-step process. The first step is a signal in response to recognized damage on the DNA, which is transmitted by a cascade of reactions towards proteins responsible for stopping DNA replication until its repair or cell death if repair proves impossible. Damage is repaired by numerous mechanisms. Among these, let us point out the mechanisms of base excision repair (BER) for small damage. For large base lesions and DNA–protein bridges, nucleotide excision repair mechanisms (NERs) are used. Double-strand breaks are repaired either by homologous recombination (HR) (faithful process) or by non-homologous end joining (NHEJ) (unfaithful process). In the latter case, the repair mechanism leads to a loss of genetic material and therefore does not faithfully repair the damage. Depending on its location, the consequences of the mutation are more or less severe and can lead to cancer, sometimes after many steps.

The reactions of water radiolysis radicals with various individual components of DNA, including the bases adenine (A), guanine (G), thymine (T) and cytosine (C), or the sugars (deoxyriboses), have been extensively studied in the past. Today, spectroscopic techniques, analytical chemistry and fine techniques from molecular biology enable the examination of damage to the DNA itself as a whole and in its cellular environment. Among the indirect effects of radicals from water radiolysis on DNA, the OH$^\cdot$ radical is the most oxidizing [BUX 08].

Some ionizing radiations can cause damage clusters, i.e. lesions grouped in multiple damage sites (MDS). This may be the case, for example, of several nearby modified bases, or a single-strand break plus an oxidized base, or two double-stranded cuts, or any other multiple combinations of these lesions located on one or both strands of the DNA. Such multiple lesions are much less efficiently repaired by enzymes and ultimately lead to mutagenesis or programmed cell death or apoptosis (Figure 7.2).

The direct effects of the absorption of ionizing radiation by DNA depend on temperature, LET (Linear Energy Transfer) and DNA conformation (Bernhard and Sevilla, 2008). Like any irradiated system, the initial ionization produces electrons and cation radicals (or break) in the DNA, which are distributed in equal proportions between the phosphate-sugar backbone of the DNA and the bases. The oxidized radicals resulting from the capture of the "breaks" are mainly those of guanine and to a lesser extent adenosine, while the electrons are captured by thymine or cytosine. The particular distribution of radicals is due to the transfer of electrons (Cyt$^-$, Thy$^-$) or of "breaks" (G$^+$) from site to site. This is explained by the ionization potentials (G<A<C, T) and electronic affinities of the bases (C, T>>A>G). The fact that the guanine base is strongly affected explains why the oxidized base (8-oxo-7,8-dihydro-2'-deoxyguanosine, 8-oxdGuo) serves as a biomarker of DNA damage (Vasseur et al. 2008). Two distinct mechanisms of charge transfer coexist: one consists of a jump from near to near, and the other occurs in a single step (super-exchange) between two bases separated by one or more higher-level bases (or "bridge") constituting a thermodynamic barrier, crossed by a tunnel effect.

DNA is permanently damaged by the attack of endogenous metabolites or by irradiation. Daily single-strand cuts are estimated for endogenous metabolites to be between 10,000 and 55,000 (for irradiation at 1,000 Gy^{-1}), oxidized bases at 3,200 (2,000 Gy^{-1}), off-base sites at 12,600 (25 Gy^{-1}) and double-stranded cuts at 8 (40 Gy^{-1}). Other lesions are more specific to irradiation, such as multiple cluster lesions (150 Gy^{-1}) and cross-links between DNA and a protein (150 Gy^{-1}).

Few studies concern marine organisms. In the mussel *Mytilus edulis*, molecular responses to ionizing radiation have been quantified by analyses of mRNA expression of a gene involved in DNA double-strand break repair (RAD51) and by the induction of DNA strand breaks using the comet assay. Samples collected from a site impacted by low levels of radionuclides with low external and internal dose rates, approximately 0.61 $\mu Gy.h^{-1}$, are significantly lower than the generic (all species) "no effect" dose rate of 10 $\mu Gy.h^{-1}$. However, DNA strand breaks and RAD51 mRNA expression were both altered [ALA 12].

7.2.2. *Induction of cytogenetic and genetic effects*

Irradiation is a mutagenic and carcinogenic agent because it causes damage to cytogenetic material. Damage can occur directly through energy transfer to critical molecules or indirectly through interaction with constituents of the internal environment. Inside the cell, the molecular material can be ionized, forming free radicals that will cause damage to the cytogenetic material. These genetic effects will be transmitted to the offspring and will affect not only the reproductive process but also the metabolism and physiological performance of the individual.

Damage to the genetic material of a population is not always harmful because the mutation may turn out to be an advantage for the carrier individual. However, this will become harmful if the mutation rate is too high, with mutations that are too severe, not eliminated by natural selection or if the DNA damage is not repaired.

The parameters measuring cytogenetic and genetic effects are numerous. For example, Jha et al. [JHA 05, JHA 06] evaluated the genotoxic effects of tritium in the adult mussel *Mytilus edulis* using the micronucleus (MN) induction and Comet assays in hemocytes of exposed individuals. The effects for both tests are clear for an acute irradiation of 96 h at dose rates of 12 to 485 $\mu Gy\ h^{-1}$. These dose rates are found to be lower than the generic dose rate of 500 $\mu Gy\ h^{-1}$ recommended by the IAEA for the protection of aquatic biota.

Recent work has begun to show that epigenetics play an important role in the adaptation of organisms to environmental stimuli. A review of the current knowledge on the possible involvement of epigenetics in the cascade of responses resulting from environmental exposure to ionizing radiation has been performed by Horemans et al. [HOR 19]. Such a biomarker could be used to detect and better understand the mechanisms of toxic action and inter-/intra-species susceptibility to radiation in the context of environmental risk assessment and management.

Assuming that the relationship between DNA mass per haploid genome and mutation rate per Gy and per locus is linear, the IAEA [IAE 76] calculates that in fish, if both parents each receive an irradiation dose of 0.005 Gy, less than one embryo in a thousand will not be viable (mutation rate estimated at 7.10^{-6} per Gy and per zygote).

Mutations in minisatellites (non-coding, repetitive areas of DNA with a high mutation rate) have also been extensively studied in mammals [DUB 93] and fish (the medaka *Oryzias latipes*) [SHI 98, SHI 04b] after irradiation. The studies showed that some mutations were transmissible to subsequent generations. The instability of minisatellites observed after exposure to non-genotoxic agents raises the question of the influence of epigenetic events on genomic functions.

7.2.3. *Effects on other biomolecules*

The radiation-induced modification of the structure of proteins leads to an alteration of their function (enzymatic, transport, structure, hormonal). These radiation-induced modifications lead to disturbances of cellular physiology and can contribute to the lethality process. In terms of lipids, lipid peroxidations are at the origin of a decrease in membrane fluidity, having as a consequence, once again, an alteration of the properties of the cell's membrane system. The attack of sugars (polysaccharides) is less well known, but reactive oxygen species attack mucopolysaccharides and in particular the proteoglycans of cartilage. In addition, glucose can be oxidized and release ketoaldehydes, H_2O_2 and OH° (Figure 7.3).

As a result, the major cellular functions that can be altered by ionizing radiation are numerous. Four major cellular functions are particularly affected. These are the regulation of oxidative stress (MT, SOD), cell cycle control (p53, p21), induction of chaperone proteins (Hsp70, Hsp90) and cellular detoxification (CYP450, GST, MXR).

The p53 protein is present in all cells of the body. In the absence of stress, p53 is present in an inactive state and is not phosphorylated. During normal cell division, p53 is not required. During the response of a normal cell to DNA damage, the "genome guard" function of p53 is induced. Cells in which p53 is absent or inactivated by mutation or by interaction with viral or cellular proteins, divide while the DNA has not been repaired. This results in mutations, aneuploidy or defects in mitosis leading to cell death. Cells that survive these alterations are likely to become malignant.

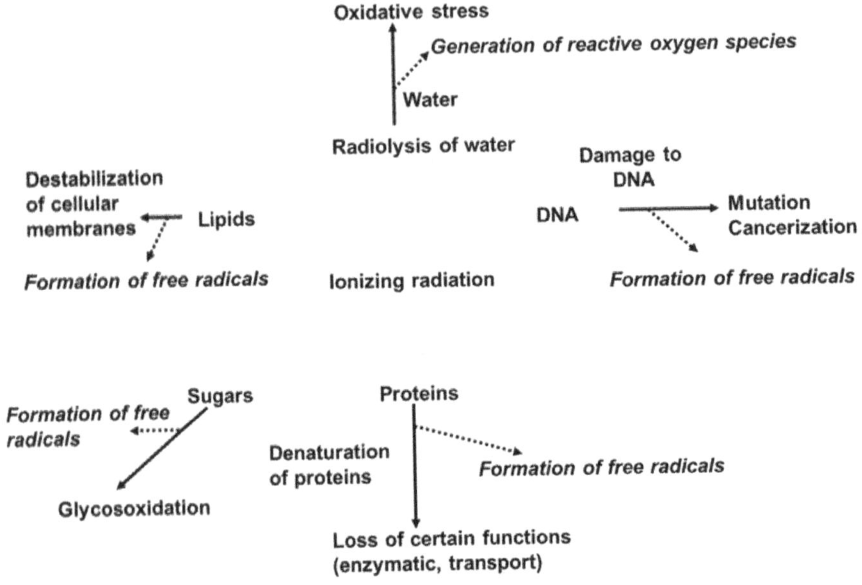

Figure 7.3. *Effects of ionizing radiation on various biomolecules*

7.3. Effects of irradiation at the subcellular and cellular levels

Cellular organelles, especially chromosomes, are subject to structural alterations under the effect of ionizing radiation.

7.3.1. *Physical alterations of chromosome structure*

Direct or indirect damage to DNA can result in irreversible damage or alterations to genetic material, and responses can be immediate or latent. Significant damage occurs when the genomic template is altered. This is particularly the case in transcriptional defects, point mutations, dominant lethal mutations, deletions or when there is no transcription.

Physical alterations of the chromosome structure can concern chromosomal aberrations (CAs), sister chromatid exchanges (SCEs), anaphase aberrations (AAs) and micronuclei formation (MNs).

CAs are formed by deletions and breaks with possible reassemblies giving rise to dicentric chromosomes, rings, acentric fragments, and so on. SCEs are formed by exchanges of chromosomal material between sister chromatids at the same locus. MNs are formed by consequent breaks of chromosomes and improper realignment. AAs are formed when there are chromosome breaks, spindle failures and chromosome viscosity.

Various criteria or biomarkers provide information on the cytogenetic and genetic effects of ionizing radiation and may be more sensitive than fertility criteria. This is the case for DNA strand breaks, which are increased with an irradiation dose of 0.3 Gy or dose rates of 0.38 $Gy.d^{-1}$. These criteria can also be used in in situ studies [HAR 01].

All levels of ionizing radiation tested from 1.8 to 6.8 Gy (180 to 680 rads) were found to cause significant chromosomal damage in *Nereis arenaceodentata*. The nature of this damage depended on the dose, the dose rate and the DNA repair capacity of the worm [PES 81].

7.3.2. *Histopathological changes*

Histopathological studies of high doses of radiation in invertebrates and lower vertebrates demonstrate that the mechanisms of induction of radiation-induced mortality are the same as in mammals. Studies have focused on the hematopoietic system, the digestive tract and the immune system. The effects observed are only at high doses of irradiation and can therefore only concern the consequences of the radioactive releases of severe accidents.

Findings of subcellular alterations due to ionizing radiation have been little researched in marine organisms. However, a comparison of radiation effects on primary cultures of rainbow trout and Dublin Bay prawn, *Nephrops norvegicus* for irradiation doses of 0.5 and 5 Gy was performed. Irradiated cultures of both organisms showed pronounced morphological damage, particularly to the nucleus and mitochondria. Cytoskeletal damage was also evident. However, the cells of the hematopoietic tissue of *Nephrops norvegicus* appeared to be considerably more radiosensitive than the zygote cells of the rainbow trout (Lyng et al. pp. 25–31, *in:* [IAE 03a]).

Primary cultures of rainbow trout skin receiving cobalt-60 gamma radiation doses ranging from 0.5 Gy to 15 Gy showed nuclear aberrations and changes in cytoplasmic organelles, particularly mitochondria (Olwell et al. pp.390, *in:* [IAE 03a]).

At the Mururoa site, in six species of serranid fish, the gonadal organization and the ultrastructural characteristics of the germ cells and somatic cells are not modified. This implies that the reproductive potential of these fish has remained intact despite environmental pollution, in particular the presence of radionuclides [BRU 98].

7.3.3. *Induction of defense and damage biomarkers*

Following the example of the effects of chemical pollutants whose impact is evaluated by the induction of biomarkers which reflect either defense or damage [AMI 08a; AMI 08b; AMI 17b], various research studies have been initiated for radionuclides. Thus, for example, within the framework of the ENVIRHOM program, Farcy [FAR 06] highlighted two potential candidate genes in the oyster *Crassostrea gigas* for the study of the effects of ionizing radiation in chronic and low-dose situations. These genes code for metallothionein (MT) and for multi-xenobiotic resistance (MXR). The transcriptional induction of these two genes has been confirmed in response to high doses of acute irradiation.

Of the metals measured in oyster tissues, Ag, Cd, Cu, Ni and Zn showed the strongest associations with pro-oxidant and oxidative stress responses (superoxide dismutase, lipid peroxidation and lysosomal membrane destabilization) and detoxification responses (glutathione and metallothionein), suggesting that current metal contamination still exerts a significant amount of stress in biota [CHA 19].

Metallothionein (MT) plays an important role in the detoxification of metals and their radionuclides in vertebrates and invertebrates. However, in case of very high contamination, this protein does not assume its role anymore [AMI 06]. This was observed by Liu and Wang [LIU 16] in two species of oysters transplanted for two months into a highly polluted Chinese estuary (Juilong).

7.4. Effects of irradiation on individuals

Research has been conducted to determine the doses received, the responses to those doses and the mechanisms of action of radioactivity on living matter. Numerous reviews of the effects of radiation on aquatic organisms have been conducted because of the introduction of radionuclides into these environments

[IAE 88, NCP 91, UNS 96]. The two main types of studies are in situ and experimental. Much of our knowledge comes from information and observations gathered following the Chernobyl accident.

Three types of experiments reproduce with more or less realism a chronic irradiation dose: the use of external sources (X-rays, sources of ^{60}Co or ^{137}Cs) simulating external irradiation, irradiation due only to incorporated radionuclides (internal irradiation) and organisms living in a contaminated environment (combining external and internal irradiation).

Responses to radiation are a function of many extrinsic factors. The first factor is the type of radiation. Radiation is diverse (X-rays, γ rays, β particles, Auger electrons and neutrons). A relationship between the type of ionizing radiation and the effect on living matter has been established [BAR 93, BAR 94, GOO 93]. Due to the different densities of the ionizations, the hierarchy of biological effects is as follows: α particles > protons = neutrons[1] > β particles = γ rays. In humans, this is taken into account by means of the EBR. This coefficient is a measure used to compare the biological effect of two radiations. To characterize the relative biological effectiveness, it is conventional to take as a reference the biological effect of X- or gamma radiation from cobalt-60 [IAE 79].

The second factor that affects the response is the dose rate. A higher dose is required at low dose rates to produce the same effect as a given dose at high dose rates. The importance of dose rate explains the reason for the separate review of acute and chronic exposures.

Responses to radiation can also be modified by environmental factors such as oxygen level, temperature, salinity and even pressure. Little information is available on the effect of oxygen level in non-mammals, whereas the radioresistance of mammals is increased by a factor of 2 to 4 in individuals in anoxic environments. This is related to oxidative stress.

Low temperatures, slowing down metabolism, reduce the effects of radiation even though the repair of molecular damage is also slowed. Salinity increases osmoregulatory efforts and makes individuals more radiosensitive.

While significant efforts have been made to develop radiation protection for human populations, their offspring and the human race against cancers and genetic diseases, little has been done for all other living species. The assumption has been that if humans were protected, all other species that were less radiosensitive would be protected [ICR 91].

1 In humans, the effect of neutrons is generally considered to be greater than that of protons.

7.4.1. *Mortality*

In acute dose studies (effects at high doses and high dose rates of radiation to individuals), the main mortality criterion used in the experiments is the lethal dose that causes death in half the group of individuals subjected to that dose (LD_{50}). In mammals, the onset time is expressed in seconds or minutes and the standard observation time is 30 or 60 days. In invertebrates, the onset of mortality may be slower but there is no standard period.

The other important parameter to consider is the dose–response curve. It is often a sigmoidal response with for cumulative effects a normal distribution. However, two species may have the same LD_{50} value and a different response curve. This is especially true for chronic effects.

7.4.1.1. *Mortality of invertebrate organisms*

Sensitivity to radiation is variable according to systematic group (Table 7.1). Bacteria appear to be highly radioresistant, with *Deinococcus radians* in particular tolerating doses as high as 30,000 Gy. Deep-sea marine bacteria seem less radioresistant. Similarly, protozoa are relatively resistant, such as *Paramecium aurelia*, whose LD_{50} is 1,000 Gy [DAL 95]. For many adult invertebrates, the LD_{50} varies from a few dozen to a few hundred Gy. However, values as low as 2.1 Gy are reported for the decapod crustacean *Palaemontes pugio* [ENG 74]. The comparison of radiosensitivity will be developed below (section 7.4.4). The crab *Callinectes sapidus* is one of the most radioresistant species, since an overall dose of 457 Gy delivered in 70 days seems to be necessary for the lethality rate to increase [ENG 67].

There are few studies of chronic effects on mortality in invertebrates. The high dose of 0.3 $Gy.h^{-1}$ for 70 days causes the death of the adult blue crab *Callinectes sapidus* [ENG 67]. The same is true for juvenile mollusks (the scallop *Argopecten irradians*) for a dose of 0.37 $Gy.h^{-1}$ for 84 days, whereas the same dose for 14 months in another bivalve mollusk (*Mercenaria mercenaria*) only leads to a decrease in reproduction [CHA 72].

The genotoxic, cytotoxic, developmental and survival impact of tritiated water (HTO) was studied in embryonic larvae of the marine mollusk *Mytilus edulis*. One-hour-old embryos were exposed to a range of concentrations (0.37–370 $kBq.\ mL^{-1}$) of HTO, which delivered a dose between 0.02 and 21.41 mGy. From the lowest dose of irradiation chromosomal aberrations appeared and the survival rate decreased [HAG 05].

Taxonomic group	DL_{50} (Gy)
Microorganisms	
Protozoa	>1,000
Bacteria	>30,000
Coelenterates	
Hydrozoans	20–120
Poriferous	
Sponges	20–120
Platelminthes	
Planarians	55
Annelids	
Adults	500–650
Juveniles	500
Mollusks	
Adults	50–500
Juveniles	11
Crustaceans	
Adults	2.1–566
Echinoderms	100

Table 7.1. *Lethal dose ranges (Gy) in invertebrates (modified from [HAR 01])*

7.4.1.2. *Mortality of vertebrats*

As for invertebrates, it is the juvenile stages of vertebrates that are the most radiosensitive. In adults of various fish species, the LDs_{50} vary from 3.75 to 100 Gy (Table 7.2). The dose rate and the interval between the various irradiation periods have an influence on the effects. If the dose rate is lower and the time interval between irradiation periods is increased, a higher radiation dose is required to observe the same effects as with high irradiation without division spread over time.

The correlation between radiosensitivity and phylogenetic position is not absolute. Thus, the LD_{50} in the amphipod crustacean *Gammarus duebeni* is 40 Gy and in *Gammarus zaddachi* 17 Gy [HOP 73]. Similarly, the crab *Callinectes sapidus*has a high radioresistance (LD_{50} of 560 Gy).

Taxonomic group	DL_{50} (Gy)
Juvenile fish	0.16–5
Adult fish	3.75–100
Juvenile amphibians	10
Adult amphibians	7–23

Table 7.2. *Lethal dose ranges (Gy) in vertebrates (modified from [HAR 01])*

Mortality following chronic radioactive contamination is variable among the various zoological groups (Table 7.3).

Organism	$DL_{50/30j}$ Gy
Mollusks	370–1,100
Crustaceans	15–510
Except young amphipods	6
Fish	11–56
Except eggs and biostoderm stage	1

Table 7.3. *Sensitivities of various zoological groups of marine organisms (from [WOO 97b])*

7.4.2. Alteration of reproduction

7.4.2.1. *Sterility and reduced fertility*

There is a wide range of responses for sterility-inducing doses in marine organisms ranging from 21 to 1,000 Gy. Sterility-inducing dose rates vary widely among species ranging from 20 to 140 $mGy.h^{-1}$ (Table 7.4). Sterility is dependent on dose rate and exposure time. Mammals appear to be more sensitive than fish and invertebrates, but the most sensitive fish, *Ameca splendens* (freshwater fish), is almost as sensitive as humans (0.6 and 0.23 mGy.h, respectively^{-1}) [HAR 01].

In the mollusk *Physa heterostropha* (a freshwater species that tolerates a certain salinity), doses of 100 and 250 $mGy.h^{-1}$ for 24 weeks cause a decrease in fecundity and lifespan and an acceleration of growth [COO 71]. In this species, an irradiation dose of 0.0065 $Gy.d^{-1}$ causes a decrease in the number of reproductive capsules maturing, but because each capsule contains a greater number of eggs, the total number of eggs laid is identical in controls and irradiated individuals [COO 70, COO 71]. There is a linear relationship between the radiation dose to the sperm of the spawners and the occurrence of eye malformations in trout embryos between 0.25 and 4 Gy [MCG 72].

	Sterility		**Fertility**	
Taxonomic group	A	B	A	B
Invertebrates				
Neanthes arenaceodentata	50	20	0.5	0.19
Artemia salina	21	1,400		
Fish				
Oryzias latipes	80	140	2.5	4.2

Table 7.4. *Induction of sterility and reduction of fertility in invertebrates and vertebrates after acute and chronic exposure (modified from [HAR 01]). A: acute exposure (dose in Gy); B: chronic exposure (dose rate in mGy.h^{-1})*

Examples will be provided below for reference species. *Daphnia* (some species withstanding low salinity) were chronically exposed to the alpha-emitting radionuclide americium-241 at concentrations of 0.4 to 40 Bq.mL^{-1} corresponding to average irradiation rates of 0.3, 1.5 and 15 mGy.h^{-1}. Am-241 amounts were determined in the medium, tissues and cuticles of the *Daphnia*. The calculated average dose rates were 0.02, 0.11 and 0.99 mGy.h^{-1} for whole organisms, and the internal radiation dose contributed 99% of the total dose rate. The increase in respiratory demand after 23 days of exposure at the highest dose rate suggested an increased metabolic cost due to counteracting radiation stress from alpha radiation. Exposure at dose rates of 0.11 and 0.99 mGy.h^{-1} caused a significant 15% reduction in organism mass. Fecundity remained unchanged over the 23-day period, but individual egg and neonate masses were significantly reduced compared to controls. Since ingestion rates were similar for all individuals, it therefore appears that the increased metabolic expenditure in the chronically irradiated *Daphnia* came from a decrease in their energy investment in their offspring. As a result, the neonates showed significantly reduced resistance to starvation at each dose rate compared to controls [ALO 06, GIL 08].

In another 70-day experiment, Alonzo et al. [ALO 08a] exposed *Daphnia* (*Daphnia magna*) to the same doses of americium-241 (0.4 to 40 Bq.mL^{-1}). Various parameters and reproduction were followed over three successive generations. This study confirmed that oxygen consumption increased significantly in the first generation (F0) after 6 days of exposure to an irradiation dose of 1.5 mGy.h^{-1}. The consequences were limited to a reduction in body length (5%) and in the dry mass of females (16%) and eggs (8%) after 23 days of exposure, while mortality and fecundity remained unchanged. For neonates of successive generations, the results were in sharp contrast to those observed in the first generation (F0). At the highest radiation dose rate (15 $mGy.h^{-1}$), an initial mortality of 38–90% affected the juveniles while the survivors had delayed reproduction and reduced fecundity for the F1 and F2 generations. At 0.3 and 1.5 $mGy.h^{-1}$, *Daphnia* mortality ranged from 31 to 38%, but only for the F1 generation started with the first breeding neonates. Reproduction was affected by a reduction in the proportion of females reproducing, for the first generation at 1.5 $mGy.h^{-1}$ (62% of total *Daphnia*) and for the second generation at 0.3 $mGy.h^{-1}$ (69% of total *Daphnia*). Oxygen consumption remained significantly higher at doses above 0.3 $mGy.h^{-1}$ than in controls for almost all generations. Organism size and mass continued to decrease in a dose-dependent manner, with a significant reduction in mass from 15% at 0.3 $mGy.h^{-1}$ to 27% at 15 $mGy.h^{-1}$ for the second-generation offspring [ALO 08a].

Alonzo et al. [ALO 08b] evaluated the potential effect of irradiation on population growth using simple population models and parameter values from chronic exposure experiments in two invertebrate species, the earthworm and the Daphnia, with contrasting life history strategies. In the earthworm *Eisenia fetida*, models predicted an increase in population growth delay with increasing gamma irradiation dose rates (up to 0.6 of a control generation time at 11 $mGy.h^{-1}$), with population extinction predicted at 43 $mGy.h^{-1}$. In the microcrustacean *Daphnia magna*, models predicted an increase in population growth delay with increasing alpha irradiation dose rates (up to 0.8 of a control generation time at 15.0 $mGy.h^{-1}$), while only two successive generations were exposed. The effects of population-level changes were estimated using different individual criteria (including survival, number of offspring produced and time to first reproduction). The two species did not respond in the same way. Fast-growing *Daphnia* were more susceptible to reduced fecundity or delayed first reproduction than slow-growing earthworms. Alonzo et al. [ALO 08b] suggest that it is not advisable to consider the sensitivity of a population to irradiation independently of the life history of the species.

7.4.2.2. Reproductive success

The reproductive success of a species is related to a number of parameters such as the inherent radiosensitivity of the tissues of the reproductive system and juvenile stages of the species, gametogenesis processes, reproductive strategy and historical exposure of the individual. These effects may be revealed with low doses and low dose rates of radiation.

Contamination of the marine amphipod *Allorchestes compressa* with uranium for approximately 10 weeks influences its growth and sex ratio. At 0.1 $mg.L^{-1}$, uranium increased growth by 23%, as measured by mean weight after 4 weeks, and at 2 $mg.L^{-1}$, growth was reduced by 28% compared to the control. The amphipod bioaccumulated uranium from seawater with a concentration factor of 10. The number of males, sex ratio and respiration rate (measured on males only) at 1 $mg.L^{-1}$ were significantly lower than in the control [AHS 86].

In fish, the UNSCEAR synthesis [UNS 08] reports no harm for dose rates ranging from 0 to 99 $\mu Gy.h^{-1}$.

The review by Egami and Jiri [GGA 79] pointed out that the radiosensitivity of gametes and embryonic cells undergoes a considerable change during these processes. The order of decreasing sensitivity of spermatogenic cells is from the most sensitive spermatogonia, through spermatocytes and primary spermatids. Bergonié and Tribondeau's classical law [BER 06] states that the sensitivity of cells varies directly with their mitotic activity and inversely with their degree of differentiation.

The long-term development of populations of two fish, European plaice (*Pleuronectes platessa*) and thornback skate (*Raja clavata*), using mathematical models confirms the relative sensitivities of the two populations, as might be predicted on the basis of their life histories and reproductive strategies, to the possible effects of radiation on individual fertility, fecundity and mortality (Woodhead, pp. 32–42, *in*: [IAE 03a]). Combining these three factors, the survival of the two species decreased in a fairly similar manner.

Comparison of the effects of chronic exposure to the same absorbed dose rates of β-radiation from tritiated water, or radiation from an external source of ^{137}Cs, on the reproductive performance of the polychaete *Ophryotrocha diadema* was performed. At 7.3 $mGy.\ h^{-1}$, β and γ radiation resulted in similar and significant reductions in larval numbers compared to controls. Irradiated worms had significant reductions in egg production [KNO 97].

7.4.2.2.1. Inherent radiosensitivity and fertility

The factors inherent in radiosensitivity are controlled by the genetic material of the species and the repair processes of its DNA. They can be affected by the developmental mechanisms of the various cell types. Because of the high sensitivity of gonadal tissues at early stages and because fertility affects reproductive success directly, fertility has been selected as a useful indicator of inherent radiosensitivity.

The effects of acute irradiation on fertility vary according to species from 0.02 to 20 Gy, i.e. three orders of magnitude (Table 7.5.). However, for most of the organisms studied, the critical limit for observing a decrease in fertility appears to be between 0.1 and 1 Gy. Doses below 0.02 Gy do not appear to affect fertility.

Effects of chronic exposures on fertility in invertebrates are observed from 0.2 to 55 $mGy.h^{-1}$, in fish from less than 0.6 to 13 $mGy.h^{-1}$ and in mammals from 0.023 to 0.07 $mGy.h^{-1}$. Dose rates below 0.02 $mGy.h^{-1}$ do not appear to affect fertility.

Data on the chronic effects of irradiation on invertebrates are few and also very limited in vertebrates for amphibians, reptiles and birds. For mammals, the UNSCEAR synthesis [UNS 93] suggests that responses depend on the stage of development of gonadal tissues and the duration of irradiation, with great variability depending on the species concerned. The fact that the values for effects in invertebrates are close to those observed in mammals seems to indicate that at the cellular and molecular levels, radiosensitivity differs little for the same stage of gametogenesis.

Groups	Dose (Gy)	Dose rate ($mGy.h^{-1}$)
Invertebrates		
Neanthes arenaceodentata	0.5	0.19
Fish		
Oncorhynchus tschawytcha	2.5	4.2
Oryzias latipes	5	2.8

Table 7.5. *Acute radiation doses (Gy) and chronic dose rates ($mGy.h^{-1}$) with an effect on fertility (modified from [HAR 01])*

Eelpouts (*Zoarces viviparus*), a marine fish, were exposed to γ radiation at a dose rate of 2.0 $mGy.h^{-1}$. There was a significant radiation-induced reduction in relative weight (gonadosomatic index) [GRE 95].

Two experiments were performed with flatfish (plaice). In the first experiment, adult male fish were exposed to average absorbed dose rates of 0.25, 0.5 or 1.2 $mGy.h^{-1}$ for 73 days (average cumulative doses of 0.43, 0.85 and 2.03 Gy, respectively). In the second experiment, the dose rates were 0.24, 0.5 or 1.0 $mGy.h^{-1}$ for 197 days (average cumulative doses of 1.07, 2.24 and 4.57 Gy). From the dose rate of 0.24 $mGy.h^{-1}$, the weight of the fish testicles decreased [KNO 99].

7.4.2.2.2. Factors affecting inherent radiosensitivity

The main parameters that contribute to the inherent radiosensitivity are the biological repair capabilities, which are sensitive to dose and dose rate, and the repopulation and specialization capabilities of cells. Biological repair includes the repair of nuclear and cytoplasmic materials. In the nucleus, the main repair is to the DNA.

The ability of cells to repair damage from ionizing radiation was noted very early on. Thus, organisms exposed to split doses of radiation are less sensitive than those receiving a single dose. The hypothesis was that this division of the dose allows the repair processes to reduce the damage. The most studied repair mechanism is the one involving DNA (see above). Information exists on DNA repair systems for fish, bivalve mollusks and annelids (*in:* [HAR 01]). One of the most remarkable examples is the repair capacity developed by the bacterium *Deinococcus radiodurans*, which is able to survive irradiations of more than 30,000 Gy, while its DNA is split into several hundred fragments [DAL 95].

Irradiation of a dividing cell usually causes the process to stop, otherwise the cell is much more sensitive to radiation at this stage. Monitoring mechanisms of the cell machinery are responsible for checking the correct state of the DNA and the sequence of events in the cell cycle. In case of problems, inhibitors are responsible for stopping the cell machinery [CAR 95, CAR 96]. The monitoring mechanisms are conserved in all eukaryotes [ELL 96].

The antioxidant defenses against the appearance of free radicals during irradiation are of three major classes: water-soluble reducers (glutathione, ascorbate, urate), fat-soluble vitamins (alpha-tocopherol, beta-carotene) and enzymes (peroxidase, catalase, superoxide dismutase). The latter are inducible under conditions of oxidative stress.

Cellular repopulation and specialization are also an inherent component of radiosensitivity, varying between species and between organs. Some species are able to regenerate tissues and organs even as adults. These species will therefore be more radioresistant than others. However, if the repopulation and specialization of the primordial cells of the reproductive organs do not have this possibility, then the reproductive success is concerned and these species will probably be radiosensitive. In humans, while the cells of the digestive tract, the hematopoietic tissues and the cells of the testicles are able to divide, those of the nervous tissues and the primary cells of the ovaries no longer divide or divide at a very low rate. We must remain cautious about this information because our knowledge of stem cells is evolving rapidly at the moment.

7.4.2.2.3. Gametogenesis strategy

The processes occurring during gametogenesis and their duration also significantly affect the reproductive success of a species, and they vary greatly among species. The most important parameters are the capacity for repopulation and repair of primordial germ cells (see above 7.4.2), the duration and synchrony of the stages of gametogenesis, the latency time between gamete production and utilization, and the time to reach sexual maturity. Thus, in the polychaete annelid *Neanthes arenaceodentata*, oocyte development is synchronous and the female dies after a single oviposition, implying that oocytes will only be exposed to chronic doses for a short period of time. Moreover, in this species, the testes are more radioresistant than the ovaries.

7.4.2.2.4. Reproductive strategy

The reproductive strategies of species are of two types, including r-strategists and K-strategists. The r-strategists with a large number of offspring have a considerable mortality under normal conditions, and this mortality may go unnoticed in the event of irradiation. For example, according to Blaylock [BLA 69], the fish *Gambusia affinis* (freshwater fish) receiving a dose of 0.109 $Gy.d^{-1}$ has a higher mortality and number of abnormal embryos than control populations, but this is compensated for by the spawning of a greater number of eggs. This may be an adjustment of the natural irradiated population to an increase in the nuisance experienced. K-strategists generally have strategies for protecting their offspring (brooding of young stages, nest guarding, viviparous development) that are ineffective against the effects of irradiation.

Another important factor in reproductive strategy besides the total number of gametes produced, the rate of division and the sensitivity of the gametes is the time between the formation of primary germ cells and the release of mature gametes. The longer this time, the higher the irradiation dose. Here also long-lived species will be more exposed to chronic irradiation

7.4.2.2.5. Historical exposure

The reproductive success of a species in a natural ecosystem is affected by changes in part of the gene pool as a result of exposure to irradiation affecting successive generations. Since exposure of organisms is usually chronic, low-level and multigenerational, it is possible that genetic drift occurs in irradiated populations leading to some resistance of exposed individuals. Data in this area are very scarce, most studies being shorter than the lifespan of the individuals of the species studied. On the other hand, over long periods of time, confounding factors in addition to irradiation are likely to cause physiological, genetic effects.

There is considerable evidence that individuals subjected to low levels of radiation exhibit changes at the cellular level that reflect the ability of organisms to adapt to the effects of radiation. This phenomenon is called the “adaptive response”. The response can last for several hours in mammals and is sometimes referred to as the stress response or the genotoxic stress response. This may lead to an overestimation of the effects at low doses of radiation [UNS 94].

7.4.3. *Influence of the vital stage*

It has long been known that the life stage is of enormous importance with regard to radiosensitivity. The early stages of embryonic development are the most radiosensitive and the embryo is the most radiosensitive stage of the entire life cycle of an organism [RUG 60]. Early work with irradiation doses, often of X-rays, verifies this statement for many marine organisms, such as annelids, echinoderms, mollusks (including cephalopods) and fish. Ova are generally more radiosensitive than spermatozoa, and the stages during the period of intense cell division are the most radiosensitive [RUG 60].

The radiosensitivity of eggs depends on the degree of radionuclide accumulation in and on the egg, as well as on the radionuclide itself, the chemical composition of the water, the epiphytes present on the egg and the fish species involved [EGA 79].

So far, the stage most radiosensitive to acute external irradiation has been observed to be the first mitosis of Pacific salmon (*Oncorynchus kisutch*) eggs with a LD_{50} of 160 mGy. For the axolotl embryo at the early gastrula stage, the LD_{50} is less than 100 mGy in 17 days (*in:* [POL 66]). Feldt [FEL 72] reports that the growth of herring is seriously impaired for an acute irradiation dose of 50 mGy, undergone at the stage of first mitosis.

Juvenile stages are generally more sensitive than adults and sensitivity decreases with age[2], as is well illustrated in Table 7.6 for salmon. Thus, the LD_{50} in *Physa acuta* is 11, 14, 49 and 110 Gy for the four-cell, trochophore, veliger and hippostage stages, respectively [RAV 67]. This is also true in Tables 7.1 and 7.2 for lethal doses in invertebrates and vertebrates.

Irradiated stage	**DL_{50} (Gy)**
1 cell	0.58 ± 0.04
32 cells	3.1 ± 0.1
Early-ring germ	4.6 ± 0.2
Late-ring germ	4.5 ± 0.2
Early "pigmented eyes" stage	4.2 ± 0.2
Late "pigmented eyes" stage	9.0 ± 0.4

Table 7.6. *Relative radiosensitivity of different embryonic stages of Salmo gairdnerii to X-ray irradiation (modified from [DON 57])*

2 This is also observed in humans.

In the case where irradiation is only external, the most conclusive experiment because of its long duration and the large number of individuals tested is that by Bonham and Donaldson (*in:* [BON 77]). In this experiment, Chinook salmon eggs (*Oncorhynchus tschawystscha)* were irradiated with 0.005, 0.05, 0.10 or 0.20 $Gy.d^{-1}$ (source: ^{60}Co) from fertilization to the stage of yolk resorption (about 80 days). The juveniles were then marked and released into the river from where they reached the ocean. After one to three years at sea, the return of mature adults was quantified and crosses between irradiated and non-irradiated individuals were made. The first harmful effects (early death, developmental and growth rate abnormalities, delayed gonad development, less frequent return of adults, etc.) appeared only for doses equal to or higher than 0.10 $Gy.d^{-1}$ (total dose ~ 8.10 Gy). The authors point out that this dose, for which harmful effects appear, is 800 times higher than the irradiation dose experienced by organisms living in the Columbia River. However, there is a shortcoming in that the authors provide us with the radiation doses delivered to salmon eggs and larvae and not the absorbed radiation doses, which must be less.

External irradiation cannot account for all the effects of radioactive contamination, because we have seen that radionuclides can accumulate preferentially in certain organs and then low-energy or high-mass radiation (conversion electrons, Auger electrons, alpha emitters) increase the absorbed dose by a factor of 10 to 20, or even more than 300.

Irradiation will only be internal in the case of contamination by food (or by peritoneal injection). Thus, Watson et al. [WAT 59] found that daily ingestion of ^{32}P for 25 weeks by *Salmo gairdnerii* at a rate of $0.22.10^3$ $Bq.g^{-1}$ had no effect, but that with $2.2.10^3$ $Bq.g^{-1}$, growth was slowed and leukopenia appeared after four months. In the same species, Nakatani and Foster [NAK 63] showed that ingestion of $18.5.10^6$ $Bq.g^{-1}.d^{-1}$ of ^{90}Sr-Y affected growth, mortality rate, damage to the digestive mucosa and the appearance of leukopenia within 10 weeks. Zinc-65 appears to be less harmful to trout [NAK 66] since with the ingestion of 37.10^3 $Bq.g^{-1}.d^{-1}$, leukopenia took 10 weeks to appear. The amount of radionuclides actually retained by the fish is small. It is, for example, about 25% in the case of ^{90}Sr-Y.

The incorporation of certain radionuclides is much more harmful. Thus, it is sufficient for phytoplanktonic algae to remain for 24 hours in water containing 1,110 $Bq.L^{-1}$ of ^{226}Ra for their photosynthetic activity to be reduced [HAV 73]. Similarly, because of the low-energy beta particles of tritium, the dose delivered to an organism living in tritium-contaminated water is largely due to the incorporated radionuclide. For eggs and larvae of several fish species (*Cyprinus carpio*, *Gasterosteus aculeatus*, *Lebistes reticulatus*, *Paralichthys olivaceus* and *Fugu niphobles*), no effects were observed at a concentration of $18.5.10^9$ $Bq.L^{-1}$ (about 1.5 $Gy.d^{-1}$) (*in:* [IAE 76]) at least for the somatic effects and physiological functions studied. In contrast, Strand et al. [STR 73] showed that tritiated water concentrations as low as 37.10^3 Bq.L^{-1} suppressed the immunity response to *Chondrococcus columnaris* in trout.

The experimental contamination of aquatic organisms by water should have improved the study of the effects of ionizing radiation, the irradiation being both internal and external. The results of the various works on the contamination of fish eggs and larvae fall into two quite distinct categories. On the one hand, Soviet work (*in:* [POL 66]) detects the appearance of anomalies in the development of fish larvae for doses of ^{90}Sr-Y as low as 3.7 Bq.L^{-1} (assuming uniform contamination by ^{90}Sr-Y and a concentration factor of 10,000, the radiation dose is estimated at 1 μGy.d^{-1} for 37.10^{-3} Bq.L^{-1} by Fontaine, [FON 64]). On the other hand, English and Japanese works report no effect on fish eggs and larvae living in water containing less than 37.10^5 Bq.L^{-1} of ^{90}Sr-Y. The explanations for these divergences must be found in the experimental protocols, the different species, and so on. Nelson (*in:* [PRU 72]) found that larvae of the oyster *Crassostrea gigas* showed a high proportion of abnormal individuals when they lived for 48 hours in a medium containing 37.10^5 Bq.L^{-1} of ^{65}Zn or 37.10^6 Bq.L^{-1} of ^{51}Cr (i.e. irradiation doses estimated by the author to be 4 and 3 mGy.d^{-1}, respectively). Japanese researchers (Mikami et al. 1956, *in:* [POL 66]) have shown that an unknown mixture of fission products from ashes collected from the fishing boat "*Dragon Bienheureux*" after the thermonuclear explosion of March 1, 1954, caused deleterious effects (abnormalities and delayed hatching) in the larvae of the fish *Zebra danio* at a concentration of 1.48 Bq.L^{-1}. The effects are manifested at very low concentrations, and it is necessary to underline here again the more or less significant effect according to the incriminated radionuclides and also the interaction of the various radionuclides.

7.4.4. *Radiosensitivity and radioresistance of organisms*

Depending on the taxonomic group, the radiosensitivity is very different (Figure 7.4.). Generally, the most advanced organisms are also the most radiosensitive [UNS 96].

Some species of aquatic invertebrates seem particularly radiosensitive. These include crustaceans and mollusks (*Daphnia magna, Pollicipes polymerus* and *Mytilus edulis*). Their radiosensitivity to tritiated water is significantly lower than the value of 10 μGy.h^{-1} usually considered as a protection criterion for ecosystems subjected to chronic irradiation by gamma radiation [GAR 09b]. Recent work by Hagger et al. [HAG 05] and Jha et al. [JHA 05] on a marine mollusk (*Mytilus edulis*) at the egg stage and at the adult stage have highlighted the great sensitivity of this species of invertebrates (compared to vertebrates) and the fact that the damage to the DNA, depending on the dose rate from 13 μG.h^{-1}, has dose-independent cytogenetic consequences, namely sister chromatid exchange (SCE) and chromosomal aberrations, significantly, as early as the dose rate of 0.03 mGy.d^{-1} (1.3 µGy.h^{-1}). The lowest exposure dose rate for which a statistically significant effect is observed relative to the biological response of the unexposed control group is 0.0002 mGy.d^{-1} in the barnacle *Pollicipes polymerus* [ABB 79] or 1.7 µGy.h^{-1} relative to normal development of 48-hour-old *Mytilus edulis* larvae after 23 hours exposure to tritiated water (HTO). In a study of five generations of *Daphnia magna* chronically exposed to tritiated water, Gudkov and Kipnis (1996) concluded that the developmental abnormalities of the larvae worsened over the generations and were statistically significant at 3 µGy.h^{-1}.

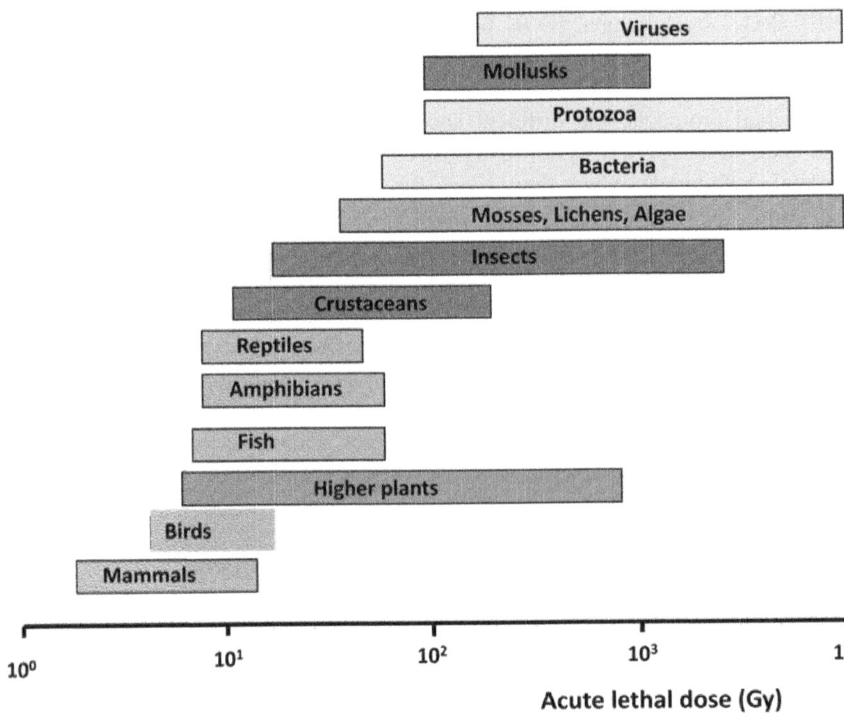

Figure 7.4. *Lethal dose and radioresistance in several taxonomic groups (modified from [UNS 96]). For a color version of this figure, see www.iste.co.uk/amiard/radioecology.zip*

Ionizing radiation produces several types of damage to nucleic acids, proteins and other crucial cellular components. Nevertheless, various microorganisms from phylogenetically distant taxa (bacteria, archaea, fungi) can withstand ionizing radiation levels several orders of magnitude above natural background. These examples suggest that resistance to chronic ionizing radiation and chemical contamination is not limited to rare, specialized strains from extreme environments but can occur among common microbial taxa, perhaps due to the overlap between ionizing radiation resistance mechanisms and mechanisms to meet other stressors [SHU 19]. Radioresistance mechanisms remain poorly understood [SHU 20].

7.4.5. *Conclusions on the effects of irradiation at the individual level*

Concerning the effects of irradiation on non-human organisms, many uncertainties and discrepancies persist, in particular because the data are very incomplete. The available information shows an extreme fluctuation of responses. For example, there is a factor of 1,000 in the worm *Neanthes arenaceodentata* between mortality (500 Gy) and the decrease in fertility (0.5 Gy). Similarly, the difference between the decrease in fertility and sterility reaches a factor of 100 in the worm *Neanthes arenaceodentata*.

Moreover, acute doses do not correspond to any reality in nature, where radiation doses are delivered to organisms in a chronic way. However, organisms are capable, to a certain extent, of restoring the damage suffered as a result of irradiation, and this is all the better if the irradiation dose is low and spread out over time. It is unfortunate that experiments on the effects of chronic radiation doses represent a small number of the total number of experiments on the effects of ionizing radiation. Moreover, the doses used are generally much higher than those encountered in situ.

7.5. Ecological community effects of irradiation

The lack of information on the effect of ionizing radiation is even more evident for ecological communities than for species. Experiments are extremely rare and always incomplete.

The effect of irradiation on ecological communities can be quantified mainly by controlled irradiations on suitable land using radioactive sources, but this type of experiment has not been carried out in the marine environment. Another possibility is the environmental monitoring of chronically exposed fauna and flora in contaminated marine sites, but this requires a good estimate of the radiation doses received. Environmental dosimetry relies on a good estimation of the exposure pathways and on the quantification of the received dose or dose rate using dosimetric models. However, the uncertainties in this approach remain very high.

7.5.1. *Monitoring of highly contaminated areas*

The highly contaminated areas are those of the first nuclear centers such as Hanford or White Oak, the sites of atomic weapon experiments or the sites of nuclear accidents such as Chernobyl or Fukushima. Only the atomic tests and the Fukushima accident concern the marine environment.

Kolesnikov et al. [KOL 70] reported the presence of mutants of the marine copepod crustacean *Pleuromamma piseki* (absence of luminescent organ, smaller size, slightly modified genital segment) around Mururoa Atoll. These mutants are confined around the Tuamotu archipelago, and the proportion of mutants is 100% at the plants situated 120 and 135 miles from Mururoa Atoll, which are also those where the plankton was the most contaminated following the explosion of October 5, 1966.

According to Templeton et al. (*in:* [NAS 71]), in the Pacific atolls, after each nuclear explosion, a large number of animals died as a result of the effects of intense irradiation, heat and blast, but they were replaced by new arrivals and the more subtle effects were excessively difficult to detect. Gorbman and James (1963, *in:* [NAS 71]), however, found abnormalities in the thyroid of fish living near Eniwetok due, according to the authors, to radioisotopes of iodine. On the other hand, the abundance and giant size of certain species of fish and lobsters at Bikini Atoll would not be due to a favorable effect of radiation but probably to the absence of human predation (*in:* [NAS 71]).

Experimental irradiation doses of 0.40 $Gy.d^{-1}$ lead to an increase in the biomass of species constituting the fouling community in marine or freshwater environments and to a change in the composition of the communities over time that is distinct from that observed for controls (*in:* [POL 66]).

Because of the disparity of these data, the authors are reduced to hypotheses. According to Polikarpov [POL 66], following a radioactive contamination, the level of which is not specified, the most radiosensitive species would disappear while bacteria, microphytes and other resistant organisms would remain. He also estimated that fisheries would cease to function if 50% of the fish stock were affected. According to this author, ionizing radiation must not affect more than 10% of the eggs, otherwise the number of catches would be affected. This argument is refuted by Templeton et al. (*in:* [NAS 71]) and by the IAEA [IAE 76] for all species with high fecundity. In this case, mortality due to irradiation would be insignificant compared to natural mortality. These opinions must be qualified for species with high reproductive power and/or high radiosensitivity, in particular coastal and aquatic birds, reptiles, amphibians, aquatic mammals and certain fish such as elasmobranchs (rays, sharks).

7.5.2. *Partial conclusions*

In conclusion, there are three possible ways to assess the potential impact of an increase in the level of radioactivity on a population:

a) the dose rate estimate can be compared with variations in natural irradiation (background);

b) comparisons between estimated dose rates and those that have been shown to cause significant deleterious effects in laboratory or field populations;

c) and when environmental dose rates reach the limits of protection for non-human species, verification of the validity of these criteria.

The first comparisons of natural radiation and anthropogenic sources were made in the 1980s. Thus, Woodhead [WOO 84] calculated the dose rates received by marine organisms due to natural radioactivity, atmospheric fallout from atomic testing and nuclear waste and compared the amounts received from natural and anthropogenic sources. Routinely, a nuclear reactor radiates significantly less than natural radioactivity. Only the Windscale (Sellafield) discharges cause a significant irradiation of a few tens of $mGy.h^{-1}$. In general, with the exception of nuclear accidents, irradiation from anthropogenic sources is low [UNS 96].

7.6. Confounding factors on the effects of irradiation

Several factors, called confounding factors, are likely to interfere by attenuating or on the contrary by accentuating the biological effects of irradiation. It is traditional to distinguish the intrinsic effects which are biological and depend on the individual and the extrinsic effects which depend on environmental factors. But often these factors interact with each other.

The biological effect of ionizing radiation will depend not only on the dose of irradiation but also on its distribution in time (intensity or dose rate in $Gy.h^{-1}$), organotropism, the radionuclide, the size and geometry of the individuals, the nature of the irradiated tissues and organs and the vital stage during which the irradiation caused its effect.

One of the biological factors showing the greatest dependence on the effects of radiation is the age of the individual. The juvenile stages of plants and animals are the most radiosensitive. This is due to the intensity of the youngest's metabolism.

The distribution or organotropism of the radionuclide will greatly influence the radiation dose to the organism. For example, the fact that cadmium-109 is adsorbed with a concentration factor of 170 on the external surface of the cells of the unicellular alga *Dunaliella bioculata* results in an electron irradiation dose of 5.50 $mGy.h^{-1}$; this dose would be limited to 0.0012 $mGy.h^{-1}$ if the radionuclide were uniformly distributed in the cell [AMI 78b]. The organotropism of cobalt-60 in marine and freshwater organisms, and of silver-110m in marine organisms, varies according to the contamination pathway: the radionuclides will be mainly on the external organs after direct contamination by water and mainly in the liver, digestive gland or hepatopancreas after contamination via food. The result is that the radiation doses in the center of the animal (where the most radiosensitive organs are generally located, such as the gonads, blood-forming organs and intestinal mucosa) are much lower in the case of contamination by water than in the case of contamination by food, although the amount of radionuclide accumulated is much greater in the first case [AMI 78a, AMI 78c, AMI 79b].

While we observe a decrease in radioactive contamination due to cobalt-60 and silver-110m along the food chain, on the contrary, the irradiation doses suffered by the organisms of the highest trophic levels are higher [AMI 78a]. This is related to the fact that, in small organisms of the first trophic levels, a significant proportion of the radiation, γ–, X- and β–rays, escapes into the surrounding environment, whereas it is much more absorbed in relatively larger consumers. This is no longer true in the case of internal contamination by alpha emitters. In this case, single-celled organisms are likely to disappear, whereas multi-celled organisms will have only a limited number of cells affected and, depending on the importance of their function, this will have more or less harmful repercussions on the individual.

All biological tissues do not react in the same way to ionizing radiation. In mammals, according to the law of Bergonié and Tribondeau [BER 06], the radiosensitivity of a cell varies proportionally to its capacity for division and inversely proportionally to its degree of differentiation. This explains the extreme radiosensitivity of gonads and blood-forming organs. This rule seems general for all organisms. Thus, the eggs of *Artemia salina* in dormancy are much less sensitive to ionizing radiation than rehydrated and developing eggs of the same species [ENG 62]. This law also applies to the various life stages since, as we have seen above, the early embryonic and larval stages, where biochemical activities are intense, are also the most radiosensitive stages.

Organisms are generally more resistant to stress when they are otherwise in optimal conditions [HUM 11] and this is valid for the effect of ionizing radiation. Thus, euryhaline species living in brackish (estuarine) waters such as *Fundulus heteroclitus* and *Palaemonetes pugio* are better able to withstand irradiation at low salinities, in contrast to *Artemia salina*, a species of hypersalineenvironments (Angelovic et al. 1966, *in:* [IAE 76]. Angelovic et al. (1969, *in:* [IAE 76]) were thus able to study the combined effects of irradiation (5, 10, 20 and 25 Gy), salinity (5, 15 and 25‰) and temperature (12, 17, 22 and 27°C) in *Fundulus heteroclitus* (Table 7.7). Poikilothermic aquatic organisms are much more resistant to ionizing radiation at low temperatures; this is the case for *Physa acuta*, *Artemia salina*, *Cyprinus carpio*, *Oryzias latipes*, *Carassius auratus*, and so on (*in:* [IAE 76]). Osmoregulatory phenomena are disturbed when animals are irradiated; this has been shown for the salmon *Oncorynchus kisutch* (*in:* [IAE 76]) and for *Callinectes sapidus* [ENG 73].

Days after irradiation	**Salinity**	**Temperature (°C)**			
		12	**17**	**22**	**27**
30	**5**	>25	>25	20.5	15.9
	15	>25	>25	15.4	7.7
	25	>25	>25	10.2	7.2
60	**5**	12.6	19.6	15.8	3.5
	15	13.9	17.2	10.5	3.3
	25	18.9	15.0	4.8	3.0

Table 7.7. *Influence of various ecological factors on the radiosensitivity of Fundulus heteroclitus (LD50 in Gy modified from Angelovic et al. 1969 in: [IAE 76])*

Researchers have long found that the same dose of radiation has a greater effect when delivered in a short time rather than over a relatively long period of time. This is largely due to the fact that irradiated tissues are able to restore some of the damage, especially at low doses of radiation. It is presumably this mechanism that minimizes the effects resulting from natural irradiation.

Among extrinsic factors, temperature is an abiotic factor of particular concern in assessing potential effects of radionuclides on marine species. Sea surface temperatures are expected to increase by 0.5 to 3.5 C in the next 30 to 100 years, so determining the interaction between elevated temperature and radiological exposure is relevant. This type of study was performed by Dallas et al. [DAL 16] for the genotoxic effect of tritium (HTO) on the mussel *Mytilus galloprovincialis*. The activity concentration used (15 $MBq.L^{-1}$) resulted in an accumulation of tritium that varied with both time and temperature, but consistently produced dose rates (calculated using the ERICA tool) below 20 $Gy.h^{-1}$, i.e. well below the recommendations of the IAEA and EURATOM. Despite this, there was a significant induction of DNA strand breaks (measured by the comet assay), also with a temperature-dependent time lag. At 15°C, DNA damage was significantly elevated only after 7 days, in contrast to 25°C, where a similar response was observed after only 3 days. From an ecological perspective, this study suggests that mussels (or similar marine species) exposed to elevated temperature and HTO may have a compromised ability to defend themselves against genotoxic stress.

7.7. Systematic under-evaluation of the biological effects of ionizing radiation

The main factors leading to an underestimation of the biological effects of ionizing radiation are six.

7.7.1. *Harmfulness according to life stage*

As with the vast majority of stresses experienced by living organisms, ionizing radiation is particularly harmful to the earliest life stages – embryos and larvae.

7.7.2. *Harmfulness by tissue or organ*

For the majority of available effects data, they are expressed on the basis of whole-body dose rates [HOW 11].

7.7.3. *Harmfulness according to emitters*

The harmfulness of emitters to organisms is not of the same intensity, and this phenomenon is practically ignored for non-human organisms.

7.7.4. *Not taking into account the bystander effect*

The bystander effect (or non-targeted effect) induced by ionizing radiation is a phenomenon by which cellular damage (sister chromatid exchanges, chromosomal aberrations, apoptosis, micronucleation, transformation, mutations and gene expression) is expressed in neighboring non-irradiated cells in the vicinity of an irradiated cell or cells. The communication between irradiated and non-irradiated cells takes place through two types of cell secretions, bridges between the two cell types (*gap-junction*) or vesicular secretions. The main signals linked to the bystander effect are cytokines, reactive oxygen species (ROS) and microRNAs.

7.7.5. *Failure to take into account the interactions between the effects of radionuclides*

The confounding factors mentioned in section 7.6 are not sufficiently taken into account. However, an individual is more radiosensitive if it is already stressed by another physical, chemical or biological stress [AMI 17a].

7.7.6. *Failure to take into account biodiversity in the marine environment*

The ICRP has selected three marine reference organisms, a macro-alga, a crab and a flatfish. This is paltry compared to the immense biodiversity of the marine environment. The number of marine species known today is 226,000, with the total number estimated to be between 700,000 and 1 million. The marine environment differs from the terrestrial environment by the number of phyla, since it has 28 animal phyla (14 of which are exclusively marine and two have been discovered since 1980) compared to 11 in the continental environment (only one of which is exclusively continental).

A very large number of taxonomic groups are thus left out of the ICRP approach, such as members of the phytoplankton and zooplankton, the basis of a large number of trophic chains, or mollusks, many of whose representatives are strong bioaccumulators of radionuclides. It is worth noting the great diversity of metal bioaccumulation in marine organisms. Thus, in crustaceans alone, Rainbow [RAI 99, RAI 02] has identified six different adaptive strategies of metal metabolism resulting in as many types of metal bioaccumulation.

7.8. Conclusion

The most serious and significant gap in radioecology is undoubtedly that relating to the sublethal effects (physiological and genetic) of low doses of chronic irradiation. For flora and fauna, the estimation of the absorbed dose remains rough and the influence of the type of tissue or organ irradiated and of the type of radiation involved are practically never taken into account. Similarly, in the environment, the pollutants that impact organisms are not only radionuclides but many other chemical, biological and even physical pollutants. The interactions between these various types of pollution are always neglected.

In a review, the NCRP [NCP 91] noted that the acute effect (LD_{50}) appeared for fish and crustaceans from 10 to 100 Gy, and for mollusks and echinoderms from 100 to 1,000 Gy. In the larval development of fish, the egg was no longer fertilized at 0.1 Gy, and aberrations appeared at the one-cell stage from 0.01 to 0.1 Gy and at the 32-cell stage from 1.0 to 10.0 Gy. In invertebrate larval development, eggs and embryos were altered at doses ranging from 10 to 100 Gy. No effect was observed for doses below 10 $mGy.d^{-1}$ during chronic exposure.

Biological effects such as genomic damage occur in organisms at dose rates well below those required to obviously impair reproduction. The question is whether such effects are detrimental to populations of the irradiated generation or subsequent generations, and if so, under what conditions? Finally, little or nothing is known about the possible interactions of various environmental stresses, such as chemical stresses, with the stress of chronic irradiation [WHI 03].

Table 7.8 provides a summary of the dose ranges and dose rates that cause mortality, infertility or decreased fertility.

	Dose (Gy)		**Dose rate ($mGy.h^{-1}$)**	
	Range	**NOEL**	**Range**	**NOEL**
Mortality	<3–> 30,000	< 1	0.1–0.48	< 0.1
Sterility	1–1,000	< 1	0.17–1,400	< 0.1
Fertility	0.06–20	< 0.05	0.023–550	< 0.02

Table 7.8. *Ranges of doses and dose rates causing mortality, sterility and reduced fertility in mammals, fish and invertebrates and recommended no observable effect levels (NOELs)*

On the other hand, according to the estimates of Blaylock and Witherspoon [BLA 75], the most critical stages from the point of view of irradiation would not be reprocessing plants but mines and ore purification plants where the estimated irradiation doses are 33 $mGy.d^{-1}$ for aquatic plants, 10 $mGy.d^{-1}$ for invertebrates and 0.6 $mGy.d^{-1}$ for fish. Moreover, we may wonder to what extent this value of 0.24 $mGy.d^{-1}$ (10 $\mu Gy.h^{-1}$), the lowest value proposed by various European radioecologists and the European Union obtained with the help of the ERICA tool, is really protective for flora and fauna? Deleterious effects (aberrant chromosomal mutations, decreased fertility, increased number of abnormal larvae, etc.) have appeared for doses below 10 $mGy.d^{-1}$ (6.5 $mGy.d^{-1}$ for *Physa heterostropha* and *Chironomus tentans*; 3 to 4 $mGy.d^{-1}$ for *Crassostrea gigas*) and also as low as 0.1 $mGy.d^{-1}$ for various fish eggs according to Polikarpov [POL 66]. The safety factor is therefore either low (less than 10) or zero.

The lack of information on the effects of low doses of irradiation does not allow us to draw conclusions on the development and growth of populations. For an edible and exploited species, such as certain aquatic species (fish, crustaceans, mollusks, cetaceans, etc.), the reduction in fertility due to a pollutant may reduce the exploitable portion of the population. According to Polikarpov [POL 66], if only 10% of the fish eggs are damaged, the stock will be halved in 29 years for mullet, in 34 years for mackerel and in 50 years for anchovy.

The disappearance of one or more species will be particularly serious for ecosystems whose natural balance is fragile. This is true for some terrestrial and freshwater ecosystems, but especially for estuarine ecosystems where specific diversity is low. This can result in a radical and complete change in community structure.

8

Characterization of Radioactive Risk in Marine Organisms

8.1. The principle of radioactive risk characterization

After the first steps of a radioactive risk assessment, the quantification of radionuclides in the organism's environment (Chapters 3 and 5), the quantification of bioaccumulation (Chapter 4), the estimation of the absorbed radiation dose (Chapter 6) and the quantification of the biological adverse effects of ionizing radiation (Chapter 7), the last step remains to be carried out: the characterization of the radioactive risk. Theoretically, this step is simple, since it is sufficient to calculate the ratio (risk quotient) between exposure and effect, a PEC (Predicted Environmental Concentrations) and a PNEC (Predicted No-Effect Concentrations) in the case of chemical pollutants. In the case of radioactive risk, the ratio values are called PEDR (Predicted Environmental Dose Rate) and PNEDR (Predicted No-Effect Dose Rate).

If the ratio is greater than unity, then there is a risk; if it is less than unity, then there is no risk. The applications are much more difficult because while the first two steps are relatively well understood, the last two steps (estimation of the dose received and the harmful effect of ionizing radiation) are poorly estimated. The reliability of the PEDR/PNEDR ratio depends on the accuracy and veracity of the two estimated values.

The estimation of the two reference values (PEDR and PNEDR) is based on the dose–response relationship. This relationship is assumed to be linear and as the dose increases, the adverse effect increases. This is true for most chemical stresses. However, in the case of ionizing radiation, the linear relationship is well established in humans for medium and high doses, but is controversial for low doses. This is a fundamental question for estimating the two reference values (Figure 8.1). For low doses, is this response linear or not? Is there or is there not a threshold below which there is no harmful effect of ionizing radiation? Is the effect always harmful or is it even beneficial at certain low doses (hormesis effect)? This question is the subject of heated debate.

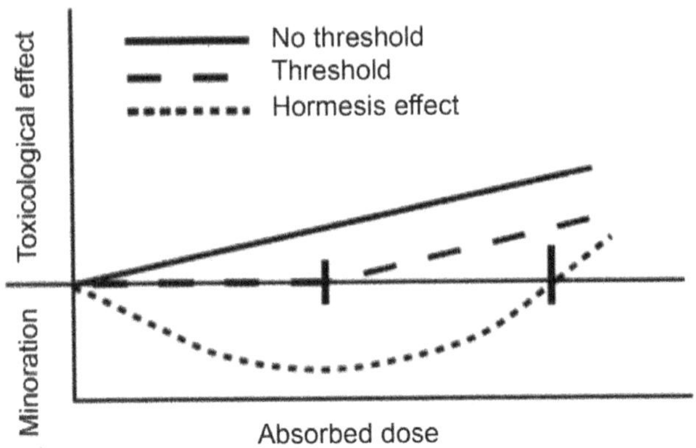

Figure 8.1. *Relationship between absorbed dose and biological effect (modified from [LAN 04a])*

8.1.1. *International organizations involved in radioactive risk assessment*

There are three international organizations working on radioactive risk assessment: UNSCEAR, ICRP and IAEA. Following the historical logic of the construction of basic standards in radiation protection, UNSCEAR, by updating in 2008 the synthesis on the effects of ionizing radiation in non-human species [UNS 08], has consolidated the scientific basis for recommendations on the subject. The ICRP [ICR 08], in its following publications, detailed its approach, which were translated into basic standards in radiation protection by the IAEA [IAE 11b]. Today, the collaboration of these organizations, especially within the Coordination Group on Environmental Protection, has become much more interactive. Their participation in this group, along with other entities such as the UIR, the OECD-NEA, and the UNEP, makes it possible to envisage better risk assessment.

Historically, the international awareness by politicians that ionizing radiation can have an impact on ecosystems and that humans are not necessarily the most radiosensitive species occurred following Rio (1992) and the Convention on Biological Diversity, a treaty signed by the United Nations in 1992. In 1996, UNSCEAR [UNS 96] proposed its first reference values. In 2008, UNSCEAR [UNS 08] proposed threshold values of 100 $\mu Gy.h^{-1}$ (2.4 $mGy.d^{-1}$) for terrestrial organisms and 400 $\mu Gy.h^{-1}$ (9.6 $mGy.d^{-1}$) for aquatic organisms. Since 1998, 16 European States have explicitly taken radioactive substances into account for the protection of an environment (OSPAR Convention) (Chapter 5).

For most states, the approaches followed are consistent with those for the protection of the environment with regard to chemical substances and for the protection of Man with regard to radioactive substances.

Some states have been quicker than others to begin thinking about this issue. This is the case in the United States with the CERCLA (Comprehensive Environmental Response, Compensation and Liability Act), more commonly known as the Superfund Act (see also the US DOE publication [USD 02]), Canada (Canadian Nuclear Safety and Control Act 2000) and the United Kingdom (The Environment Permitting Regulation, Schedule 23 of 2010). Canada highlights the toxicity of human activities releasing uranium.

8.1.2. *European research*

In Europe, research projects supported by the Euratom domain have helped to improve knowledge and develop methods and tools. The main past scientific projects have been FASSET (2000–2003) [LAR 04], ERICA (2004–2007) [LAR 08], PROTECT (2007–2008) [HOW 10], the STAR Network of Excellence (2011–2015) [HIN 13], the European Radioecology Research Platform Alliance (2012-) and the COMET project (2013–2017). The European Commission organized in November 2012, in cooperation with the Group of Experts of Article 31 of the Euratom Treaty, a scientific seminar on the topic of environmental protection from ionizing radiation.

The coordination of research in the field of radiation protection in the broad sense is now carried out by four European associations that have signed an agreement of understanding between them, each one being dedicated to a subfield of radiation protection: concerning crisis and post-accident (NERIS), concerning radioecology (ALLIANCE), concerning low doses (MELODI) and concerning dosimetry (EURADOS).

European efforts have focused on the creation of computerized databases accessible to all. For example, the database collecting radionuclide concentration factors of non-human organisms has gathered 90,000 values (Wildlife Transfer Database, 2011, http://www.wildlifetransferdatabase.org/). However, there is a strong distortion of the amount of data according to the environment (high for terrestrial, low for marine), as well as according to taxonomic group (some have no data), as well as the radionuclides (high for cesium-137, low for the majority).

Similarly, the FREDERICA database that aggregates the effects of ionizing radiation for non-human organisms included in 2016 about 32,000 data with the same distortions as the concentration factor data [COP 08, BEA 16].

8.2. Methods for selecting reference values

Three methods for deriving reference values are the safety factors (1 to 10,000) for the smallest value of an adverse effect, the statistical extrapolation of the HR_5 (5% Hazard Rate) and the "expert judgment" method based on a literature review.

8.2.1. *The ICRP approach*

The consideration of radioactive risk to the environment only began in 2000. The evolution of the concepts, doctrines and recommendations of this organization has been analyzed and detailed by Clarke and Valentin [CLA 09]. The ethical and philosophical foundations of environmental protection were presented in ICRP Publication 91 [ICR 03]. The system, to date, details the biology of RAPs (Publication 108, [ICR 08]); presents transfer factors for estimating internal concentrations of radionuclides of environmental significance in different situations (Publication 114, [ICR 09]); provides additional information (Publication 108) on dosimetry, biological effects and derived reference levels (environmental dose rate bands where potential adverse effects may warrant attention); and provides information on the application of the system in planned, emergency and existing exposure situations (Publication 124, [ICR 14, LAR 15]). The ICRP has since provided dose coefficients (Publication 136, [ICR 17]) and determined relative biological effectiveness, to guide the calculation of specific weighting factors for use in environmental radiation protection (Publication 148, [ICR 21]).

8.2.2. *The ERICA approach*

The "ERICA" approach is an integrated approach that includes aspects of risk assessment, management and communication [LAR 08].

From the FREDERICA database, the calculation of ED_{50} (doses inducing 50% of the effect in case of acute exposure) or EDR10 (dose rates inducing 10% of the effect in case of chronic exposure) is performed. These critical ecotoxicity results are in turn used to plot species sensitivity distributions (SSDs) and estimate doses (or dose rates) below which 95% of the species in the ecosystem are expected to be protected (HD_5 or HDR_5) [ERI 06].

The choice of the reference value is the protection threshold for 95% of the species in an ecosystem during exposure to chronic external gamma irradiation. This value (equal to 10 $\mu Gy.h^{-1}$, or 0.24 $mGy.d^{-1}$) was obtained by applying a safety factor to the value of the fifth percentile of the distribution of the radiosensitivity of species (SSD). This procedure aims to take into account the uncertainty linked to the extended use of effects data obtained by external gamma irradiation only in the field of internal irradiation by all types of emitters [BEA 16].

The IRSN considers that the ERICA graded approach, which is more operational than the ICRP approach while remaining compatible with it, is the basis for explicitly demonstrating environmental protection in the context of assessing and controlling the radiological impact associated with planned environmental exposure situations [BEA 16].

8.3. Tools for assessing radioactive risk to non-human organisms

Due to the increasing interest in environmental protection against ionizing radiation on the international scene, the need for operational tools for radiological environmental risk assessment has become critical. Beaugelin-Seiller et al. [BEA 06b] propose a fast and user-friendly computerized method allowing the transformation of the exposure (expressed in Bq per unit mass or volume) to the dose (i.e. energy deposited in the body, expressed in Gy) received by any non-human species, for any radionuclide for a unit of time and a unit of "concentration" of the radionuclide in the source of radiation.

Three tools are currently operational, two American tools (RESRAD-Biota and SADA) one European tool (ERICA Assessment tool and Doses3D), and two approaches are in the process of being developed, one Canadian and the other Korean associated with a tool (K-Biota) [BEA 16].

A comparison of three tools: RESRAD-Biota, ERICA and SENES Risk Model was performed [GAR 10a]. Little difference was observed, except in the use of aggregate transfers by RESRAD and ERICA, which does not make the evaluation transparent. On the other hand, in a comparison of eight tools in the IAEA EMRAS program, strong variability was found between tools [BER 08a, BER 08b].

8.3.1. *The ERICA tool*

The ERICA tool (Environmental Risks from Ionising Radiation in Environment: Assessment and Management) is a computer application dedicated to the assessment of radiological risk to aquatic and terrestrial organisms. This integrated approach, developed in European countries, uses mainly two concepts: the concentration ratio *CR* and the distribution coefficient K_d between aqueous and sedimentary phases. The novelties of the ERICA tool consist of a probabilistic approach and relatively exhaustive lists of organisms and radionuclides. Many old data having been acquired for the purpose of human radiation protection, only the consumable part of the organism has been taken into account (often the muscles). In order to extrapolate the results obtained in muscles to the whole individual, conversion factors have been proposed and have been collected by Hosseini et al. [HOS 08]. When data are missing, the authors also propose to take results for taxonomic groups or analogous radionuclides. Two tables with *CR* summaries for all marine and freshwater organisms are provided.

Organ mass is variable according to the zoological group. Thus, bones represent 10% in mammals and 7% in birds, while muscle mass represents 25% in birds. These proportions vary with the age of the individuals. Similarly for organotropism, in mammals, 90% of strontium, 70% of lead, 60% of polonium, 45% of plutonium and 45% of americium will be found in the bones. In birds, 55% of the plutonium will be in the bones.

Erica's integrated approach requires that a risk assessment examining the dose rate be defined for risk characterization in steps 1 and 2 (hazard identification and exposure quantification). In step 3 (dose–effect relationship), no dose rate quantification is used and the methods leading to the risk characterization are based on the possible effects of irradiation on reproduction, mortality and morbidity. The species sensitivity distribution was used to perform the predicted no-effect dose rate (PNEDR) ERICA risk assessment. The method used was based on mathematical processing of FRED data (see below, FASSET and EPIC databases are merged to form FREDERICA) and resulted in a PNEDR of 10 $\mu Gy.h^{-1}$. This rate provides sufficient protection from adverse effects on the structure and function of all ecosystems undergoing chronic exposures [GAR 08].

Andersson et al. [AND 09], using European guidelines on the derivation of predicted chemical no-effect concentrations, applied appropriate radiation sensitivity data. They also lead to the generation of a no-effect dose rate of 10 $\mu Gy.h^{-1}$, following a distribution of sensitivity data for the various species.

8.3.2. *The FASSET database*

The FASSET project (Framework for Assessment of Environmental Impact, [LAR 04]) initiated the development of the methodology now known as ERICA. The scenarios were chosen to cover all the components of all types of ecosystems and all the situations in which the method is applied.

The FRED database (FASSET Radiation Effects Database) is a unique structured resource of the biological effects of ionizing radiation on non-human species, primarily from temperate ecosystems, encompassing 26,000 primary data entries. Data considered to be of high quality were extracted from FRED and effect–dose relationships were constructed to provide estimates of EDs_{50} (inducing a 50% change in observed effect for an acute exposure) and EDR_{10} (dose rate giving a 10% change in observed effect). These assessments are doses (or dose rates) related to the percent change in mean endpoint level for a particular effect (50% or 10% for acute or chronic exposure regimes, respectively). Acute and chronic species sensitivity distributions (SSDs) were constructed from these data sets, and the assessment factor method (AFM) was applied when data were too sparse.

Garnier-Laplace et al. [GAR 04], inspecting the FASSET database on the effects of irradiation on non-human biota (FRED), consider that one of the major difficulties in implementing ecological risk assessments for radioactive pollutants is in the lack of data for chronic low-level exposure. The critical review of a number of publications on this topic points out that a large number of extrapolations are made to perform an ecological risk assessment (acute exposure versus chronic exposure regime, quality of radiation type including relative biological effectiveness and weighting factors, individual and population biological effects including radiosensitivity and lifestyle variations throughout the life cycle, single- or multi-pollutant radionuclide). Of all the parameters, the reproductive ones that drive the population demography in a given ecosystem and, consequently, the structure and functioning of that ecosystem are the most important. Garnier-Laplace et al. [GAR 04] consider that the uncertainties of extrapolation from one species to another are the highest, followed by acute to chronic, external to internal and single pollutant to a mixture of stressors, then extrapolation from the individual to the population, and finally extrapolation from ecosystem structure to ecosystem functioning. To extrapolate from the individual to the population, the IAEA launched the MODARIA (I and II) program. Eight technical meetings were held during the two phases of this program.

8.4. Recommendations of radiation doses for marine organisms

The reference values vary according to the international organizations and the states concerned. In its synthesis, UNSCEAR [UNS 08] limits the absorbed doses to 400 $\mu Gy.h^{-1}$. The ICRP [ICR 08] defines the DCRL (Derived Consideration Reference Levels) in plants and reference animals (RAP, Reference Animal or Plant). For flatfish and algae, the recommended values are 42 to 420 $\mu Gy.h^{-1}$, and for crabs, 420 to 4,200 $\mu Gy.h^{-1}$. The ICRP daily values are 1–10 $mGy.d^{-1}$ for fish and 10–100 $mGy.d^{-1}$ for algae and crabs.

Using the ERICA tool [ERI 21], several European countries propose a conservative PNEC (Predicted No Effect Dose-Rate) of 10 μGy.h^{-1} or 87.6 mGy.yr^{-1} (derived from SSD, Species Sensitivity Distributions). The work carried out in the framework of the European project PROTECT (Protection of the Environment from Ionising Radiation in a Regulatory Context) [HOW 10] has made it possible to consolidate the reference value of the dose rate known as no effect, for all or part of an ecosystem chronically exposed to ionizing radiation, produced in the framework of ERICA (i.e. 10 μGy.h^{-1} or 24 mGy.d^{-1}) [GAR 08, LAR 08]. This same project uses a value of 4.8 mGy.d^{-1} for invertebrates [GAR 10b].

In the United States, the USDOE considers that exposure levels for natural populations of aquatic biota (10 mGy.d^{-1}) are safe for radiological protection (Jones et al. *in*: [IAE 03b]). In Canada, Environment Canada uses values of 20 μGy.h^{-1} for fish, 110 μGy.h^{-1} for algae and macrophytes and 200 μGy.h^{-1} for invertebrates [BIR 03]. In the United Kingdom, the values adopted for populations of freshwater and coastal organisms are chronic dose rates lower than 400 μGyh^{-1}, and for populations of deep-sea organisms are chronic dose rates lower than 1,000 μGyh^{-1} [COP 01]. Russian authors [SAZ 02] propose to protect marine fauna by retaining as a concentration of various radionuclides the value that represents 1% of the acute lethal dose.

Table 8.1 presents the estimated no-effects values (ENEVs) from Health Canada and enables comparison of radiation causing deleterious effects.

Taxa	ENEV (Gy.yr^{-1})
Fish	0.2
Benthic invertebrates	2
Algae	1
Macrophytes	1

Table 8.1. *Summary of estimated no-effect values (ENEVs) by Health Canada (modified from [UNS 08])*

From these reference values, the assumption is made that no potential ecological risk can exist; the following assumptions of the effects of irradiation on biota are stated:

– Doses and dose rates that cause no effect on fertility in most radiosensitive species will not decrease the reproductive success of other invertebrate, fish and mammal species.

– Doses and dose rates lower than those causing mortality and sterility in the most radiosensitive species will not cause loss of endangered species, loss of populations or reduced biodiversity in radionuclide-contaminated ecosystems.

On the other hand, this value of 0.24 $mGy.d^{-1}$ supposed to protect the majority of organisms has been, or is being, exceeded in several regions of the world. This was the case at Windscale (Sellafield) for benthic organisms (0.8–4.8 $mGy.d^{-1}$), in the Columbia River for benthic organisms (6 $mGy.d^{-1}$) or for zooplankton, respectively, 1.65 and 3.6 $mGy.d^{-1}$ at WindscaleSellafield and in the Columbia River. This is the case in large areas around Chernobyl and Fukushima. These areas are obviously small in surface area (at Windscale (Sellafield), 1,500 km^2 of sediment delivering a dose greater than the natural irradiation according to [PEN 73]), but they multiply after each accident rather rapidly.

8.5. Applications and feedback

ICRP Publication 124 [ICR 14] describes how to implement the concepts previously discussed in the environmental radiation protection demonstration framework and for different environmental exposure situations (planned, emergency and existing).

For a tritium release into the marine environment and using only the "lowest observed effect" and "highest no effect" dose rates, applying a criterion for protection of 95% of species, absorbed doses were estimated to be 15 $\mu Gy.h^{-1}$ (with a lower 95% uncertainty limit of 3 $\mu Gy.h^{-1}$). Compared to the proposed Canadian criteria for aquatic biota, applying a safety factor of $1/10^{th}$, the no-effect value is 5 to 10 $\mu Gy.h^{-1}$, and thus the radioactive risk was low (Twining et al. pp. 3–6, *in*: [IAE 03b]).

During the MARINA II program (1986–2001), dose rates up to 0.1 $mGy.d^{-1}$ were estimated in the OSPAR region but were apparently without effect [EC 03b]. In Natura, 2,000 areas in England and Wales where the permissible limit is 1 $mGy.d^{-1}$, 10 areas have dose rates of 0.5 to 1, two areas have dose rates exceeding 12 $mGy.d^{-1}$, but using the ERICA tool, the dose rates are below 1 $mGy.d^{-1}$ [COP 01]. Over the period 2000–2004, the European project EPIC (Environmental Protection from Ionising Contaminants, EC Inco-Copernicus Program) had the objective of examining the protection of the Arctic environment against the effects of ionizing radiation. For marine ecosystems, the rates are of the order of 10^{-2} to 10^{-1} mGy over 20 years and the effects limited [BRO 03]. The radioactive risk has been estimated to be low for all nuclear fuel sites [CHA 08]. For example, at the spent fuel reprocessing plant at La Hague, the estimated dose rates were low and less than 10^{-3} $mGy.d^{-1}$ [CHA 05]. The demonstration of conformity of post-closure safety studies of radioactive waste repositories with protection objectives for non-human organisms was discussed at the BIOPROTA forum [SMI 12].

Exercises of inter-comparison of approaches, methods and tools devoted to environmental radioprotection are regularly carried out within the framework of the IAEA programs (EMRAS I, 2003-2007; EMRAS II, 2009-2011; MODARIA, 2012-2015, appendix p.137). On this occasion, a comparison between radiotoxicity data acquired in situ and under controlled conditions was undertaken [GAR 13]. This showed an apparent greater radiosensitivity (by a factor of 4 to 6) of organisms in the field than in the laboratory.

Various scientific publications have presented results of radiological risk assessment for the environment. For the marine environment, the Fukushima accident ([GAR 11] with ERICA), in Australia a STEP with release of ^{131}I (nuclear medicine) [VEL 11], in the Gulf of Mannar (India) ([FER 12] with ERICA), in the Barents Sea (Norway) [IOS 12] with NRPA), in Japan 100 days after the accident ([KEU 14] with K-BIOTA) or 4 months after the Fukushima accident ([FIS 13] with EDEN). In the case of oil and gas operations, the review clearly showed that only a limited number of studies have examined the possible impacts on biota of naturally occurring radionuclides in produced water. The substantial uncertainties involved in the impact assessments made it difficult to reach a conclusion [HOS 12].

A case that also leads to an ecological risk situation due to exposure to ionizing radiation is the theoretical case of a nuclear submarine accident in the Barents Sea [IOS 12].

In Fukushima, on the basis of the first partial information available on the contamination of the forest and marine environment, the dose rates received by representative organisms of various taxonomic groups, during the first 30 days following the accident, were evaluated generically with the ERICA tool. The comparison between the results obtained with the very conservative approach and the ICRP DCRLs showed a risk to the marine ecosystem [GAR 11]. Controversies have arisen over the potential radiological impact on wildlife. In 2011, UNSCEAR established an international expert panel to analyze the dosimetric impact of the Fukushima disaster, and the health implications, for workers, the public and non-human species. This study was published in a paper released in April 2014 [UNS 14]. The conclusions of this study are as follows for marine species: the doses calculated for the short-term phase after the releases do not indicate the possibility of acute effects to populations except for macro-algae near the release outfall; in the medium and long-terms, the exposure levels for freshwater and marine species are well below those characteristic of significant effects. Strand et al. [STR 14] estimated that only macroalgae (cumulative dose of 7 Gy) greatly exceeded their corresponding reference. They pointed out that if individual plants and animals could have had their health degraded in the weeks directly following the accident in relatively contaminated areas, impacts on population integrity would have been unlikely because of the short duration of the highest exposures.

8.6. Gaps in the characterization of radioactive risk

8.6.1. *Position of international organizations with respect to the ICRP approach*

The IAEA agrees with the ICRP approach to the assessment and control of radiological impact associated with planned environmental exposure situations, but considers that for existing and emergency situations, there is a need for further reflection. The IAEA has published several guides contributing to environmental protection [IAE 12a, IAE 14, IAE 15b]. UNEP finds the definitions of "environment" and "environmental protection" more relevant than the narrow ICRP approach. For its part, the UIR [UIR 12] proposes a more ecological approach called "ecosystems". The arguments and the various steps to be taken in favor of an ecosystem approach, to complete the evaluation criteria at the level of organisms currently used in radiation protection, are summarized by Bradshaw et al. [BRA 14]. This approach already exists in support of various conventions, such as the OSPAR Convention [OSP 92], the Convention on Biological Diversity [CBD 14] or the European Marine Strategy Framework Directive [EC 08].

8.6.2. *Transgenerational effects*

Differences in radiation sensitivity between species and phyla are poorly understood but have significant implications for understanding global radiation effects and radiation protection. These differences in sensitivity between species also lead to global effects at higher levels (community, ecosystem), as interactions between species may be altered. Animal and plant studies suggest that gamma irradiation may cause observable effects in the F1 generation that are not attributable to the inheritance of a rare and stable DNA mutation. Several studies have demonstrated an increase in genomic instability detected in germ cells or somatic cells of F1 organisms from exposed F0 organisms. The possible role of this phenomenon may be due to epigenetic modifications, such as methylation, in the genome. The possible role of this phenomenon may come from epigenetic modifications, such as DNA methylation, histone modifications or non-coding RNA expression related to radiosensitivity, as well as from adaptation effects [ADA 18].

8.6.3. *Ignorance of biodiversity*

Based on a shared vision of the need to simplify the diversity of species within the animal and plant kingdoms, the concept of reference organisms varies in quality and number: 12 for the ICRP, 31 for Europe, 4 (3 animals and 1 plant) for the United States (US DOE), 40 for the United Kingdom, etc. This underlines the weakness of this approach. The ICRP argues that reference species (RAP, [ICR 14]) are almost equivalent to representative organisms (ROs). A representative organism is a particular species or group of organisms selected as part of a site-specific impact assessment, taking into account its geographical location relative to the radiation source(s). Given the considerable biodiversity of marine organisms, the selection of only three marine species is a challenge. Thompson [THO 88] emphasized the high radiosensitivity of large marine predators,

To fill gaps in bioaccumulation data, one approach is based on the assumption that an underlying taxonomic and/or phylogenetic relationship exists for radionuclide transfer: a residual maximum likelihood (REML) mixed model regression modeling approach. In the marine environment, unlike the terrestrial and freshwater environments, this model only works to a limited extent. Other factors can likely affect transfer to a much greater degree in marine systems, such as quite diverse life histories and different diets, which can confound any phylogenetic model [BRO 19].

8.6.4. *The choice of the threshold adverse effect*

For environmental radiation protection, the biological effects of interest are those with population-level consequences, whether deterministic or stochastic[1]. The biological effects thus considered most relevant are early mortality, certain forms of morbidity (which would impact the health of individuals), alterations in reproduction and the induction of chromosomal damage. Unfortunately, the number of data concerning these various effects is extremely limited in the marine environment.

Several methods exist (see section 8.2). The method used by the ICRP ("expert judgment") is only relevant if the data on the subject are numerous and reliable and if the expert has no conflict of interest in the field. In this case, the conditions are not met. Garnier-Laplace et al. [GAR 10b] proposed an adaptation of a method commonly used for chemicals. The predicted no-effect dose rates (PNEDR) were derived by applying, as far as possible, the European guidelines developed for chemicals. To derive the PNEDR at the ecosystem (or generic) level, EDR10

1 Stochastic effects are rarely taken into account by environmental radiation protection.

radiotoxicity data (dose rates giving a 10% effect relative to the control) were used to fit a species sensitivity distribution (SSD) and estimate the HDR5 (the hazardous dose rate affecting 5% of species with a 10% effect). Next, a multi-criteria approach was developed to justify the use of an assessment factor (AF) to be applied to HDR5 to estimate a PNEDR

8.7. Conclusion

The characterization of the radioactive risk is obtained from a ratio of two terms: the dose received and the harmful effect. However, these two terms are very poorly evaluated and are affected by strong uncertainties, an underestimation of the absorbed dose (Chapter 6) and an underestimation of the harmful effect. The result is a final estimate of radioactive risk that is very uncertain and very small.

The method most commonly used in chemical risk assessment is not used for radioactive risk. However, because of the number of data, it would be relevant. It is true that it is the most penalizing, especially for the marine environment, since precautionary factors are necessary and that for the marine environment they can frequently lead to dividing the PNEC value by a factor of 10,000. The method used by the ICRP is not relevant; the one proposed by Europe is correct but suffers, especially for the marine environment, from an insufficient number of relevant data on adverse effects.

In assessing radioactive risk, politicians have for too long made the mistake of believing that humans are the most radiosensitive species. Similarly, it is a mistake to believe that the risk due to natural radioactivity is zero; natural is not synonymous with harmless.

It is a serious omission to include only three marine species (an alga, a crab and a fish) among the reference species. It is aberrant that there are no mollusks, no mammals and no polychaetes. This ignores the great biodiversity of the living marine world, and the great diversity of metabolisms of marine organisms.

The end point of protection for humans relates to stochastic effects on individuals, while for wildlife it relates to deterministic effects on species and ecosystems [VAN 18].

Conclusion

Two of the objectives of radioecology directly concern the assessment of radioactive risk to non-human organisms: the first is to understand the processes of transfer and concentration of radionuclides in ecosystems, the second is to assess the impact of natural and artificial radioactivity on the environment (radioecological impact). Only the first is adequately understood, while the second is very deficient.

C.1. Behavior and fate of radionuclides in the marine environment

The fate of radionuclides in the marine environment is well understood and fairly well modeled. The behavior is explained mainly by chemical speciation when the radionuclide is introduced into seawater and the changes of this speciation over time. Dispersion is mainly dependent on whether the radionuclide is conservative or not. The adsorption of the radionuclide on the sedimentary phase is quantified by the distribution coefficient K_d. This depends on many factors (particle size, carbonate and stable element contents of the particles, salinity, etc).

C.2. Bioaccumulation of radionuclides in marine organisms

The pathways and mechanisms of radionuclide bioaccumulation are now well known for the major radionuclides and the most studied taxonomic groups. Many ecological factors (biotic and abiotic) influence bioaccumulation such as chemical speciation, salinity, temperature, age and size of individuals, sex and sexual maturity, but the two most important factors are taxonomic group and species membership. Indeed, between neighboring species, bioaccumulation can vary considerably. A phenomenon present in the majority of organisms is the non-homogeneity of the bioaccumulation of radionuclides because they bioaccumulate preferentially in certain organs (organotropism) according to their chemical characteristics and their possible physiological function. The quantification of bioaccumulation is expressed by the concentration factor (CF) or the concentration ratio (CR). The elimination of radionuclides by marine organisms is often a relatively slow and always incomplete process. Transfers of radionuclides to organisms can occur via seawater, sediment or their diet. Trophic transfers can be biomagnified along the various links of the trophic chain.

C.3. Estimation of absorbed dose to marine organisms

The radiation dose absorbed by marine organisms can be natural or anthropogenic, the former being significant. Anthropogenic doses are especially significant for benthic organisms (those living in contact with the sediment) or in contaminated sites following atomic bomb tests, or near the discharges of spent nuclear fuel reprocessing plants, or following a nuclear accident such as that of Fukushima. The estimation of the absorbed dose is only possible by calculation. The latter has biases that systematically underestimate the dose value. The main biases are due to a simplification of the calculation hypotheses, such as considering that the organisms are homogeneously contaminated and that they have regular geometric shapes. But above all, it is the failure to take into account certain emitters with short paths in living matter, such as alpha emitters or low-energy beta emitters, Auger electrons, and the fact that secondary radiation due to the interaction of certain radionuclides with the target is ignored.

C.4. Estimation of the effect of ionizing radiation on marine organisms

The effects of irradiation in organisms occur at all levels of biological organization from the molecule (especially DNA) to the ecosystem. Deterministic effects occur at high radiation doses and with notable effects (mortality), while stochastic effects are observed at lower doses or for chronic dose rates. The effects are earlier for the first vital stages (embryonic) and at the lowest levels of biological organization (molecular).

There are serious gaps in our knowledge of the effects of irradiation on marine organisms. Indeed, the number of scientific studies are few and concern adult stages and rarely the juvenile stages. Moreover, the effects due to alpha and beta emitters are little studied, the bystander effect (effect of proximity of irradiated cells) is ignored and the number of species studied is ridiculously reduced.

The essential question is "what dose rate has no harmful effect on living organisms?" Garnier-Laplace et al. [GAR 04] propose a dose rate of 10 $\mu Gy.h^{-1}$, which is supposed to protect 95% of living species. A first remark is in order: do we have the right to condemn 5% of species? Second, various dose rates well below 10 $\mu Gy.h^{-1}$ trigger harmful effects. For example, in the case of tritium (tritiated water), one-third of the 24 data listed have values (LDR_{10}) below this limit, and these data are all related to effects observed on three species of aquatic invertebrates (*Daphnia magna, Mytilus edulis* and *Pollicipes polymerus*) [ABB 79, HAG 05, JHA 05, JHA 06].

Recent work confirms the high radiosensitivity of some marine organisms to tritium irradiation. AlAmri et al. [ALA 12] and Jha [JHA 08] emphasize the importance of the comet assay in determining genetic effects in wildlife. The micronucleus (MN) test in sentinel species appears to be a good biomarker of individual health status. However, further validation studies are needed to better characterize the different types of nuclear changes and clarify the role of interspecific biotic and abiotic factors, as well as interindividual variability [BOL 11].

C.5. Towards an improved characterization of the radioactive risk to marine organisms

The approach adopted by the ICRP is not at all scientific in the choice of reference values ("expert judgment"). It is also too simplistic and has many shortcomings. It is therefore necessary to try to improve the characterization of the radioactive risk for non-human organisms.

C.5.1. *The shortcomings of the ICRP approach*

The ICRP approach is based on RAPs. However, the enormous diversity of non-human biota is a specific challenge. The dosimetric models, adopted by the International Commission on Radiological Protection (ICRP), provide dose conversion coefficients for a small variety of biota, including reference animals and plants. In addition, the models use a number of simplified approaches, such as: simple body shapes, uniform composition and density, homogeneous internal contamination, a limited set of external radiation sources for non-human organisms and truncation of radioactive decay chains. The ICRP has corrected this somewhat in its publication 107 with a more systematic consideration of internal exposures to radionuclides concentrated in specific organs and the choice of decay chains (Ulanovsky and Prohl, pp.218–232, *in:* [IAE 12b]). Consideration of weighting factors related to the type of radiation is essential (Higley et al. pp.233–245, *in:* [IAE 12b]). The ICRP effort on this point is paltry. To correct the oversimplification of the complexity of the morphology, internal structure and density of the body, dose rates could be calculated using voxelized phantoms [HIG 15].

C.5.2. *A consequent deepening of scientific knowledge*

Scientific knowledge for the natural and anthropogenic radionuclides most present in the marine environment is sufficient for their fate in the marine environment and in organisms. On the other hand, there is a clear lack of assessment of absorbed doses and adverse effects of ionizing radiation on marine organisms.

In particular, the effects of low doses are poorly understood. At these low doses, the potentially relevant effects are genomic instability, bystander effects, hypersensitivity at low doses and adaptive or inducible responses. They appear to have an epigenetic basis. Radioecologists therefore need a simple test that detects effects at low doses, is preferably non-lethal to the test organism (i.e. collection is not essential) and works for a wide range of species, and where field sampling is possible. The proximity effect meets these criteria (Mothersill and Seymour, pp. 197–207, *in*: [CIG 06]). Crustaceans appear to be very sensitive to radiation, and in particular to the bystander effect, in contrast to terrestrial arthropods and some other invertebrates, which are generally considered radioresistant [MOT 01].

C.5.3. *Consideration of weighting factors*

Human radiation protection uses two weighting factors to estimate the biological dose (expressed in sievert) from the physical dose (expressed in gray). The first weighting factor takes into account the type of ionizing radiation involved, and the second weighting factor takes into account the radiosensitivity of the tissue (or organ) affected by the dose. In the dose assessment of animals and plants, only the first factor is considered.

C.5.3.1. *The weighting factor of the type of radiation*

It is well known that the potential for radiation effects on living organisms depends not only on the absorbed dose and dose rate in the tissue or organ of interest but also on the type and energy of the radiation causing the dose. This can be translated into the relative biological effectiveness (or RBE). Other authors use the terms quality factor, radiation weighting factor, and for non-human biota, ecodosimetric weighting factor (eR). Overall, a nominal eR of about 10 for deterministic parameters relevant to the population, with a range of about 5 to 20, is recommended by Chambers et al. (*in*: [IAE 03a]).

Experimental results suggest that RBE values tend to be higher at lower dose rates and when the irradiation is from tissue-embedded radionuclides rather than external radionuclides. Studies using endpoints such as mortality, reproduction and immune function have provided widely varying RBE values such as 40–50 for life shortening in mice, 47–377 (best estimate 80) for oocyte mortality, 130–180 for hemopoiesis (acute gamma reference) and 250–360 for hemopoiesis (chronic gamma reference). Based on the available data, Thompson et al. (pp. 93–100, in; IAEA, 2003a) recommended an RBE value of 40 for the ecological risk assessment of biota chronically exposed to alpha-emitting radionuclides.

The ICRP has initiated studies to estimate the weighting factors to be applied for the various types of radiation. For tritium, the reported values were centered around 1.5 to 2 for X-rays and 2 to 2.5 for gamma rays; the values for alpha particles were generally significantly higher. The ICRP proposes the use of a weighted absorbed dose RBE for protective purposes, with an RBE weighting for biota of 1 for all low TEL radiation and 10 for alpha particles [ICR 21]. The ICRP position for RBE values is unacceptable.

This situation leads to aberrations. For example, the annual exposure limit for humans is 1 $mSv.yr^{-1}$, and for non-human mammals (including marine mammals), the chronic exposure limits are 1 $mGy.d^{-1}$, i.e. three orders of magnitude higher (Bréchignac and Barescut, pp.119–128, *in*: [IAE 03a]). This situation is aberrant, even if the comparison sensu stricto is not possible, since the annual effective dose limits for humans are established with respect to stochastic risk, whereas for non-human species, the reference values for chronic dose rate are established with respect to deterministic effects.

C.5.3.2. *The weighting factor for the type of tissue or organ*

This weighting factor is rarely mentioned by radioecology experts for non-human organisms. The problem is complex, because it requires knowledge of the distribution of each radionuclide in the various tissues of the species studied. However, it is not insoluble because if there is a great biodiversity, there is also a uniqueness of life at the molecular and cellular levels and many metabolisms are comparable through the evolution of living organisms. Our knowledge on humanity can be easily extrapolated to other mammals including marine mammals and with some nuances to a majority of other phyla.

C.5.4. *Taking interactions into account*

The interactions between the effects of ionizing radiation and another agent are poorly understood, but they could be synergistic in their harmful effects. Two hypotheses can be considered. The first hypothesis is that the mechanism of the synergistic interaction of ionizing radiation with another effective agent can be attributed to a reduced cellular capacity to repair damage after their combined action. The second possible mechanism to explain the synergistic interaction may be based on the assumption that the synergy could be related to additional lethal or potentially lethal damage resulting from the interaction of lesions induced by the two agents (Kim and Petin, pp. 383–388, *in*: [IAE 03a]). Similarly, additional stress (irradiation, Cd overload) can cause rapid reallocation of radionuclides in an ecosystem with drastic changes in transfer factors (Kutakhmedov et al. pp. 175–185, *in:* [CIG 06]).

C.6. Future of radioecology

Radioecology has gone through three phases: an anthropocentric phase, a biocentric phase and an ecocentric phase [OUG 16].

C.6.1. *Level of biological organization of the estimation of the harmful effect of ionizing radiation*

The harmful effects of ionizing radiation can be estimated at various levels of biological organization from the molecular level to the ecosystem level through the individual level. The opinions of scientists differ. The ICRP favors the individual level, while the IUR recommends the ecosystem level and other experts propose the molecular level with ecological biomarkers [AMI 13, AMI 17b]. Strand et al. [STR 17] suggest that a combination of laboratory and field studies of ecosystem functioning and effects may offer the best opportunities for consistency in interpreting the results of studies of the environmental impacts of ionizing radiation. According to Kautsky and Gilek (pp. 77, *in*: [IAE 03a]), there is a too great gap between the goal of protecting the environment, i.e. ecosystem function and population maintenance, and estimates of effects that are typically studied at the individual, organ, cellular or subcellular levels. The impact of individual changes at the population and ecosystem levels depends on population size, total biomass, life history, reproductive investment, generation time, inter- and intraspecific competition, predation and resource exploitation. The authors (Kausky and Gilek) addressing effects on individuals may be a very pessimistic approach.

It is necessary to place this debate within the evolution of concepts in radioecology.

The anthropocentric vision of radioecology considering humans as the most radiosensitive species was only abandoned in the 2000s. The approach became biocentric. This shift is noticeable in ICRP policy. In 2003 and 2007 [ICR 03, ICR 07], this international organization continued to believe that this anthropocentric approach was "enveloping" but proposed the development of tools to assess the impact on the environment. In 2008, the ICRP [ICR 08] proposed the concept and use of reference animals and plants (RAPs) to assess the impact on the environment. A dozen reference organisms, supposedly representative of all living organisms, were selected on which an effort to quantify the activity, exposure pathways, radiation doses and resulting adverse effects is required from organizations conducting research in radioecology. In addition, radioecology adopts the risk assessment tools commonly used for chemical toxicants [AMI 17a].

But this biocentric (or "reference organism") approach is highly reductionist. It is based entirely on the responses of individuals to the effects of radiation, without consideration of higher levels of biological organization. The methodology is out of step with the protection objectives it is supposed to serve, which are the protection of populations and, beyond that, of communities and ecosystems. Furthermore, it ignores the interactions between species that govern impacts at the system level. It cannot therefore account for certain effects that occur at the ecosystem scale, such as indirect effects and the transgenerational propagation of effects from individuals to the population and the ecosystem.

It is therefore necessary to develop an ecosystem approach for tomorrow. The ecosystem approach is applied in many fields, outside the field of radiation, and is thus recommended in the marine environment for the management of fisheries [FAO 03, NOA 03] and marine coasts [LAF 04]. This approach is also recommended in international agreements and conventions such as the Convention on Biological Diversity [UNE 07], the Water Framework Directive [EC 00] or OSPAR [OSP 10] (Bergen statement). In the context of the radiation field, the IUR [IUR 12] has written its own reflections.

This ecosystemic approach, which takes an ecocentric view, takes into account indirect effects and in particular possible interactions. It takes into account the higher levels of organization and should be able to account for the concept of resilience, i.e. the capacity of the ecosystem to absorb the pressure of a disturbance without apparent damage.

However, the effects on ecosystems are delayed and are often too significant for remediation to be considered. It would be important and even essential to have early and sensitive warning signals of the effects of weak radiation on living organisms. To do this, tools exist. These are biomarkers. They can be divided into two main categories: defense biomarkers, which react to the presence of stress, and damage biomarkers, which indicate that the individual has not only been stressed but also that this stress has caused harmful effects. For radionuclides, the most common defense biomarker is metallothionein [CAI 99, AMI 06]; for damage biomarkers, the main ones are antioxidant and genetic markers [AMI 08a]. To counteract the disadvantage that the biomarker response is not followed by an effect at the population or ecosystem level, the use of ecologically oriented biomarkers seems to be a good future solution [AMI 08a, AMI 15, AMI 17b]. The biomarker approach in radioecology can be inspired by medical research on the impact of ionizing radiation following palliative cancer care. This research shows that various proteins or biochemical mechanisms (histone phosphorylation) [BON 08] could advantageously serve as biomarkers. Indeed, some protein targets have been shown to be promising for use as radiation biomarkers in future molecular epidemiology studies [TAR 13].

C.6.2. *Major research programs*

In the field of radioecology, the European Community is very active and supports many coordination and research projects.

Thus, the COMET project (COordination and iMplementation of a pan-European instrumenT for radioecology) is a combined collaborative project and coordination and support action under the 7th Framework Program of the European Commission and Euratom. This project has contributed to the integration and sustainability of radioecology in Europe through its support and interaction with the European Radioecology ALLIANCE. In the short-term, the work to promote radioecology will continue in the framework of the H2020 EJP-CONCERT project (European Joint Programme for the Integration of Research in Radiation Protection). The EJP-CONCERT project (2015–2020) aims to develop a sustainable structure to promote and administer joint programming and open research calls in the field of radiation protection research for Europe. In the longer term, radioecological research will be facilitated by ALLIANCE [MUI 18].

The European Radioecology Alliance project is a transnational organization aiming at the integration of the research programs of the partners involved in radioecology, more precisely in the field of the assessment of the impact of radioactive substances on humans and the environment. ALLIANCE was born in the orbit of COMET [MUI 13].

The Strategic Research Agenda (SRA) initiated by the STAR network of excellence (STrategy for Allied Radioecology, program completed in 2015) and integrated into the research strategy implemented by the COMET consortium defines a long-term vision (20 years) of the needs and implementation of radioecology research. Five-year roadmaps are being developed in the field of marine radioecology. For example, two programs are underway, FRAME (Impact of recent releases from the Fukushima nuclear accident on the marine environment) and RATE (RAdioactive particle Transformation procEss). For example, kinetic transfer models have been improved to better capture dynamic changes in activity concentration in marine biota and sediments during post-accident situations, and improved trophic transfer models have been developed. The FRAME science campaigns provided data for the improvement and validation of marine models, while the experimental work on human food chain modeling and particle behavior provided data on the dynamic parameters of their respective models.

The Radioecology Exchange website was established in 2011 as part of the FP7 STAR and COMET [VAN 17].

C.6.3. *Gaps in marine radioecology*

Within ALLIANCE, the Marine Radioecology Working Group has published a position paper [VIV 18] that proposes priorities for marine radioecology. These include the combination of modeling and experimentation (both laboratory and field), the effect of speciation on radionuclide circulation and bioavailability, and improving risk assessment with validated tools for decision-makers.

The role of microorganisms with respect to radionuclides is not well known. However, the fate of radionuclides in the natural environment cannot be considered without examining the role that microbial activity plays either in improving their mobility or in immobilizing them [KON 02].

In conclusion, there is a need for a holistic assessment of the environmental impacts of ionizing radiation that considers not only the direct health consequences for humans and non-human species but also the more complex social, ethical and economic consequences of both human and non-human exposures [OUG 16].

With regard to the characterization of the radioactive risk to non-human organisms, the ICRP has accumulated gaps and errors. Historically, it made the mistake of believing that humans were the most radiosensitive species. It has partially reversed this position. Another error of the ICRP has been to minimize the precautionary factors due to the type of radiation. Finally, the great marine biodiversity is underestimated and it is a mistake to retain only three marine species among the reference species (an alga, a crab and a fish). No mollusks, no mammals, no polychaetes. This is an aberration. The result is a very uncertain final estimate of the radioactive risk. Indeed, the risk is the ratio between two terms: the dose received and the harmful effect. However, these two terms are very poorly evaluated with strong uncertainties.

More generally, most radioecologists make the mistake of estimating the risk due to natural radioactivity as zero. However, a natural phenomenon is not synonymous with harmlessness.

The ultimate goal of radioecology would be to have an estimation of the radioactive risk of non-human organisms as effective as that applied to humans, and thus to estimate the absorbed radiation dose expressed in an ecological equivalent of the sievert and no longer in gray. This assumes a reliable estimate of the thresholds for cancer induction by ionizing radiation for non-human organisms and a quantification over several generations of the genetic effects due to irradiation of their gonads. It would therefore be necessary to define the long-term risks for the various living species, hoping to find sufficient inter-species commonalities to define the elements of this sievert equivalent.

List of Acronyms

AA	Anaphase Aberration
AFM	Assessment Factor Method
AMAP	Arctic Monitoring and Assessment Programme
ASN	*Autorité de Sûreté Nucléaire* (French Nuclear Safety Authority)
BER	Base Excision Repair
BNFL	British Nuclear Fuels Ltd.
BNI	Basic Nuclear Installations
CA	Chromosomal Aberration
CEA	*Commissariat à l'Énergie Atomique et aux énergies alternatives* (French Atomic Energy and Alternative Energies Commission)
CEFAS	Centre for Environment, Fisheries and Aquaculture Science
CERCLA	Comprehensive Environmental Response, Compensation and Liability Act
CFDT	*Confédération Française Démocratique du Travail* (French Democratic Confederation of Labor)
COMET	COordination and iMplementation of a pan-European instrumenT for radioecology

DCF	Dose Conversion Factors
DCRL	Derived Consideration Reference Levels
DNA	DeoxyriboNucleic Acid
EDEN	Elementary Dose Evaluation for Natural environment
EMRAS	Environmental Modelling for RAdiation Safety
ENEV	Estimated No-Effects Value
EPIC	Environmental Protection from Ionizing Contaminants in the Arctic
ERICA	Environmental Risk for Ionizing Contaminants: Assessment and management
FASSET	Framework for ASSessment of Environmental ImpacT
FDC	Fast Distribution Coefficient
FRED	FASSET Radiation Effects Database
HC5	Hazardous Concentration 5%
HELCOM	HELsinki COMmission
HR	Homologous Recombination
IAEA	International Atomic Energy Agency
IASAP	International Arctic Seas Assessment Project
ICRP	International Commission on Radiological Protection
IRSN	*Institut de Radioprotection et de Sûreté Nucléaire* (French Institute for Radiation Protection and Nuclear Safety)
ITER	International Thermonuclear Experimental Reactor
ITF	Indonesian Through Flow

JCAE	Joint Committee on Atomic Energy
JRNC	Joint Russian–Norwegian Collaboration
LET	Linear Energy Transfer
MAFF	Ministry of Agriculture, Fisheries and Food
MEL	Marine Environment Laboratory
MMBI	Murmansk Marine Biological Institute of the Russian Academy of Sciences
MN	MicroNuclei formation
MT	MetalloThionein
MXR	Multi-Xenobiotic Resistance
NAS	National Academy of Sciences
NER	Nucleotide Excision Repair
NHEJ	Non-Homologous End Joining
NOEC	No Observed Effect Concentration
NOEL	No Observed Effect Level
OSPAR	OSlo and PARis convention
PEC	Predicted Effect Concentration
PEDR	Predicted Environmental Dose Rate
PNEC	Predicted No-Effect Concentration
PNEDR	Predicted No-Effect Dose Rate
PROTECT	PROTECTion of the environment from ionizing radiation in a regulatory context

rad	Radiation Absorbed Dose
RAP	Reference Animal or Plant
RATE	RAdioactive particle Transformation procEss
RBE	Relative Biological Efficiency
REML	REsidual Maximum Likelihood
RNA	RiboNucleic Acid
RNM	*Réseau National de Mesures* (French National Measurement Network)
RNO	*Réseau National d'Observation de la qualité du milieu marin* (French National Marine Environmental Quality Monitoring Network)
RO	Representative Organism
ROCCH	*Réseau d'Observation de la Contamination CHimique du littoral* (French Coastal Chemical Contamination Monitoring Network)
SCE	Sister Chromatid Exchange
SRA	Strategic Research Agenda
SSD	Species Sensitivity Distribution
STAR	STrategy for Allied Radioecology
UNSCEAR	United Nations Scientific Committee on the Effects of Atomic Radiation
USAEC	United States Atomic Energy Commission
USDOE	US Department Of Energy
WOMARS	WOrldwide MArine Radioactivity Studies

References

[AAR 70] AARKROG A., "Radioecological investigations", *Nucl. Saf.*, 65, pp. 74–79, 1970.

[AAR 98] AARKROG A., "A retrospect of anthropogenic radioactivity in the global marine environment", *Radiat. Prot. Dosimetry*, vol. 75, nos 1–4, pp. 23–31, 1998.

[AAR 03] AARKROG A., "Input of anthropogenic radionuclides into the world ocean", *Deep-Sea Res. II*, vol. 50, pp. 2597–2606, 2003.

[ABB 79] ABBOTT D.T., MIX M.C., "Radiation effects of tritiated seawater on development of the goose barnacle, *Pollicipes polymerus*", *Health Phys.*, vol. 36, pp. 283–287, 1979.

[ADA 18] ADAM-GUILLERMIN C., HERTAL-AAS T., OUGHTON D. et al., "Radiosensitivity and transgenerational effects in non-human species", *Proceedings of the Fourth International Symposium on the System of Radiological Protection, Annals of the ICRP*, vol. 47, nos 3–4, pp. 327–341, 2018.

[AHS 86] AHSANULLAH M., WILLIAMS A.R., "Effect of uranium on growth and reproduction of the amphipod, Alorchestes compressa", *Mar. Biol.*, vol. 93, pp. 459–464, 1986.

[ALA 12] ALAMRI O.D., CUNDY A.B., DI Y. et al., "Ionizing radiation-induced DNA damage response identified in marine mussels, *Mytilus* sp.", *Environ. Pollut.*, vol. 168, pp. 107–112, 2012.

[ALE 21] ALEXIS-MARTIN B., BOLTON M.B., HAWKINS D. et al., "Addressing the humanitarian and environmental consequences of atmospheric nuclear weapon tests: A case study of UK and US test programs at Kiritimati (Christmas) and Malden islands, Republic of Kiribati", *Glob. Policy*, vol. 12, no. 1, pp. 106–121, 2021.

[ALI 06] ALIEV R., KALMYKOV S., SAPOZHNIKOV Y., "Artificial radionuclides in Black Sea sediments", *IAEA-CSP-26*, IAEA, Vienna, pp. 263–264, 2006.

[ALO 04] ALONGI D.M., SASEKUMAR A., CHONG V.C. et al., "Sediment accumulation and organic material flux in a managed mangrove ecosystem: Estimates of land-ocean-atmosphere exchange in peninsular Malaysia", *Mar. Geol.*, vol. 208, pp. 383–402, 2004.

[ALO 06] ALONZO F., GILBIN R., BOURRACHOT S. et al., "Effects of chronic internal alpha irradiation on physiology, growth and reproductive success of *Daphnia magna*", *Aquat. Toxicol.*, vol. 80, pp. 228–236, 2006.

[ALO 08a] ALONZO F., GILBIN R., ZEMAN F.A. et al., "Increased effects of internal alpha irradiation in *Daphnia magna* after chronic exposure over three successive generations", *Aquat. Toxicol.*, vol. 87, pp. 146–156, 2008.

[ALO 08b] ALONZO F., HERTEL-AAS T., GILEK M. et al., "Modelling the propagation of effects of chronic exposure to ionising radiation from individuals to populations", *J. Environ. Radioact.*, vol. 99, pp. 1464–1473, 2008.

[AMA 16] AMAP, *AMAP Assessment 2015: Radioactivity in the Arctic*, Arctic Monitoring and Assessment Programme, Oslo, 2016.

[AMI 73] AMIARD J.C., "Contribution à l'étude du métabolisme du strontium et du calcium chez deux Téléostéens marins : le Mulet (*Mugil auratus* Risso) et la Plie (*Pleuronectes platessa* L.)", *Bull. Un. Océanogr. Fr.*, vol. 5, pp. 1–6, 1973.

[AMI 74] AMIARD-TRIQUET C., "Etude expérimentale de la contamination par le cérium 144 et le fer 59 d'un sédiment à *Arenicola marina*", *Cah. Biol. Mar. Roscoff*, vol. 15, pp. 483–494, 1974.

[AMI 75a] AMIARD J.C., Influence de divers facteurs écologiques sur l'accumulation du radiostrontium chez deux Téléostéens euryhalins : *Mugil auratus* Risso et *Pleuronectes platessa* L., Specialty thesis, CEA-R-4706 report, 1975.

[AMI 75b] AMIARD J.C., AMIARD-TRIQUET C., "Expérience préliminaire a l'utilisation d'une chaine trophique marine dans l'étude d'une pollution par le cobalt 60 : bilan après une ingestion unique", *Water Air Soil Pollut.*, vol. 5, pp. 221–229, 1975.

[AMI 75c] AMIARD-TRIQUET C., Etude du transfert des radionucléides entre le milieu sédimentaire et les invertébrés qui y vivent, PhD thesis, Université de Nantes, CEA-R-4705 report, 1975.

[AMI 75d] AMIARD TRIQUET C., "Etude comparative de la contamination par le cobalt 60 et le caesium 137 de quelques invertébrés marins fouisseurs", *Vie et Milieu*, vol. 25, no. 1, ser. A, pp. 19–40, 1975.

[AMI 75e] AMIARD-TRIQUET C., AMIARD J.C., Etude expérimentale du transfert du cobalt 60 entre une Annélide marine (*Arenicola marina* l.) et ses prédateurs : le crabe enragé (*Carcinus maenas* l.) et la plie (*Pleuronectes platessa* l.), CEA-R-4658 report, 1975.

[AMI 76a] AMIARD-TRIQUET C., AMIARD J.C., "La pollution radioactive du milieu aquatique et ses conséquences écologiques", *Bull. Ecol.*, vol. 7, pp. 3–32, 1976.

[AMI 76b] AMIARD-TRIQUET C., AMIARD J.C., "Organotropisme du cobalt 60 chez un Lamellibranche (*Scrobicularia plana* da Costa) et un Crustacé (*Carcinus maenas* L.) en fonction du vecteur de contamination", *Oïkos*, vol. 27, pp. 122–126, 1976.

[AMI 76c] AMIARD-TRIQUET C., AMIARD J.C., "Influence du mode de contamination sur l'élimination du cobalt 60 par *Scrobicularia plana* (Mollusque, Bivalve) et *Carcinus maenas* (Crustacé, Décapode)", *Cah. Biol. Mar.*, vol. 17, pp. 1–8, 1976.

[AMI 77] AMIARD J.C., AMIARD-TRIQUET C., "Health and ecological aspects of cobalt 60 transfer in a seawater food chain typical of an intertidal mud-flat", *Intern. J. Environm. Studies*, vol. 10, pp. 113–118, 1977.

[AMI 78a] AMIARD J.C., Contribution à l'étude de l'accumulation et de la toxicité de quelques polluants stables et radioactifs chez des organismes marins : l'antimoine, l'argent, le cobalt et le strontium chez des mollusques, des crustacés et des téléostéens, PhD thesis, Université Pierre et Marie Curie (Paris VI), CEA-R-4928 report, 1978.

[AMI 78b] AMIARD J.C., "Exemples de calculs de doses d'irradiation reçues par des organismes aquatiques en milieu expérimental contaminé par divers radionucléides I. Cas de cultures de *Dunaliella bioculata* contaminées par ^{51}Cr et ^{109}Cd", *Rev. Intern. Océanogr. Méd.*, vol. 51, no. 52, pp. 139–151, 1978.

[AMI 78c] AMIARD J.C., "Modalités de la contamination d'une chaîne trophique marine benthique par l'argent 110m 3. Influence du mode de contamination sur l'organotropisme du radionucléide", *Helgölander wiss. Meerest.*, vol. 31, pp. 444–456, 1978.

[AMI 79a] AMIARD J.C., AMIARD-TRIQUET C., "Interférences du cycle d'intermue et de la mue avec la radioécologie des crustacés aquatiques", *Vie et Milieu*, vols 28–29, no. 3, ser. AB, pp. 371–392, 1979.

[AMI 79b] AMIARD J.C., AMIARD-TRIQUET C., "Distribution of cobalt 60 in a mollusc, a crustacean, and a freshwater Teleost: Variations as a function of the source of pollution and during elimination", *Environ. Pollut.*, vol. 20, pp. 199–214, 1979.

[AMI 79c] AMIARD-TRIQUET C., "Synthèse bibliographique sur la radioécologie du plutonium et des autres transuraniens dans les milieux aquatiques", *Bull. Ecol.*, vol. 10, pp. 241–277, 1979.

[AMI 80] AMIARD-TRIQUET C., AMIARD J.C., *Radioécologie des milieux aquatiques. La pollution radioactive et ses conséquences écologiques*, Masson, Paris, 1980.

[AMI 81] AMIARD J.C., KHALANSKI M., "Conséquences de l'élévation de la température de l'eau sur la radiocontamination des organismes aquatiques", *Année Biologique*, vol. 20, pp. 115–130, 1981.

[AMI 06] AMIARD J.C., AMIARD-TRIQUET C., BARKA S. et al., "Metallothioneins in aquatic invertebrates: Their role in metal detoxification and their use as biomarkers", *Aquatic Toxicology*, vol. 76, pp. 160–202, 2006.

[AMI 08a] AMIARD J.C., AMIARD-TRIQUET C., *Les biomarqueurs dans l'évaluation de l'état écologique des milieux aquatiques*, Lavoisier, Tec&Doc, Paris, 2008.

[AMI 08b] AMIARD-TRIQUET C., COSSU-LEGUILLE C., MOUNEYRAC C., "Les biomarqueurs de défense, la tolérance et ses conséquences écologiques", in AMIARD J.C., AMIARD-TRIQUET C., (eds), *Les biomarqueurs dans l'évaluation de l'état écologique des milieux aquatiques*, Lavoisier, Tec&Doc, Paris, 2008.

[AMI 09] AMIARD-TRIQUET C., BERTHE T., CRÉACH A. et al., "Tolerance in organisms chronically exposed to estuarine pollution", in AMIARD-TRIQUET C., RAINBOW P.S. (eds), *Environmental Assessment of Estuarine Ecosystem: A Case Study*, CRC Press, Boca Raton, 2009.

[AMI 10] AMIARD J.C., GAZAL S., "Tritium et environnement", in GAZAL S.S., AMIARD J.C., (eds), *Le tritium. Actualité d'aujourd'hui et de demain*, Lavoisier, Tec&Doc, Paris, 2010.

[AMI 11] AMIARD-TRIQUET C., RAINBOW P.S., ROMEO M., *Tolerance to Environmental Contaminants*, CRC Press, Boca Raton, 2011.

[AMI 13] AMIARD J.C., *Le risque radioactif. Devenir des radionucléides dans l'environnement et impacts sur la santé*, Lavoisier, Tec&Doc, Paris, 2013.

[AMI 15] AMIARD-TRIQUET C., AMIARD J.C., MOUNEYRAC C., *Aquatic Ecotoxicology. Advancing Tools for Dealing with Emerging Risks*, Elsevier, London, 2015.

[AMI 17a] AMIARD J.C., *Les risques chimiques environnementaux. Méthodes d'évaluation et impacts sur les organismes*, 2nd edition, Lavoisier, Tec&Doc, Paris, 2017.

[AMI 17b] AMIARD J.C., AMIARD-TRIQUET C., *Biomarqueurs en écotoxicologie aquatique*. Lavoisier, Tec&Doc, Paris, 2017.

[AMI 21] AMIARD J.C., *Management of Nuclear Wastes*, vol. 5. ISTE, London and John Wiley & Sons, Hoboken, 2021.

[AMO 63] AMOUREUX L., "Notes biologiques et écologiques sur l'Arénicole", *C. R. Soc. Biol.*, vol. 157, pp. 1711–1715, 1963.

[AND 04a] ANDRAL B., STANISIÈRE J.Y., SAUZADE D. et al., "Monitoring chemical contamination levels in the Mediterranean based on the use of mussel caging", *Mar. Pollut. Bull.*, vol. 49, pp. 704–712, 2004.

[AND 04b] ANDRIANOV K.N., VYSKREBENTSEV V.V., DUBASOV Yu.V. et al., "North test site: Basic information on nuclear tests (1955–1992)", *Nuclear Explosions in the USSR: The North Test Site Reference Material*, IAEA, Vienna, pp. 4–34, 2004.

[AND 09] ANDERSSON P., GARNIER-LAPLACE J., BERESFORD, N.A. et al., "Protection of the environment from ionising radiation in a regulatory context (protect): Proposed numerical benchmark values", *J. Environ. Radioact.*, vol. 100, no. 12, pp. 1100–1108, 2009.

[ANO 81] ANONYMOUS, Marshall Islands: A chronology 1944–1981, Micromesia Support Committee, Honolulu, 1981.

[ANO 12] ANONYMOUS, An ecological assessment of Johnston Atoll, Washington Group International, New Jersey, 2012.

[ANS 03] ANSOBORLO E., CHIAPPINI R., MOULIN V., "La spéciation dans l'environnement et en milieu biologique", *Clefs CEA*, no. 48, 2003.

[ARE 11] AREVA, Rapport de sûreté nucléaire et de radioprotection 2010, AREVA, Report, 2011.

[ASS 99] ASSINDER D.J., "A review of the occurrence and behaviour of neptunium in the Irish Sea", *J. Environ. Radioact.*, vol. 44, pp. 335–347, 1999.

[AUF 71] AUFFRET J.P., GERMAIN P., GUEGUENIAT P. et al., "Etude expérimentale de la fixation du caesium 137 par certains sédiments de la Manche", *Cah. Océanogr.*, vol. 10, pp. 935–955, 1971.

[AVA 64] AVARGUES M., "Problèmes posés par le rejet de résidus radioactifs en milieu marin", *Colloque de Radioécologie marine*, Université de Caen et CEA, 1964.

[AVA 71] AVARGUES M., "La contamination radioactive de l'environnement", in LE PRAT G. (ed.), *Précis général des nuisances*, vol. 2, *Bibliothèque de l'environnement*, Paris, 1971.

[BAI 95] BAILLY DU BOIS P., SALOMON J.C., GANDON R. et al., "A quantitative estimate of English Channel water fluxes into the North Sea from 1987 to 1992 based on radiotracer distribution", *J. Mar. Syst.*, vol. 6, pp. 457–481, 1995.

[BAI 99] BAILLY DU BOIS P., GUÉGUÉNIAT P., "Quantitative assessment of dissolved radiotracers in the English Channel: Sources, average impact of La Hague reprocessing plant and conservative behaviour (1983, 1986, 1988 and 1994)", *Cont. Shelf Res.*, vol. 19, pp. 1977–2002, 1999.

[BAI 05a] BAILLY DU BOIS P., DUMAS F., "Fast hydrodynamic model for of medium- and long-term dispersion in seawater in the English Channel and southern North Sea, qualitative and quantitative validation by radionuclide tracers", *Ocean Modell.*, vol. 9, pp. 169–210, 2005.

[BAI 05b] BAILLY DU BOIS P., DUMAS F., "TRANSMER, hydrodynamic model for medium- and long-term simulation of radionuclides transfers in the English Channel and southern North Sea", *Radioprotection*, vol. 40, pp. S575–S580, 2005.

[BAI 05c] BAILLY DU BOIS P., DUMAS F., SOLIER L., "Hydrodynamic modelling of short-term dispersion in a macro-tidal sea, validation by high-resolution radionuclide tracer measurements", *Radioprotection*, vol. 40, no. 1, pp. S563–S568, 2005.

[BAI 08] BAINES S.B., FISHER N.S., "Modeling the effect of temperature on bioaccumulation of metals by a marine bioindicator organism, *Mytilus edulis*", *Environ. Sci. Technol.*, vol. 42, pp. 3277–3282, 2008.

[BAL 85] BALASZ G.H., "Status and ecology of marine turtles at Johnston Atoll", *Atoll Res. Bull.*, vol. 285, pp. 1–46, 1985.

[BAR 93] BARENDSEN G.W., "Subletahal damage and DNA double strand breaks have similar RBE-LET relationships: Evidence and implications", *Int. J. Radiat. Biol.*, vol. 63, pp. 325–330, 1993.

[BAR 94] BARENDSEN G.W., "The relationships between RBE and LET for different types of lethal damage in mammalian cells: Biophysical and molecular mechanisms", *Radiat. Res.*, vol. 139, pp. 257–270, 1994.

[BAS 11] BASKARAN M., "Po-210 and Pb-210 as atmospheric tracers and global atmospheric Pb-210 fallout: A review", *J. Environ. Radioact.*, vol. 102, pp. 500–513, 2011.

[BAX 95] BAXTER M.S., BALLESTRA S., GASTAUD J. et al., "Marine radioactivity studies in the vicinity of sites with potential radionuclide releases", *Environmental Impact of Radioactive Releases Symposium*, IAEA, Vienna, pp. 125–141, 1995.

[BEA 78] BEASLEY T.M., HEYRAUD M., HIGG J.J.W. et al., "^{210}Po and ^{210}Pb in *Zooplankton Fecal Pellets*", *Mar. Biol.*, vol. 44, pp. 325–328, 1978.

[BEA 06a] BEAUGELIN-SEILLER K., FIEVET B., GARNIER-LAPLACE J., "Theoretical dose reconstruction along life-cycle for plaices (*Pleuronectes platessa*) living in the Northern Cotentin waters after 1966", *IAEA-CSP-26*, IAEA, Vienna, pp. 243–247, 2006.

[BEA 06b] BEAUGELIN-SEILLER K., JASSERAND F., GARNIER-LAPLACE J. et al., "Modelling the radiological dose in non-human species: Principles, computerization and application", *Health Phys.*, vol. 90, pp. 485–493, 2006.

[BEA 16] BEAUGELIN-SEILER K., GARNIER-LAPLACE J., Radioprotection de l'environnement. État de l'art et recommandations de l'IRSN, IRSN report 2016–03, 2016.

[BEN 74] BENAYOUN G., FOWLER S.W., OREGIONI B., "Flux of cadmium through euphausiids", *Mar. Biol.*, vol. 27, pp. 205–212, 1974.

[BER 68] BERNHARD M., "Research on the metabolism of some radioactive elements in the marine environment", *Rev. Intern. Océanogr. Méd.*, vol. 11, pp. 11–59, 1968.

[BER 70] BERNHARD M., "The role of vertebrates on in the uptake and loss of radioisotopes", *Rev. Intern. Océanogr. Méd.*, vol. 20, pp. 101–123, 1970.

[BER 04] BERESFORD N.A., BROADLEY M.R., HOWARD B.J. et al., "Estimating radionuclide transfer to wild species-data requirements and availability for terrestrial ecosystems", *J. Radiol. Prot.*, vol. 24, pp. A89–A103, 2004.

[BER 06] BERGONIE J., TRIBONDEAU L., "Interprétation de quelques résultats de la radiothérapie et essai de fixation d'une technique rationnelle", *Comptes-rendus de l'Académie des Sciences (Paris)*, vol. 143, pp. 983–984, 1906.

[BER 07] BERGEY L.L., WEIS J.S., "Molting as a mechanism of depuration of metals in the fiddler crab, *Uca pugnax*", *Mar. Environ. Res.*, vol. 64, pp. 556–562, 2007.

[BER 08a] BERESFORD, N.A., BARNETT, C.L., BROWN, J. et al., "Inter-comparison of models to estimate radionuclide activity concentrations in non-human biota", *Radiat. Environ. Biophys.*, vol. 47, pp. 419–514, 2008.

[BER 08b] BERESFORD N.A., BALONOV, M., BEAUGELIN-SEILLER, K. et al., "An international comparison of models and approaches for the estimation of the radiological exposure of non-human biota", *Appl. Radiat. Isot.*, vol. 66, pp. 1745–1749, 2008.

[BER 10] BERESFORD N.A., BARNETT C.L., BROWN J.E. et al., "Predicting the radiation exposure of terrestrial wildlife in the Chernobyl exclusion zone: An international comparison of approaches", *J. Radiol. Prot.*, vol. 30, pp. 341–373, 2010.

[BIR 03] BIRD, G.P., THOMPSON, P., MACDONALD, C.R. et al., "Assessment of the impact of radionuclide releases from Canadian nuclear facilities on non-human biota", *SPEIR 3*, Darwin, 2003.

[BLA 69] BLAYLOCK B.G., "The fecundity of a *Gambusia affinis affinis* population exposed to chronic environmental radiation", *Radiat. Res.*, vol. 37, pp. 108–117, 1969.

[BLA 73] BLANC D., *Physique nucléaire*, Masson, Paris, 1973.

[BLA 75] BLAYLOCK B.G., WITHERSPOON J.P., "Dose estimation and prediction of radiation effect on aquatic biota resulting from radioactive releases from the nuclear fuel cycle", *Impacts of Nuclear Releases into the Aquatic Environment*, IAEA, Vienna, pp. 377–393, 1975.

[BLU 01] BLUST R., "Radionuclide accumulation in freshwater organisms", in VAN DER STRICHT E., KIRCHMANN R. (eds), *Radioecology. Radioactivity and Ecosystems*, UIR, Liège, 2001.

[BOI 97] BOISSON F., HUTCHINS D.A., FOWLER S.W. et al., "Influence of temperature on the accumulation and retention of 11 radionuclides by the marine alga *Fucus vesiculosus* (L.)", *Mar. Pollut. Bull.*, vol. 35, pp. 313–321, 1997.

[BOL 11] BOLOGNESI C., HAYASHI M., "Micronucleus assay in aquatic animals", *Mutagenesis*, vol. 26, pp. 205–213, 2011.

[BON 77] BONHAM K., DONALDSON L.R., HERSHBERGER W.K., Low level chronic irradiation of salmon, Final report, RLO-2225–T2–8, 1977.

[BON 08] BONNER W.M., REDON C.E., DICKEY J.S. et al., "γH2AX and cancer", *Nat. Rev. Cancer*, vol. 8, no. 12, pp. 957–967, 2008.

[BOR 16] BORDNER A.S., "Measurement of background gamma radiation in the northern Marshall Islands", *Proc. Natl. Acad. Sci.*, vol. 113, no. 25, pp. 6833–6838, 2016.

[BOU 92] BOURLAT Y., MARTIN G., "Precise determination of the concentration of radiocaesium in the water of Mururoa lagoon", *J. Environ. Radioact.*, vol. 17, pp. 13–29, 1992.

[BOU 95] BOURLAT Y., MILLIES-LACROIX J.C., NAZARD R., "Determination of plutonium radioactivity in Mururoa lagoon water", *J. Radioana. Nucl. Chem.*, vol. 197, pp. 387–408, 1995.

[BOU 96] BOUST D., MITCHELL P.I., GARCIA K. et al., "A comparative study of the speciation and behaviour of plutonium in the marine environment of two reprocessing plants", *Radiochim. Acta*, vol. 74, pp. 203–210, 1996.

[BOU 97a] BOURLAT Y., MILLIÈS-LACROIX J.C., Le PETIT G. et al., "^{90}Sr, ^{137}Cs and $^{239+240}$Pu in world ocean water samples collected from 1992 to 1994", in GUÉGUÉNIAT P., GERMAIN P.P., MÉTIVIER H. (eds), *Radionuclides in the Oceans, Inputs and Inventories*, Les Editions de Physique, Paris, 1997.

[BOU 97b] BOUST D., COLIN C., LECLERC G. et al., "Distribution and transit times of plutonium-bearing particles throughout the Channel", in GERMAIN P., GUARY J.-C., GUÉGUÉNIAT P. et al. (eds), *Radioprotection, special edition, "Radionuclides in the Oceans, RADOC 96–97, Proceedings Part 1 "Inventories, behaviour and processes"*, Cherbourg-Octeville, 7–11 October 1996, April 1997, vol. 32, C2, pp. 123–128, 1997.

[BOU 99] BOUST D., "Distribution and inventories of some artificial and naturally occurring radionuclides in medium to coarse-grained sediments of the channel", *Cont. Shelf Res.*, vol. 19, pp. 1959–1975, 1999.

[BOU 06] BOUDJENOUN R., TEYSSIE J.L., COTRET O. et al., "Bioaccumulation and retention of ^{210}Pb in the Mediterranean mussel *Mytilus galloprovincialis*", *IAEA-CSP-26*, IAEA, Vienna, pp. 321–322, 2006.

[BOV 75] BOVARD P., GRAUBY A., SAAS A. et al., "Incidence de la charge polluante des eaux sur le comportement des radionucléides", *Combined Effects of Radioactive, Chemical and Thermal Releases to the Environment*, AIEA, Vienna, pp. 179–190, 1975.

[BOW 75] BOWEN V.T., Distributions of transuranic nuclides in the oceans: Possibilities for their cycling, COO-3563-40, 1975.

[BRA 06] BRADSHAW C., KUMBLAD L., FAGRELL A., "The use of tracers to evaluate the importance of bioturbation in remobilising contaminants in Baltic sediments", *Estuar. Coast. Shelf Sci.*, vol. 66, pp. 123–134, 2006.

[BRA 14] BRADSHAW C., KAPUSTKA L., BARNTHOUSE, L. et al., "Using an ecosystem approach to complement protection schemes based on organism level endpoints", *J. Environ. Radioact.*, vol. 136, pp. 98–104, 2014.

[BRA 19] BRAINARD R.E., ACOBA T., ASHER M.A.M. et al., "Coral reef ecosystem monitoring report for the Pacific Remote Islands Marine National Monument 2000–2017", *PIFSC Special Publication SP-19–006f*, Pacific Islands Fisheries Science Center, pp. 1–89, 2019.

[BRE 17] BRESSAC M., LEVY I., CHAMIZO E. et al., "Temporal evolution of Cs-137, Np-237, and Pu239+240 and estimated vertical Pu239+240 export in the northwestern Mediterranean Sea", *Sci. Total Environ.*, vol. 595, pp. 178–190, 2017.

[BRO 02] BROWN J.E., NIKITIN A., VALETOVA N.K. et al., "Radioactive contamination in the marine environment adjacent to the outfall of the radioactive waste treatment plant at ATOMFLOT, northern Russia", *J. Environ. Radioact.*, vol. 61, pp. 111–131, 2002.

[BRO 03] BROWN, J.E., THØRRING, H., HOSSEINI, A. et al., Deliverable D6: The "EPIC" impact assessment framework: Towards the protection of the Arctic environment from the effects of ionizing radiation, Project ICA2–CT-2000–10032, 2003.

[BRO 04] BROWN J.E., JONES J.R., SAXÉN R. et al., "Radiation doses to aquatic organisms from natural radionuclides", *J. Radiol. Prot.*, vol. 24, no. 4A, pp. A63–A77, 2004.

[BRO 05] BROWN J.E., GAFVERT T., "Oil and gas platforms case study report", in BERESFORD N.A., HOWARD B.J. (eds), *Application of FASSET Framework at Case Study Sites*, Deliverable Report D9 to the ERICA Project, Swedish Radiation Protection Authority, Stockholm, pp. 43–58, 2005.

[BRO 19] BROWN J.E., BERESFORD N.A., HEVRØY T.H., "Exploring taxonomic and phylogenetic relationships to predict radiocaesium transfer to marine biota", *Sci. Total Environ.*, vol. 649, pp. 916–928, 2019.

[BRU 73] BRUNER H.D., "Distribution of tritium between the hydrosphere and invertebrates", in MOGHISSI A.A., CARTER M.W. (eds), *Tritium*, Messinger Graphies, Phoenix, 1973.

[BRU 98] BRUSLÉ J., BRUSLÉ-SICARD S., FOURCAUD B., "Gonads of groupers (Serranidae) from Mururoa and Hao (French Polynesia) used as indicators of the quality of the coral reef environment: Histological and ultrastructural aspects", *Cybium*, vol. 12, no. 4, pp. 371–382, 1998.

[BUE 14] BUESSELER K.O., "Fukushima and ocean radioactivity", *Oceanography*, vol. 27, no. 1, pp. 92–105, 2014.

[BUE 17] BUESSELER K., DAI M., AOYAMA M. et al., "Fukushima Daiichi-derived radionuclides in the ocean: Transport, fate and impacts", *Ann. Rev. Mar. Sci.*, vol. 9, pp. 173–203, 2017.

[BUS 02] BUSTAMANTE P., GERMAIN P., LECLERC G. et al., "Concentration and distribution of ^{210}Po in the tissues of the scallop *Chlamys varia* and the mussel *Mytilus edulis* from the coasts of Charente-Maritime (France)", *Mar. Pollut. Bull.*, vol. 44, pp. 997–1002, 2002.

[BUX 08] BUXTON G.V., "An overview of the radiation chemistry of liquids", in SPOTHEIM-MAURIZOT M. et al. (eds), *Radiation Chemistry: From Basics to Application in Material and Life Sciences*, EDP Sciences/L'actualité Chimique, 2008.

[CAI 99] CAI L., SATOH M., TOHYAMA C. et al., "Metallothionein in radiation exposure: Its induction and protective role", *Toxicology*, vol. 132, pp. 85–98, 1999.

[CAL 86] CALMET D., Synthèse radioécologique des différents compartiments de l'environnement marin du cotentin, Thesis, Université d'Aix-Marseille II, 28 March 1986.

[CAQ 01] CAQUET T., RAMADE F., "The ecological dimensions of ecotoxicological research", in BRÉCHIGNAC F., HOWARD B.J. (eds), *Radioactive Pollutants: Impact on the Environment*, Les Ulis, EDP Sciences, 2001.

[CAR 66] CAREY JR. A.G., PEARCY W.G., OSTERBERG C.L., "Artificial radionuclides in marine organisms in the northeast Pacific Ocean off Oregon, U.S.A", *Symposium for the Disposal of Radioactive Waste into Seas, Oceans, and Surface Waters*, Vienna, 17 May 1966.

[CAR 95] CARR A.M., HOEKSTRA M.F., "The cellular responses to DNA damage", *Trends Cell Biol.*, vol. 5, pp. 32–40, 1995.

[CAR 96] CARR A.M., "Checkpoints take the next step", *Science*, vol. 271, pp. 314–315, 1996.

[CAR 97a] CARPENTER R., "Anthropogenic radionuclides in the marine environment: Case studies", in IAEA (ed.), *Strategies and Methodologies for Applied Marine Radioactivity Studies*, IAEA Training Courses Series no. 7, Vienna, 1997.

[CAR 97b] CARVALHO F.P., "Distribution, cycling and mean residence time of Ra-226, Pb-210 and Po-210 in the Tagus estuary", *Sci. Tot. Environ.*, vol. 196, no. 2, pp. 151–161, 1997.

[CAR 11] CARVALHO F.P., OLIVEIRA J.M., MALTA M., "Radionuclides in deep-sea fish and other organisms from the North Atlantic Ocean", *ICES J. Mar. Sci.*, vol. 68, no. 2, pp. 333–340, 2011.

[CAR 18] CARVALHO F.P., "Radionuclide concentration processes in marine organisms: A comprehensive review", *J. Environ. Radioact.*, vol. 186, pp. 124–130, 2018.

[CBD 14] CBD, "COP 5 Decision V/6," Annex B: Ecosystem Approach, CBD, available at: http://www.cbd.int/decision/cop/?id=7148, 2000.

[CEF 20] CEFAS, "Radioactivity in food and the environment", 25th edition, *CEFAS*, 2020.

[CFD 75] CFDT, *L'électronucléaire en France*, Le Seuil, Paris, 1975.

[CHA 72] CHAPMAN W.A., "Ionizing radiation", in KINNE O. (ed.), *Marine Ecology*, vol. I, *Environmental Factors*, Part 3, Wiley, London, 1972.

[CHA 98] CHARMASSON S., GERMAIN P., LECLERC G., "^{210}Po as a tracer of variations of trophic input to deep-sea benthic ecosystems: A study of the deep-sea amphipod *Eurythenes Gryllus* from the tropical Atlantic", *Radiat. Prot. Dosim.*, vol. 75, nos 1–4, pp. 131–138, 1998.

[CHA 03] CHARMASSON S., "(CS)-Cs-137 inventory in sediment near the Rhone mouth: Role played by different sources", *Oceanologica Acta*, vol. 26, pp. 435–441, 2003.

[CHA 05] CHAMBERS, D.B., MULLER, E., SAINT-PIERRE, S. et al., "Assessment of marine biota doses arising from the radioactive sea discharges of the COGEMA La Hague facility. A comprehensive case study (Consensus Appraisal)", *IRPA11 – International Conference on the Protection of the Environment from the Effects of Ionizing Radiation: Protection of the Environment from the Effects of Ionizing Radiation*, pp. 159–174, 2005.

[CHA 06] CHAMBERS D.B., OSBORNE R.V., GARVA A.L., "Choosing an alpha radiation weighting factor for doses to non-human biota", *J. Environ. Radioact.*, vol. 87, pp. 1–14, 2006.

[CHA 08] CHAMBERS, D.B., FERNANDES, S.L., PHILLIPS, H.A. "Overview of representative ecological risk assessments conducted for sites with enhanced radioactivity", *IRPA12, International Conference on the Protection of the Environment from the Effects of Ionizing Radiation: Strengthening Radiation Protection Worldwide*, 2008.

[CHA 11] CHARMASSON S., LE FAOUDER A., LOYEN J. et al., "^{210}Po and ^{210}Pb in the tissues of the deep-sea hydrothermal vent mussel *Bathymodiolus azoricus* from the Menez Gwen field (Mid-Atlantic Ridge)", *Sci. Total Environ.*, vol. 409, no. 4, pp. 771–777, 2011.

[CHA 19] CHAN C.Y., WANG W.X., "Biomarker responses in oysters Crassostrea hongkongensis in relation to metal contamination patterns in the Pearl River Estuary, southern China", *Environ. Pollut.*, vol. 251, pp. 264–276, 2019.

[CHE 81] CHERRY R.D., HEYRAUD M., "Polonium-210 content of marine shrimp: Variation with biological and environmental factors", *Mar. Biol.*, vol. 65, pp. 167–175, 1981.

[CHE 82] CHERRY R.D., HEYRAUD M., "Evidence of high natural radiation doses in certain mid-water oceanic organisms", *Science*, vol. 218, pp. 54–56, 1 October 1982.

[CHE 86] CHEN J.H., LAURENCE E.R., "^{238}U, ^{234}U and ^{232}Th in seawater", *Earth Planet. Sci. Lett.*, vol. 80, nos 3–4, pp. 241–251, 1986.

[CHE 87] CHERRY M.I., CHERRY R.D., HEYRAUD M., "Polonium-210 and lead-210 in Antarctic marine biota and sea water", *Mar. Biol.*, vol. 96, pp. 441–449, 1987.

[CHE 91] CHERRY R.D., HEYRAUD M., "Pownium·210 and Lead·210 in marine organisms: Allometric relationships and their significance", in KERSHAW P.J., Woodhead, D.S. (eds), *Radionuclides in the Study of Marine Processes*, Elsevier Science Publishers Ltd, Dordrecht, 1991.

[CIG 06] CIGNA A.A., DURANTE M. (eds), *Radiation Risk Estimates in Normal and Emergency Situations*, Springer, Berlin, 2006.

[CLA 09] CLARKE R.H., VALENTIN J., "The history of ICRP and the evolution of its policies", *Ann. ICRP 109*, vol. 9, no. 1, pp. 75–110, 2009.

[COM 55] COMAR C.L., *Radioisotopes in Biology and Agriculture: Principles and Practices*, McGrawth-Hill, New York, 1955.

[COO 70] COOLEY J.L., NELSON D.J., Effects of chronic irradiation and temperature on population of the aquatic snail *Physa heterostropha*, Report ORNL-4612, Oak Ridge National Laboratory, 1970.

[COO 71] COOLEY J.L., MILLER F.L., "Effects of chronic irradiation on laboratory populations of the aquatic snail *Physa heterostropha*", *Radiat. Res.*, vol. 47, pp. 716–724, 1971.

[COO 79] COOPER M.B., HARTLEY B.M., Residual radioactive contamination of the Montebello Islands from nuclear weapons tests conducted in 1952 and 1956, Australian Radiation Laboratory, ARL/TR010, April 1979.

[COO 83] COOPER M.B., LOKAN K.H., WILLIAMS G.A., The radiological status of the Montebello Islands; May 1983, Australian Radiation Laboratory, ARL/TR062, October 1983.

[COO 90] COOPER M.B., MARTIN L.J., WILKS M.J. et al., Radiological hazard assessment at the Montebello Islands, Australian Radiation Laboratory, ARL/TR96, December 1990.

[COO 98] COOPER L.W., BEASLEY T.M., ZHAO X.L. et al., "Iodine-129 and plutonium isotopes in Arctic kelp as historical indicators of transport of nuclear fuel-reprocessing wastes from mid-to-high latitudes in the Atlantic Ocean", *Mar. Biol.*, vol. 131, pp. 391–399, 1998.

[COP 01] COPPLESTONE D., BIELBY S., JONES S.R. et al., Impact assessment of ionising radiation on wildlife, R&D publication 128, Environment Agency, Bristol, 2001.

[COP 08] COPPLESTONE, D., HINGSTON J., REAL A., "The development and purpose of the FREDERICA radiation effects database", *J. Environ. Radioact.*, vol. 99, no. 9, pp. 1456–1463, 2008.

[COS 20] COSSA D., FOWLER S.W., MIGON C. et al., "A biogeochemical approach to contamination of the Ligurian Sea", in MIGON C., NIVAL P., SCIANDRA A. (eds), *The Mediterranean Sea in the Era of Global Change 2: 30 Years of Multidisciplinary Study of the Ligurian Sea*, ISTE, London, and John Wiley & Sons, Hoboken, New York, 2020.

[CRE 21] CRESSWELL T., PRENTICE E., HOWELL N. et al., "Bioaccumulation kinetics and internal distribution of the fission products radiocaesium and radiostrontium in an estuarine crab", *J. Hazard. Mater.*, vol. 408, 124453, 2021.

[CRO 05] CROTEAU M.N., LUOMA S.N., STEWART A.R., "Trophic transfer of metals along freshwater food webs: Evidence of cadmium biomagnification in nature", *Limnol. Oceanogr.*, vol. 50, pp. 1511–1519, 2005.

[CRO 07] CROTEAU M.N., LUOMA S.N., PELLET B., "Determining metal assimilation efficiency in aquatic invertebrates using enriched stable metal isotope tracers", *Aquat. Toxicol.*, vol. 83, pp. 116–125, 2007.

[DAH 95a] DAHLGAARD H., "Transfer of European coastal pollution to the Arctic: Radioactive tracers", *Mar. Pollut. Bull.*, 31, pp. 3–7, 1995.

[DAH 95b] DAHLGAARD H., CHEN Q., HERMANN J. et al., "On the background level of ^{99}Tc' ^{90}Sr and ^{137}Cs in the North Atlantic", *J. Mar. Syst.*, vol. 6, pp. 571– 578, 1995.

[DAH 96a] DAHLGAARD H., "Polonium-210 in mussels and fish from the Baltic-North Sea Estuary", *J. Environ. Radioact.*, vol. 31, nos 1–2, pp. 91–96, 1996.

[DAH 96b] DAHLGAARD H., AARKROG A., HALLSTADIUS L. et al., "Radiocaesium transport from the Irish Sea via the North Sea and the Norwegian coastal current to East Greenland", *Rapp. P. – v. Reun. Cons. Int. Explor. Mer*, vol. 186, pp. 70–79, 1996.

[DAH 97] DAHLGAARD H., BERGAN T.D.S., CHRISTENSEN G.C., "Technetium-99 and caesium-137 time series at the Norwegian coast monitored by brown algae *Fucus vesiculosus*", in GERMAIN P., GUARY J.C., GUÉGUÉNIAT P. et al. (eds), *Radionuclides in the Oceans – RADOC 96–97", Cherbourg-Octeville (Fr), 7–11 Oct. 1996. Radioprotection – Symposium*, vol. 32, pp. 353–358, 1997.

[DAL 95] DALEY M.J., MINTON K.W., "Resistance to radiation", *Science*, vol. 270, p. 1318, 1995.

[DAL 16] DALLAS L.J., BEAN T.P., TURNER A. et al., "Exposure to tritiated water at an elevated temperature: Genotoxic and transcriptomic effects in marine mussels (*M. galloprovincialis*)", *J. Environ. Radioact.*, vol. 164, pp. 325–336, 2016.

[DAN 02] DANGE C., Etude du comportement des éléments traces en milieu estuarien : approche expérimentale et modélisation. Application à l'étude de la spéciation de Cd, Co et Cs dans les estuaires majeurs français (Seine, Loire, Gironde, Rhône), Thesis, Université de Reims Champagne-Ardenne, 2002.

[DEB 20] DEBLONDE G.J.P., KERSTING A.B., ZAVARIN M., "Open questions on the environmental chemistry of radionuclides", *Commun. Chem.*, vol. 3, article 167, 2020.

[DEI 91] DEINES A.C., GOLDMAN D.L., HARRIS R.R. et al., Marshall Islands chronology 1944 to 1990, Draft, History Associates Incorporated, USDOE, Contract N. DE-AC08-87NV10594, January 11, 1991.

[DEL 06] DELFANTI, R., TSABARIS C., PAPUCCI C. et al., "Re-distribution of ^{137}Cs Chernobyl signal in the Aegean Sea", *IAEA-CSP-26*, Vienna, pp. 89–92, 2006.

[DEL 08] DELISTRATY D., "Radioprotection of nonhuman biota", *J. Environ. Radioact.*, vol. 99, pp. 1863–1869, 2008.

[DEL 09] DELOFFRE J., LAFITE R., "Sedimentary processes on estuarine mudflats: Examples of the Seine and Authie Estuaries", in AMIARD-TRIQUET C., RAINBOW P.S. (eds), *Environmental Assessment of Estuarine Ecosystems. A Case Study*, CRC Press, Boca Raton, 2009.

[DES 91] DESMET G.M., VAN LOON L.R., HOWARD B.J., "Chemical speciation and bioavailability of elements in the environment and their relevance to radioecology", *Sci. Total Environ.*, vol. 100, pp. 105–124, 1991.

[DOU 08] DOUKI T., CADET J. "Radiation-induced damage to DNA: From model compounds to cell", in SPOTHEIM-MAURIZOT M. et al. (eds), *Radiation Chemistry from Basics to Application in Material and Life Sciences*, EDP Sciences/L'actualité Chimique, 2008.

[DUB 93] DUBROVA Y.E., JEFFREYS A.J., MALASHENKO A.M., "Mouse minisatellite mutations induced by ionizing radiation", *Nat. Genet.*, vol. 5, pp. 92–94, 1993.

[DUF 01] DUFFUS J.H., "Risk assessment terminology", *Chem. Int.*, vol. 23, no. 2, pp. 40–42, 2001.

[DUR 99] DURAND J.P., CARVALHO F.P., GOUDARD F. et al., "^{210}Po binding to metallothioneins and ferritin in the liver of teleost marine fish", *Mar. Ecol. – Prog. Ser.*, vol. 177, pp. 189–196, 1999.

[EC 96] EC, Technical guidance document in support of commission directive 93/67/ EEC on risk assessment for new notified substances and commission regulation (EC) No 1488/94 on risk assessment for existing chemicals, part I, II, III & IV, Office des publications officielles des communautés européennes, Luxembourg, available at: http://ecb.jrc.it/existing-chemicals/, 1996.

[EC 00] EC, "Directive 2000/60/CE du Parlement européen et du Conseil du 23 octobre 2000 établissant un cadre pour une politique communautaire dans le domaine de l'eau", *Journal officiel des Communautés européennes*, 22.12.2000, L 327, 2000.

[EC 03a] EC, "Technical guidance document in support of Commission Directive 93/67/EEC on risk assessment for new notified substances and Commission Regulation (EC) No 1488/94 on risk assessment for existing substances", Directive 98/8/EC of the European Parliament and of the Council concerning the placing of biocidal products on the market. Part II. Luxembourg, Office for Official Publication of the European Communities, 2003.

[EC 03b] EC, "MARINA II – Update of the MARINA Project on the radiological exposure of the European Community from radioactivity in North European marine waters", *Radiation Protection*, 132, vol. II, annexe F, 2003.

[EC 08] EC, "Directive 2008/56/EC of the European Parliament and of the Council of 17 June 2008 establishing a framework for community action in the field of marine environmental policy", *Official J. Eur. Union*, 25 June, L164/19–40, 2008.

[EDF 04] EDF, *Centrales nucléaires et environnement. Prélèvements d'eau et rejets*, EDP Sciences, Les Ulis, 2004.

[EDF 20] EDF, *Centrales nucléaires et environnement. Prélèvements d'eau et rejets*, EDP Sciences, Les Ulis, 2020.

[EDG 76] EDGINGTON D.N., ALBERTS J.J., WAHLGREN M.A. et al., "Plutonium and americium in Lake Michigan sediments", *Transuranium Nuclides in the Environment*, IAEA, Vienna, pp. 499–516, 1976.

[EDG 80] EDGINGTON D.N., "A review of the persistence of long-lived radionuclides in the marine environment – Sediment/water interactions", *International Symposium on the Impacts of Radionuclide Releases into the Marine Environment*, IAEA-SM-248/148, IAEA, Vienna, 6–10 October 1980.

[EGA 79] EGAMI N., IJIRI K.I., "Effects of irradiation on germ cells and embryonic development in teleosts", *Int. Rev. Cytol.*, vol. 59, pp. 195–248, 1979.

[EIS 73] EISENBUD M., *Environment Radioactivity*, 2nd edition, Academic Press, New York, 1973.

[ELL 96] ELLEDGE S.J., "Cell cycle checkpoints: Preventing an identity crisis", *Science*, vol. 274, pp. 1664–16672, 1996.

[ENG 62] ENGEL D.W., FLUKE D.J., "The effect of water content and post-irradiation storage on radiation sensibility of brine shrimp cystis (eggs)", *Radiat. Res.*, vol. 16, pp. 173–181, 1962.

[ENG 67] ENGEL D.W., "Effects of single and continuous exposure of gamma radiation on the survival and growth of the blue crab, *Callinectes sapidus*", *Radiat. Res.*, vol. 32, pp. 685–691, 1967.

[ENG 73] ENGEL D.W., DAVIES E.M., ANGELOVIC J.W. et al., "Effect of radiation salinity and temperature on the ionic regulation of the blue crab, *Callinectes sapidus*", in NELSON D.J., (ed.), *Radionuclides in Ecosystems*, USAEC, 1973.

[ENG 74] ENGEL D.W., SHELTON M.G., WHITE JR. J.C., "The effect of ionizing radiation and salinity on the grass shrimp, *Palaemontes pugio*", *Radiat. Res.*, vol. 59, pp. 72, 1974.

[ERE 95] EREMEEV V.N., IVANOV L.M., KIRWAN JR. A.D. et al., "Amount of ^{137}Cs and ^{134}Cs radionuclides in the Black Sea produced by the Chernobyl accident", *J. Environ. Radioact.*, vol. 27, pp. 49–63, 1995.

[ERI 05] ERICA, Erica, deliverable D5: Derivation of Predicted No Effect Dose Rate values for ecosystems (and their suborganisational levels) exposed to radioactive substances, Report, European Commission, Contract number FI6R-CT-2004–508847, 2005.

[ERI 06] ERICA, Derivation of Predicted No Effect Dose Rate values for ecosystems (and their sub-organisational levels) exposed to radioactive substances, Deliverable D5, European Commission, 6th framework, Contract No. FI6R-CT-2003–508847, 2006.

[ERI 09] ERIKSEN1 D.Ø., SIDHU R., RAMSØY T. et al., "Radioactivity in produced water from Norwegian oil and gas installations – Concentrations, bioavailability, and doses to marine biota", *Radioprotection*, vol. 44, no. 5, pp. 869–874, 2009.

[ERI 21] ERICA, "ERICA tool", *CEH*, available at: https://wiki.ceh.ac.uk/display/rpemain/ERICA+Tool, 2021.

[EYR 04] EYROLLE E., CHARMASSON S., "Importance of colloids in the transport within the dissolved phase (<450 nm) of artificial radionuclides from the Rhône River towards the Gulf of Lion (Mediterranean Sea)", *J. Environ. Radioact.*, vol. 72, pp. 273–286, 2004.

[FAO 03] FAO, "The ecosystem approach to fisheries", *FAO Technical Guidelines for Responsible Fisheries*, Food and Agriculture Organization of the United Nations, no. 4, suppl. 2, Rome, 2003.

[FAR 06] FARCY E., Etude de l'impact des radionucléides rejetés par les installations nucléaires du Nord Cotentin sur l'huître creuse *Crassostrea gigas*. Analyse de l'expression de marqueurs moléculaires de stress, Thesis, Université de Caen, 2006.

[FEL 72] FELDT W., "Behaviour of radioactivity in the marine environment", *ipp. n: Disposal of Radioactive Waste*, OCDE, Paris, 1972.

[FER 12] FEROZ KHAN M., GODWIN WESLEY S., "Radionuclide monitoring in molluscs inhabiting intertidal region near a nuclear installation, Gulf of Mannar, India", *Mar. Pollut. Bull.*, vol. 64, no. 2, pp. 436–444, 2012.

[FES 11] FESENKO S., FESENKO J., SANZHAROVA N. et al., "Radionuclide transfer to freshwater biota species: Review of Russian language studies", *J. Environ. Radioact.*, vol. 102, pp. 8–25, 2011.

[FIÉ 20] FIÉVET B., BAILLY DU BOIS P., VOISEUX C. et al., "A comprehensive assessment of two-decade radioactivity monitoring around the Channel Islands", *J. Environ. Radioact.*, pp. 223–224, 106381, 2020.

[FIS 13] FISCHER N.S., BEAUGELIN-SEILLER, K., HINTON, T.G. et al., "Evaluation of radiation doses and associated risk from the Fukushima nuclear accident to marine biota and human consumers of seafood", *PNAS*, vol. 110, no. 26, pp. 10670–10675, 2013.

[FLA 63] FLAVIER H., Mesure des retombées radioactives dans les plantes sauvages et quelques poissons des étangs des Landes près de Bordeaux, Doctor-Engineer thesis, Bordeaux, 1963.

[FOL 75] FOLSOM T.R., HODGE V.F., GURNEY M.E., "Plutonium observed on algal surfaces in the ocean", *Mar. Sci. Comm.*, vol. 1, pp. 39–49, 1975.

[FON 60] FONTAINE Y., La contamination radioactive des milieux aquatiques, CEA-R-1588 report, 1960.

[FOR 73] FORSTER W.O., WOLFE D.A., LOWMAN F.G. et al., "Trace elements interactions between river water and seawater", in NELSON D.J. (ed.), *Radionuclides in Ecosystems*, USAEC, 1973.

[FOW 76] FOWLER S.W., HEYRAUD M., BEASLEY T.M., "Experimental studies on plutonium kinetics in marine biota", *Impacts of Nuclear Releases into the Aquatic Environment*, STI/PUB/407, IAEA, Vienna, TecDoc no. 187, pp. 157–176, 1976.

[FOW 88] FOWLER S.W., HOLM E., MEE L.D., "Monitoring the marine environment", *IAEA Bull.*, vol. 2, pp. 40–42, 1988.

[FOW 89] FOWLER S.W., "Transport and redistribution of trace metals and radionuclides in the marine environment by biogenic particles", in KLEKOWSKI R.Z., STYCZYNSKA-JUREWICZ E., FALKOWSKI L. (eds), *Proceedings of the 21st EMBS*, Gdansk, Poland, 14–19 September 1986, Polish Academy of Sciences, Wroclaw, 1989.

[FOW 97] FOWLER S.W., "Marine biogeochemistry of radionuclides. Strategies and methodologies for applied marine radioactivity studies", *IAEA-TCS-7*, 82, 1997.

[FOW 00] FOWLER S.W., NOSHKIN V.E., LA ROSA J. et al., "Temporal variations in plutonium and americium inventories and their relation to vertical transport in the northwestern Mediterranean Sea", *Limnology and Oceanography*, vol. 45, pp. 446–458, 2000.

[FOW 05] FOWLER S.W., FISHER N.S., "Radionuclides in the biosphere", in LIVINGSTON H.D. (ed.), *Marine Radioactivity*, Elsevier, Amsterdam, 2005.

[FOW 11] FOWLER S.W., "^{210}Po in the marine environment with emphasis on its behavior within the biosphere", *J. Environ. Radioact.*, vol. 102, pp. 448–461, 2011.

[FRA 74] FRAIZIER A., "Relation entre la forme physico-chimique du ruthénium 106 et les phénomènes de contamination observés expérimentalement chez diverses espèces marines", *Comparative Studies of Food and Environmental Contamination*, IAEA, Vienna, pp. 135–148, 1974.

[FRA 06] FRANIÆ Z., PETRINEC B., "Marine radioecology and waste management in the adriatic", *Arh. Hig. Rada. Toksikol.*, vol. 57, pp. 347–352, 2006.

[FRI 05] FRIEDLANDER B.R., GOCHFELD M., BURGER J. et al., "Radionuclides in the marine environment", *A Cresp Science Review*, Draft, 2005.

[FUK 82] FUKAI R., YOKOYAMA Y., "Natural radionuclides in the environment", in HUTZINGER O. (ed.), *The Handbook of Environmental Chemistry*, Springer-Verlag, Berlin, The Natural Environment and the Biogeochemical Cycles, vol. 1B, 1982.

[FUL 15] FULLER N., LEREBOURS A., SMITH J.T. et al., "The biological effects of ionising radiation on crustaceans: A review", *Aquat. Toxicol.*, vol. 167, pp. 55–67, 2015.

[FUR 06] FURUKAWA M., "Data synthesis for terrestrial gamma radiation level measured on the Japanese Islands and their adjacent ocean floor", *IAEA-CSP-26*, IAEA, Vienna, pp. 248–251, 2006.

[GAR 96] GARCIA K., BOUST D., MOULIN V. et al., "Multiparametric investigation of the reactions of plutonium under estuarine conditions", *Radiochim. Acta*, vol. 74, pp. 165–170, 1996.

[GAR 97] GARCIA K., Distribution et comportement du plutonium dans les eaux de la Manche et de l'estuaire de la Seine, Chemistry thesis, Université de Paris Sud Orsay, CEA-R-5752 report, 1997.

[GAR 00a] GARNIER-LAPLACE J., ADAM C., BAUDIN J.P., "Experimental kinetic rates of food-chain and waterborne radionuclide transfer to freshwater fish: A basis for the construction of fish contamination charts", *Arch. Environ. Contam. Toxicol.*, vol. 39, pp. 133–144, 2000.

[GAR 00b] GARNIER-LAPLACE J., ADAM C., LATHUILLIÈRE T. et al., "A simple fish physiological model for radioecologists exemplified for ^{54}Mn direct transfer and rainbow trout (*Oncorhynchus mykiss* W.)", *J. Environ. Radioact.*, vol. 49, pp. 35–53, 2000.

[GAR 00c] GARNIER-LAPLACE J., BAUDIN J.P., ADAM C., "Milieux aquatiques", in ROBEAU D., DABURON F.H., METIVIER H. (eds), *Le césium de l'environnement à l'Homme*, EDP Sciences, Les Ulis, 2000.

[GAR 04] GARNIER-LAPLACE J., GILEK M., SUNDELL-BERGMAN S. et al., "Assessing ecological effects of radionuclides: Data gaps and extrapolation issues", *J. Radiol. Prot.*, vol. 24, no. 4A, pp. A139–A155, 2004.

[GAR 06] GARIEL J.C., ADAM C., BEAUGELIN-SEILLER K. et al., Radioprotection de l'environnement. Synthèse et perspectives, IRSN 2005–56 report, 2006.

[GAR 08] GARNIER-LAPLACE J., COPPLESTONE D., GILBIN R. et al., "Issues and practices in the use of effects data from FREDERICA in the ERICA integrated approach", *J. Environ. Radioact.*, vol. 99, pp. 1474–1483, 2008.

[GAR 09a] GARCIA-ORELLANA J., PATES J.M., MASQUE P. et al., "Distribution of artificial radionuclides in deep sediments of the Mediterranean Sea", *Sci. Total Environ.*, vol. 407, pp. 887–898, 2009.

[GAR 09b] GARIEL J.C., Le tritium dans l'environnement Synthèse des connaissances, DEI 2009–05 report, 2009.

[GAR 10a] GARISTO, N., KOVACS, R., JANES, A., "Comparison of dose estimate results from radioecological risk assessment models RESRAD-BIOTA, ERICA tool and SENES risk model using a case study", *31st Annual Conference of the Canadian Nuclear Society and 34th CNS/CNA Student Conference 2010*, Canadian Nuclear Society, vol. 2, pp. 1688–1699, 2010.

[GAR 10b] GARNIER-LAPLACE, J., DELLA-VEDOVA, C., ANDERSSON, P. et al., "A multi-criteria weight of evidence approach for deriving ecological benchmarks for radioactive substances", *J. Radiol. Prot.*, vol. 30, no. 2, pp. 215–233, 2010.

[GAR 11] GARNIER-LAPLACE, J., BEAUGELIN-SEILLER, K., HINTON, T., "Fukushima wildlife dose reconstruction signals ecological consequences", *Environ. Sci. Technol.*, vol. 45, pp. 5077–5078, 2011.

[GAR 13] GARNIER-LAPLACE, J., GERAS'KIN, S., DELLA-VEDOVA, C. et al., "Are radiosensitivity data derived from natural field conditions consistent with data from controlled exposures? A case study of Chernobyl wildlife chronically exposed to low dose rates", *J. Environ. Radioact.*, vol. 121, pp. 12–21, 2013.

[GEC 19] GECKEIS H., ZAVARIN M., SALBU B. et al., "Environmental chemistry of plutonium", in CLARK D.L., GEESON D.A., HANRAHAN JR. R.J. (eds), *Plutonium Handbook*, 2nd edition, vol. 4, American Nuclear Society, 2019.

[GER 79] GERMAIN P., MASSON M., BARON Y., Etude de la répartition de radionucléides émetteurs gamma chez des indicateurs biologiques littoraux des côtes de la Manche et de la mer du Nord de février 1976 à février 1978, CEA-R-5017 report, 1979.

[GER 84] GERMAIN P., MIRAMAND P., "Distribution and behaviour of transuranic elements in the physical and biological compartment of the Channel French shore", *Nuclear Instruments and Methods in Physic Research*, vol. 223, pp. 502–509, 1984.

[GER 95] GERMAIN P., LECLERC G., SIMON S., "Transfer of polonium-210 into *Mytilus Edulis* (L.) and *Fucus vesiculosus* (L.) from the baie de Seine (Channel coast of France)", *Sci. Total Environ.*, vol. 164, pp. 109–123, 1995.

[GER 00] GERMAIN P., LECLERC G., LE CAVELIER S et al., "Evolution spatiotemporelle des concentrations, des rapports isotopiques et des facteurs de concentration du plutonium dans une espèce d'algue et deux espèces de mollusques en Manche", *Radioprotection,* vol. 35, no. 2, pp. 175–200, 2000.

[GIL 08] GILBIN R., ALONZO F., GARNIER-LAPLACE J., "Effects of chronic external gamma irradiation on growth and reproductive success of *Daphnia magna*", *J. Environ. Radioact.*, vol. 99, pp. 134–145, 2008.

[GIL 09] GIL-GARCI C., TAGAMI K., UCHIDA S. et al., "New best estimates for radionuclide solid–distribution coefficients in soils. Part 3: Miscellany of radionuclides (Cd, Co, Ni, Zn, I, Se, Sb, Pu, Am, and others)", *J. Environ. Radioact.*, vol. 100, pp. 704–715, 2009.

[GON 20] GONZALEZ S.A., SILVA T.A., CONTI C.C. et al., "ENVIRAD – An integrated database of environmental radiological monitoring information", *Braz. J. Radiat. Sci.*, vol. 8, no. 2, pp. 01–16, 2020.

[GOO 93] GOODHEAD D.T., THACKER J., COX R., "Effects of radiations of different qualities on cells: Molecular mechanisms of damage and repair", *Int. J. Radiat. Biol.*, vol. 63, pp. 543–556, 1993.

[GOR 19] GORYACHENKOVA A., BORISOV A.P., SOLOVEVA G.Y. et al., "Content of technogenic radionuclides in water, bottom sediments and benthos of The Kara Sea and Shallow Bays of the Novaya Zemlya Archipelago", *Geokhimia*, vol. 64, no. 12, pp. 1261–1268, 2019.

[GOU 91] GOUDARD E., DURAND J.P., PIERI J. et al., "Subcellular localisation and identification of 95mTc and 241Am binding ligands in the hepatopancreas of the lobster *Homarus Gammarus*", *Mar. Biol.*, vol. 108, pp. 411–417, 1991.

[GOU 98] GOUDARD F., PAQUET F., DURAND J.P. et al., "Subcellular and molecular localisation of different radionuclides (Caesium, Americium, Plutonium and Technetium) in aquatic organisms", *Radiat. Prot. Dosim.*, vol. 75, nos 1–4, pp. 117–124, 1998.

[GOU 04] GOUZY A., Etude du comportement du plutonium au cours de la diagénèse précoce des sédiments mains : applications à deux environnements marins marqués par les rejets issus d'usines de retraitement de combustibles usés, Thesis, Université de Caen, 2004.

[GOU 05a] GOUZY A., BOUST D., CONNAN O. et al., "Diagenetic reactivity of the plutonium in marine anoxic sediments (Cumbrian mud patch – eastern Irish Sea)", *Radioprotection,* vol. 40, Suppl. 1, pp. S541–S547, 2005.

[GOU 05b] GOUZY A., BOUST D., CONNAN O. et al., "Post-depositional reactivity of the plutonium in different sediment facies from the English Channel – An experimental approach", *Radioprotection*, vol. 40, Suppl. 1, pp. S613–S619, 2005.

[GRA 95] GRAY J., JONES S.R., SMITH A.D., "Discharges to the environment from the Sellafield site, 1951–1992", *J. Radiol. Prot.*, vol. 15, pp. 99–131, 1995.

[GRE 95] GREENWOOD, L.N., KNOWLES J.F., "Effect of chronic irradiation on the humoral immune response of a marine fish, the eelpout (*Zoarces viviparus* L.)", *Int. J. Radiat. Biol.*, vol. 67, no. 1, pp. 71–77, 1995.

[GRN 99a] GRNC (Groupe Radioécologie Nord Cotentin), Estimation des niveaux d'exposition aux rayonnements ionisants et des risques de leucémies associés de populations du Nord Cotentin, Synthesis report, IRSN, Fontenay aux Roses, 1999.

[GRN 99b] GRNC (Groupe Radioécologie Nord Cotentin), Inventaire des rejets radioactifs des installations nucléaires, vol. 1, July 1999.

[GRN 99c] GRNC (Groupe Radioécologie Nord Cotentin), Modélisation des transferts dans l'environnement, validation modèles – mesures, GT3, Final report, vol. 3, IRSN, Fontenay aux Roses, 1999.

[GUA 76] GUARY J.C., MASSON M., FRAIZIER A., "Etude préliminaire, *in situ*, de la distribution du plutonium dans différents tissus et organes de *Cancer pagurus* (Crustacea: Decapoda) et de *Pleuronectes platessa* (Pisce Pleuronectidae)", *Mar. Biol.*, vol. 36, pp. 13–17, 1976.

[GUA 77a] GUARY J.C., FRAIZIER A., "Etude comparée des teneurs en plutonium chez divers mollusques de quelques sites littoraux français", *Mar. Biol.*, vol. 41, pp. 263–267, 1977.

[GUA 77b] GUARY J.C., FRAIZIER A., "Influence of trophic level and calcification on the uptake of plutonium observed, *in situ*, in marine organisms", *Health Phys.*, vol. 32, pp. 21–28, 1977.

[GUA 81] GUARY J.C., NEGREL R., "Calcium phosphate granules: A trap for transuranics and iron in crab hepatopancreas", *Comp. Biochem. Phys.*, vol. 68A, pp. 423–427, 1981.

[GUA 82] GUARY J.C., FOWLER S.W., BEASLEY T.M., "Routes of plutonium uptake and their relation to biomagnification in starfish", *Mar. Pollut. Bull.*, vol. 13, pp. 99–103, 1982.

[GUE 75] GUEGUENIAT P., Comportement physico-chimique du ruthénium de fission dans le milieu marin, PhD thesis, Université de Strasbourg, CEA-R-4644 report, 1975.

[GUE 76] GUEGUENIAT P., BARON Y., AUFFRET J.P., Note sur l'évolution de la radioactivité artificielle dans les sédiments pendant les années 1971–1974, CEA-R-4739 report, 1976.

[GUE 79] GUEGUENIAT P., AUFFRET J.P., BARON Y., "Evolution de la radioactivité artificielle gamma dans les sédiments littoraux de la Manche pendant les années 1976–1977–1978", *Oceanol. Acta*, vol. 2, pp. 165–180, 1979.

[GUE 86] GUEGUENIAT P., "Comportement chimique et géochimique des éléments à l'état de traces dans l'estuaire de la Seine : une synthèse des résultats", *La Baie de Seine (GRECO-MANCHE)*, Université de Caen, 24–26 April 1985, IFREMER, Proceedings no. 4, pp. 247–272, 1986.

[GUÉ 03] GUÉGUÉNIAT P., GERMAIN P., MÉTIVIER H. (eds), *Radionuclides in the Oceans. Inputs and Inventories*, EDP Sciences, Les Ulis, 2003.

[GUI 12] GUILLEMETTE A., ZERBIB J.C., "Étude de l'incidence des rejets de carbone 14 sur l'environnement terrestre proche (1 km à 10 km) et sur l'environnement marin du site AREVA NC de La Hague", *La G@zette Nucléaire*, vol. 266, November 2012.

[GUI 13] GUILLEMETTE A., "Iode 129, rejets associés aux opérations de retraitement des combustibles irradiés et données environnementales", *La G@zette Nucléaire*, no. 267, February 2013.

[GUO 13] GUO F., YANG Y., WANG W.X., "Metal bioavailability from different natural prey to a marine predator *Nassarius siquijorensis*", *Aquat. Toxicol.*, vol. 126, pp. 266–273, 2013.

[GWY 04] GWYNN J.P., DOWDALL M., DAVIDS C. et al., "The radiological environment of Svalbard", *Polar Res.*, vol. 23, no. 2, pp. 167–180, 2004.

[GWY 12] GWYNN J.P., HELDAL H.E., GÄFVERT T. et al., "Radiological status of the marine environment in the Barents Sea", *J. Environ. Radioact.*, vol. 113, pp. 155–162, 2012.

[GWY 16] GWYNN J.P., NIKITIN A., SHERSHAKOV V. et al., "Main results of the 2012 joint Norwegian-Russian expedition to the dumping sites of the nuclear submarine K-27 and solid radioactive waste in Stepovogo Fjord, Novaya Zemlya", *J. Environ. Radioact.*, vol. 151, Pt. 2, pp. 417–26, 2016.

[HAG 05] HAGGER J.A., ATIENZAR F.A., JHA A.N., "Genotoxic, cytotoxic, developmental and survival effects of tritiated water in the early life stages of the marine mollusc, *Mytilus edulis*", *Aquat. Toxicol.*, vol. 74, pp. 205–217, 2005.

[HAM 03] HAMILTON T.F., MILLIES-LACROIX J.C., HONG G.H., "137Cs (90Sr) and Pu isotopes in the Pacific Ocean: Sources and trends", in GUÉGUÉNIAT P., GERMAIN P., MÉTIVIER H. (eds), *Radionuclides in the Oceans. Inputs and Inventories*. EDP Sciences, Les Ulis, 2003.

[HAM 05] HAMILTON T.F., "Linking legacies of the Cold War to arrival of anthropogenic radionuclides in the oceans through the 20th century", in LIVINGSTON H.D., (ed.), *Marine Radioactivity*, Elsevier, Amsterdam, 2005.

[HAR 74] HARDY E.P., Worldwide distribution and biomedical effects, Report WASH-1359, pp. 115–130, 1974.

[HAR 92] HARGRAVE B.T., GERMAIN P., PHILIPPOT J.C. et al., "Stable elements and ^{210}Po in the deep-sea amphipod *Eurythenes gryllus*", *Deep-Sea Res.*, vol. 39, no. 1, pp. 37–44, 1992.

[HAR 01] HARRISON F.L., KNEZOVICH J.P., "Effects of radiation on aquatic and terrestrial organisms", in VAN DER STRICHT E., KIRCHMANN R. (eds), *Radioecology. Radioactivity and Ecosystems*. UIR, Liège, 2001.

[HAR 12] HARMELIN-VIVIEN M., BODIGUEL X., CHARMASSON S. et al., "Differential biomagnification of PCB, PBDE, Hg and radiocesium in the food web of the European hake from the NW Mediterranean", *Mar. Pollut. Bull.*, vol. 64, no. 5, pp. 974–983, 2012.

[HAT 05] HATON C., Effets des rayonnements ionisants sur la structure et la fonction de la cellule épithéliale intestinale, Thesis, Université Pierre et Marie Curie, Paris 6, 2005.

[HAV 73] HAVLIK B., ROBERTSON E., "Radium uptake by fresh-water algae", *J. Hyg. Epidemiol. Microbiol. Immunol.*, vol. 17, no. 4, pp. 393–405, 1973.

[HEL 18a] HELCOM, "State of the Baltic Sea. Second HELCOM holistic assessment 2011–2016", available at: http://www.helcom.fi/baltic-sea-trends/holistic-assessments/state-of-the-baltic-sea-2018/reports-and-materials, 2018.

[HEL 18b] HELCOM, "HELCOM thematic assessment of hazardous substances 2011–2016", *Baltic Sea Environment Proceedings*, no. 157, 2018.

[HEL 18c] HELCOM, "Radioactive substances HELCOM core indicator 2018", HELCOM core indicator report, available at: http://www.helcom.fi/baltic-sea-trends/holistic-assessments/state-of-the-baltic-sea-2018/reports-and-materials, 2018.

[HET 75] HETHERINGTON J.A., JEFFERIES D.F., LOVETT M.B., "Some investigations into the behaviour of plutonium in the marine environment", *Impacts of Nuclear Releases into the Aquatic Environment*, AIEA, Vienna, pp. 193–212, 1975.

[HET 76] HETHERINGTON J.A., JEFFERIES D.F., MITCHELL N.T. et al., "Environmental and public health consequences of the controlled disposal of transuranic elements to the marine environments", *Transuranium Nuclides in the Environment*, IAEA, Vienna, pp. 139–154, 1976.

[HEY 76] HEYRAUD M., FOWLER S.W., BEASLEY T.M. et al., "Polonium-210 in Euphausiids: A detailed study", *Mar. Biol.*, vol. 14, pp. 127–136, 1976.

[HEY 79] HEYRAUD M., CHERRY R.D., "Polonium-210 and Lead-210 in marine food chains", *Mar. Biol.*, vol. 52, pp. 227–236, 1979.

[HEY 88] HEYRAUD M., DOMANSKI P., CHERRY R.D. et al., "Natural tracers in dietary studies: Data for ^{210}Po and ^{210}Pb in decapod shrimp and other pelagic organisms in the Northeast Atlantic Ocean", *Mar. Biol.*, vol. 97, pp. 507–519, 1988.

[HIG 07] HIGLEY K., BYTWERK D.P., "Generic approaches to transfer", *J. Environ. Radioact.*, vol. 98, pp. 4–23, 2007.

[HIG 15] HIGLEY K., RUEDIG E., GOMEZ-FERNANDEZ M. et al., "Creation and application of voxelised dosimetric models, and a comparison with the current methodology as used for the International Commission on Radiological Protection's Reference Animals and Plants", *Proceedings of the Second International Symposium on the System of Radiological Protection, Annals of the ICRP*, vol. 44, no. 1S, pp. 313–330, 2015.

[HIN 13] HINTON T.G., GARNIER-LAPLACE J., VANDENHOVE H. et al., "An invitation to contribute to a strategic research agenda in radioecology", *J. Environ. Radioact.*, vol. 115, pp. 73–82, 2013.

[HOD 73] HODGE V.F., FOLSOM T.R., YOUNG D.R., "Retention of fall-out constituents in upper layers of the Pacific Ocean as estimated from studies of a tuna population", *Radioactive Contamination of the Marine Environment*, IAEA, Vienna, pp. 263–276, 1973.

[HOL 80] HOLM E., PERSSON R.B.R., "Behavior of natural (Th, U) and artificial (Pu, Am) actinides in coastal waters", *Marine Radioecology, Proceedings of the 3rd NEA Seminar*, Tokyo, 1979, OECD, Paris, 1980.

[HOL 86a] HOLM E., RIOSECO J., AARKROG A. et al., "Technetium-99 in Algae from temperate and Arctic waters of the North Atlantic", in DESMET G., MYTENNAERE C., (eds), *Technetium in the Environment*, Elsevier, London, 1986.

[HOL 86b] HOLM E., RIOSECO J., MATTSSON S., "Technetium-99 in the Baltic Sea", in DESMET G., MYTENNAERE C. (eds), *Technetium in the Environment*, Elsevier, London, 1986.

[HOP 73] HOPPENHEIT M., "Effects on fecundity and fertility of single sublethal X-irradiation of *Gammarus duebeni* females", *Proc. Symp. Radioactive Contamination of the Marine Environment*, IAEA, Vienna, pp. 479–486, 1973.

[HOR 19] HOREMANS N., SPURGEON D.J., LECOMTE-PRADINES C. et al., "Current evidence for a role of epigenetic mechanisms in response to ionizing radiation in an ecotoxicological context", *Environ. Pollut.*, vol. 251, pp. 469–483, 2019.

[HOS 08] HOSSEINI A., THØRRING H., BROWN J.E. et al., "Transfer of radionuclides in aquatic ecosystems – Default concentration ratios for aquatic biota in the ERICA tool", *J. Environ. Radioact.*, vol. 99, pp. 1408–1429, 2008.

[HOS 10] HOSSEINI A., BERESFORD N.A., BROWN J.E. et al., "Background dose-rates to reference animals and plants arising from exposure to naturally occurring radionuclides in aquatic environments", *J. Radiol. Prot.*, vol. 30, no. 2, pp. 235–264, 2010.

[HOS 12] HOSSEINI A., BROWN J.E., GWYNN J.P. et al., "Review of research on impacts to biota of discharges of naturally occurring radionuclides in produced water to the marine environment", *Sci. Total Environ.*, vol. 438, pp. 325–333, 2012.

[HOU 02] HOU X.L., DAHLGAARD H., NIELSEN S.P. et al., "Level and origin of iodine-129 in the Baltic Sea", *Radioact. Environ. J.*, vol. 61, no. 3, pp. 331–343, 2002.

[HOU 13] HOU X., POVINEC P.P., ZHANG L., "Iodine-129 in seawater offshore Fukushima: Distribution, inorganic speciation, sources, and budget", *Environ. Sci. Technol.*, vol. 47, pp. 3091–3098, 2013.

[HOW 10] HOWARD, B.J., BERESFORD, N.A., ANDERSSON, P. et al., "Protection of the environment from ionizing radiation in a regulatory context: An overview of the PROTECT coordinated action project", *J. Radiol. Prot.*, vol. 30, no. 2, pp. 195–214, 2010.

[HOW 11] HOWARD B.J., BERESFORD N.A., "Radiation protection of the environment: A summary of current approaches for assessment of radionuclides in terrestrial ecosystems", in HESTER R.E., HARRISON R.M. (eds), *Nuclear Power and the Environment*, Issues in Environmental Science and Technology, no. 32, RSC Publishing, Cambridge, 2011.

[HU 10] HU Q.H., WENG J.Q., WANG J.S., "Sources of anthropic radionuclides in the environment: A review", *J. Environ. Radioact.*, vol. 101, pp. 426–437, 2010.

[HUG 19] HUGHES E.W., MOLINA M.R., ABELLA M.K.I.L. et al., "Radiation maps of ocean sediment from the Castle Bravo crater", *PNAS*, vol. 116, no. 31, pp. 15420–15424, 2019.

[HUM 11] HUMMEL H., SOLOLOWSKI A., HUMMEL C. et al., "Tolerance to natural environment change and the effect of added chemical stress", in AMIARD-TRIQUET C., RAINBOW P.S., ROMÉO M. (eds), *Tolerance to Environmental Contaminants*, CRC Press, Boca Raton, 2011.

[HUN 98] HUNT G.I., "Radioactivity studies in lowestoft: The first 50 years", *Radiat. Prot. Dosimetry*, vol. 75, nos 1–4, pp. 1–14, 1998.

[HUS 99] HUSSON R., "Risques de pollution radioactive liés aux rejets des eaux de refroidissement des sous-marins d'attaque de la marine nationale", Question écrite no. 19582 lors de la 11e législature publiée dans le JO Sénat du 21/10/1999, page 3440 et réponse du ministère de la Défense publiée dans le JO Sénat du 04/11/1999, page 3630, 1999.

[IAE 76] IAEA, Effects of ionizing radiation on aquatic organisms and ecosystems, Tech. Rep. Series no. 172, IAEA, Vienna, 1976.

[IAE 78] IAEA, The radiological basis of the IAEA revised definition and recommendations concerning high-level radioactive waste unsuitable for dumping at sea, IAEA Publication 211, IAEA, Vienna, 1978.

[IAE 85] IAEA, Sediment Kd's and concentration factors for radionuclides in the marine environment, International Atomic Energy Agency Technical Report Series no. 247, Vienna, 1985.

[IAE 88] IAEA, Assessing the impact of deep-sea disposal of low level radiative waste on living marine resources, Technical Report Series no. 288, IAEA, Vienna, 1988.

[IAE 91] IAEA, Inventory of radioactive material entering the marine environment: Sea disposal of radioactive waste, IAEA-TECDOC-588, IAEA, Vienna, 1991.

[IAE 92] IAEA, Effects of ionizing radiation on plants and animals at levels implied by current radiation protection standards, Technical Reports Series no. 332, IAEA, Vienna, 1992.

[IAE 94] IAEA, Handbook of parameter values for the prediction of radionuclide transfer in temperate environments, Technical Report Series no. 364, IAEA, Vienna, 1994.

[IAE 98a] IAEA, Conditions at Bikini Atoll. Radiological conditions at Bikini Atoll and the prospects of resettlement, Radiological Assessment Reports Series 2, STI/PUB/1054, IAEA, Vienna, 1998.

[IAE 98b] IAEA, The situation at Mururoa and Fangataufa Atolls, Radiological Assessment Reports Series 1, STI/PUB/1028, IAEA, Vienna, 1998.

[IAE 98c] IAEA, Radiological conditions of the Western Kara Sea. Assessment of the radiological impact of the dumping of radioactive waste in the Arctic Seas, Radiological Assessment Reports Series 4, STI/PUB/1068, IAEA, Vienna, 1998.

[IAE 99] IAEA, Inventory of radioactive waste disposal at sea, IAEA RECDOC no. 1105, IAEA, Vienna, 1999.

[IAE 00] IAEA, Global marine radioactivity database (GLOMARD), IAEA-TECDOC-1146, IAEA, Vienna, 2000.

[IAE 01a] IAEA, Inventory of accidents and losses at sea involving radioactive material, IAEA TECDOC no. 1242, 2001.

[IAE 01b] IAEA, Severity, probability and risk of accidents during maritime transport of radioactive material, IAEA-TECDOC-1231, IAEA, Vienna, 2001.

[IAE 03a] IAEA, Protection of the environment from ionising radiation. The development and application of a system of radiation protection for the environment, IAEA-CSP-17, IAEA, Vienna, 2003.

[IAE 03b] IAEA, International Conference on the Protection of the Environment from the Effects of Ionizing Radiation, Stockholm, 6–10 October 2003, IAEA-CN-109, IAEA, Vienna, 2003.

[IAE 04a] IAEA, Nuclear explosions in the USSR: The north test site reference material, IAEA, Vienna, Version 4, December 2004.

[IAE 04b] IAEA, Sediment distribution coefficients and concentration factors for biota in the marine environment, IAEA Technical Reports Series TRS-422, International Atomic Energy Agency, Vienna, 2004.

[IAE 05] IAEA, Worldwide Marine Radioactivity Studies (WOMARS). Radionuclide levels in oceans and seas, IAEA-TECDOC-1429, IAEA, Vienna, 2005.

[IAE 10] IAEA, Handbook of parameter values for the prediction of radionuclide transfer in terrestrial and freshwater environments, Technical Reports Series no. 472, IAEA, Vienna, 2010.

[IAE 11a] IAEA, Radioactive particles in the environment: Sources, particle characterization and analytical techniques, IAEA-TECDOC-1663, IAEA, Vienna, 2011.

[IAE 11b] IAEA, Radiation protection and safety of radiation sources: International basic safety standards sources. Interim edition. General safety requirements, Part 3, no. GSR, Part 3 (Interim), IAEA, Vienna, 2011.

[IAE 12a] IAEA, Modelling radiation exposure and radionuclide transfer for non-human species, Report of the Biota Working Group of EMRAS I Theme 3, IAEA-TECHDOC no. 1678, IAEA, Vienna, 2012.

[IAE 12b] IAEA, *Proceedings of the First ICRP Symposium on the International System of Radiological Protection, Annals of the ICRP*, vol. 41, nos 3–4, 2012.

[IAE 14] IAEA, Handbook of parameter values for the prediction of radionuclide transfer to wildlife, IAEA Technical Report Series no. 479, IAEA, Vienna, 2014.

[IAE 15a] IAEA, Inventory of radioactive material resulting from historical dumping, accidents and losses at sea for the purposes of the London Convention 1972 and London Protocol 1996, IAEA-TECDOC-1776, IAEA, Vienna, 2015.

[IAE 15b] IAEA, Determining the suitability of materials for disposal at sea under the London Convention 1972 and London Protocol 1996: A radiological assessment procedure, IAEA-TECDOC-1759, IAEA, Vienna, 2015.

[IAE 20] IAEA, Environmental transfer of radionuclides in Japan following the accident at the Fukushima Daiichi Nuclear power plant, IAEA-TECDOC-1927, IAEA, Vienna, 2020.

[ICR 91] ICRP, "1990 recommendations of the International Commission on Radiological Protection", *Ann. ICRP*, Publication 60, vol. 21, nos 1–3, 1991.

[ICR 01] ICRU, "Quantities, units and terms in radioecology", *J. ICRU*, vol. 1, pp. 1–44, 2001.

[ICR 03] ICRP, "A framework for assessing the impact of ionising radiation on non-human species", *Annals of the ICRP*, vol. 33, no. 3, pp. 207–266, 2003.

[ICR 07] ICRP, "Human alimentary tract model for radiological protection", *Ann. ICRP*, vol. 26, nos 1–2, Publication 100, Pergamon Press, Oxford, 2007.

[ICR 08] ICRP, "Environmental protection – The concept and use of reference animals and plants", *Ann. ICRP*, Publication 108, vol. 38, nos 4–6, pp. 1–242, 2008.

[ICR 09] ICRP, "Environmental protection: Transfer parameters for reference animals and plants", *Ann. ICRP*, Publication 114, vol. 39, no. 6, pp. 1–111, 2009.

[ICR 14] ICRP, "Protection of the environment under different exposure situations", *Ann. ICRP*, Publication 124, vol. 43, no. 1, pp. 1–58, 2014.

[ICR 17] ICRP, "Dose coefficients for non-human biota environmentally exposed to radiation", *Ann. ICRP*, Publication 136, vol. 46, no. 2, pp. 1–136, 2017.

[ICR 21] ICRP, "Radiation weighting for reference animals and plants", *Ann. ICRP*, Publication 148, vol. 50, no. 2, 2021.

[ILI 19] IL'IN G.V., USYAGINA L.S., "Radioecological studies in the Murmansk Littoral of the Barents Sea as part of environmental monitoring in radwaste management", *Atom. Energy*, vol. 126, pp. 326–331, 2019.

[ILU 06] ILUS E., VARTTI V.P., IKAHEIMONEN T.K. et al., "Recent observations of ^{99}Tc in Finnish coastal waters", IAEA-CSP-26, IAEA, Vienna, pp. 303–306, 2006.

[IOS 12] IOSJPE M., LILAND A., "Evaluation of environmental sensitivity of the marine regions", *J. Environ. Radioact.*, vol. 108, pp. 2–8, 2012.

[IRS 15] IRSN, "Retombées des essais nucléaires", available at: https://www.irsn.fr/FR/connaissances/Environnement/surveillance-environnement/resultats/retombees-tirs-armes-nucleaires/Pages/sommaire.aspx#.YKFBBaE6_Qx, 2015.

[IRS 18] IRSN, Bilan de l'état radiologique de l'environnement français de 2015 à 2017, IRSN DG/2018–00006 report, 2018.

[IUR 12] IUR, Towards an ecosystem approach for environment protection with emphasis on radiological hazards, Findings of the IUR Ecosystem Approach Task Group, IUR report no. 7, 2nd edition, 2012.

[JAC 00] JACKSON D., LAMBERS B., GRAY J., "Radiation doses to members of the public near to Sellafield, Cumbria, from liquid discharges 1952–98", *J. Radiol. Prot.*, vol. 20, pp. 139–167, 2000.

[JAI 11] JAIN N., BHATIA A.L., "Tritium in the environment and its impact assessment against the existing radiation protection framework revisited", *Radioprotection*, vol. 46, no. 6, pp. S385–S391, 2011.

[JCA 59] JCAE, "Industrial radioactive waste disposal", *Summary-analysis of Hearings, Congress of the United States*, Washington, 1959.

[JHA 05] JHA, A.N., DOGRA Y., TURNER A. et al., "Impact of low doses of tritium on the marine mussel, *Mytilus edulis*: Genotoxic effects and tissue-specific bioconcentration", *Mutat. Res.*, vol. 586, no. 1, pp. 47–57, 2005.

[JHA 06] JHA A.N., DOGRA Y., TURNER A. et al., "Are low doses of tritium genotoxic to *Mytilus edulis*?", *Mar. Environ. Res.*, vol. 62, pp. S297–S300, 2006.

[JHA 08] JHA A.N., "Ecotoxicological applications and significance of the comet assay", *Mutagen*, vol. 23, pp. 207–221, 2008.

[JOH 10] JOHANNESSEN O.M., VOLKOV V.A., PETTERSSON L.H. et al., *Radioactivity and Pollution in the Nordic Seas and Arctic Region. Observations, Modeling and Simulations*, Springer-Verlag, Berlin, 2010.

[JOH 15] JOHANSEN M.P., RUEDIG E., TAGAMI K. et al., "Radiological dose rates to marine fish from the Fukushima Daiichi accident the first three years across the North Pacific", *Environmental Sciences and Technology*, vol. 49, no. 3, pp. 1277–1285, 2015.

[JOH 19] JOHANSEN M.P., CHILD D.P., CRESSWELL T. et al., "Plutonium and other radionuclides persist across marine-to-terrestrial ecotopes in the Montebello Islands sixty years after nuclear tests", *Sci. Total Environ.*, vol. 691, pp. 572–583, 2019.

[JOH 20] JOHANSEN M.P., CHILDE D.P., HOTCHKIS M.A.C. et al., "Radionuclides in sea turtles at the Montebello Islands former nuclear test sites: Current and historical dose rates for adults and embryos", *Mar. Pollut. Bull.*, vol. 158, no. 111390, September 2020.

[JOI 96] JOINT RUSSIAN-NORWEGIAN EXPERT GROUP FOR INVESTIGATION OF RADIOACTIVE CONTAMINATION IN THE NORTHERN AREAS, Dumping of radioactive waste and radioactive contamination in the Kara Sea, Results from 3 years of investigations (1992–1994) performed by the Joint Norwegian-Russian Expert Group, NEI- no. 728, March 1996.

[JON 82] JONES C.B., DOYLE M.K., BERKHOUSE L.H. et al., Operation Argus 1958, United States atmospheric nuclear weapons teats, nuclear test personnel review, Technical report KT-82–004(R), 1982.

[JON 03] JONES D., DOMOTOR S., HIGLEY K. et al., "Principles and issues in radiological ecological risk assessment", *J. Environ. Radioact.*, vol. 66, nos 1–2, pp. 19–39, 2003.

[KAE 17] KAERIYAMA H., "Oceanic dispersion of Fukushima-derived radioactive cesium: A review", *Fish. Oceanogr.*, vol. 26, no. 2, pp. 99–113, 2017.

[KAN 99] KANIVETS V.V., VOITSEKHOVITCH O.V., SIMOV V.G. et al., "The post-Chernobyl budget of 137Cs and 90Sr in the Black Sea", *J. Environ. Radioact.*, vol. 43, pp. 121–135, 1999.

[KAR 12] KARCHER M., SMITH J.N., KAUKER F. et al., "Recent changes in Arctic Ocean circulation revealed by iodine-129 observations and modeling", *J. Geophys. Res.*, vol. 117, C08007, pp. 1–17, 2012.

[KAY 73] KAYE S.V., "Assessing potential radiological impacts to aquatic biota in response to the national environmental policy act (NEPA) of 1969", *Environmental Behaviour of Radionuclides Released in the Nuclear Industry*, IAEA, Vienna, pp. 649–661, 1973.

[KE 01] KE C.H., WANG W.X., "Bioaccumulation of Cd, Se and Zn in an estuarine oyster (*Crassostrea rivularis*) and a coastal oyster (*Saccostrea glomerata*)", *Aquat. Toxicol.*, vol. 56, pp. 33–51, 2001.

[KER 95] KERSHAW P., BAXTER A., "The transfer of reprocessing wastes from north-west Europe to the Arctic", *Deep-Sea Res. II*, vol. 42, no. 6, pp. 1413–1448, 1995.

[KER 97] KERSHAW P., GURBUTT P., WOODHEAD D. et al., "Estimates of fluxes of ^{137}Cs in northern waters from recent measurements", *Sci. Total Environ.*, vol. 202, nos 1–3, pp. 211–223, 1997.

[KER 99] KERSHAW, P.J., MCCUBBIN, D., LEONARD, K.S., "Continuing contamination of north Atlantic and Arctic waters by Sellafield radionuclides", *Sci. Total Environ.*, vol. 237–238, pp. 119–132, 1999.

[KEU 14] KEUM, D.W., KIM, B.H., LIM, K.M. et al., "Radiation exposure to marine biota round the Fukushima Daichii NPP", *Environ. Monit. Assess.*, vol. 186, pp. 2949–2956, 2014.

[KIM 12] KIM C.K., BYUN J.I., CHAE J.S. et al., "Radiological impact in Korea following the Fukushima nuclear accident", *J. Environ. Radioact.*, vol. 111, pp. 70–82, 2012.

[KIR 77] KIRCHMANN R., CASALE G., "Energie et environnement", *Consensus*, vol. 1, pp. 29–38, 1977.

[KNO 97] KNOWLES J.F., GREENWOOD L.N., "A comparison of the effects of long-term beta and gamma irradiation on the reproductive performance of a marine invertebrate *Ophryotrocha diadema* (Polychaeta, Dorvilleidae)", *J. Environ. Radioact.*, vol. 34, no. 1, pp. 1–7, 1997.

[KNO 99] KNOWLES J.F., "Long-term irradiation of a marine fish, the plaice Pleuronectes platessa: An assessment of the effects on size and composition of the testes and of possible genotoxic changes in peripheral erythrocytes", *Int. J. Radiat. Biol.*, vol. 75, no. 6, pp. 773–782, 1999.

[KOL 70] KOLESNIKOV A.N., BARINOV G.V., ZESENKO A.Y., Radioecology studies in the Mururoa Atoll (Tuoamotu Isl.) region, USSR reports on natural and fallout radioactivity, AEC-Tr-7128, 62–65, 1970.

[KOL 19] KOLAR M.V., GUGLETA M., "The consequences of disposal and leakage of radioactive materials on various species of marine and freshwater fish", *Int. J. Fish. Aquat. Stud.*, vol. 7, no. 6, pp. 185–189, 2019.

[KON 02] KONHAUSER K.O., MORTIMER R.J.G., MORRIS K. et al., "The role of microorganisms during sediment diagenesis: Implications for radionuclide mobility", in KEITH-ROACH M.J., LIVEN F.R. (eds), *Interactions of Microorganisms with Radionuclides*, Elsevier, Amsterdam, 2002.

[KRA 03] KRAJEWSKI P., SUPLIŃSKA M., "Doses to selected seawater organisms from ^{137}Cs, ^{226}Ra and ^{239,240}Pu", *International Conference on the Protection of the Environment from the Effects of Ionizing Radiation*, Stockholm, Sweden, 6–10 October 2003, IAEA-CN-109, IAEA, Vienna, pp. 216–219, 2003.

[KU 77] KU T.L., KNAUSS K.G., MATHIEU G.G., "Uranium in open oceans: Concentrations and isotopic composition", *Deep Sea Res.*, vol. 24, pp. 1005–1017, 1977.

[KUD 00] KUDO A., "Plutonium in the Environment", *Radioactivity in the Environment*, vol. 1, 2000.

[KUM 20] KUMAR A., ROUT S., PULLIANI V. et al., "A review on distribution coefficient (Kd) of some selected radionuclides in soil/sediment over the last three decades", *J. Radioanal. Nuclear Chem.*, vol. 323, pp. 13–26, 2020.

[LAC 10] LACOUE-LABARTHE T., WARNAU M., OBERHANSLI F. et al., "Contrasting accumulation biokinetics and distribution of ^{241}Am, Co, Cs, Mn and Zn during the whole development time of the eggs of the common cuttlefish, *Sepia officinalis*", *J. Exp. Mar. Biol. Ecol.*, vol. 382, pp. 131–138, 2010.

[LAF 04] LAFFOLEY d., MALTBY E., VINCENT M.A. et al., The ecosystem approach. coherent actions for marine and coastal environments, A report to the UK Government, English Nature, Peterborough, 2004.

[LAG 12] LAGUIONIE P., BAILLY DU BOIS P., BOUST D. et al., "Consequences of radioactive releases into the sea resulting from the accident at the Fukushima Dai-ichi nuclear power plant – Evolution of expert investigation according to the data available", *Radioprotection*, vol. 47, no. 3, pp. 423–447, 2012.

[LAI 99] LAI D., "An overview of five decades of studies of cosmic ray produced nuclides in oceans", *Sci. Total Environ.*, vol. 238, pp. 3–13, 1999.

[LAN 04a] LANDIS W.G., YU M.H., *Introduction to Environmental Toxicology. Impacts of Chemicals Upon Ecological Systems*, 3rd edition, Lewis Publishers, Boca Raton, 2004.

[LAN 04b] LANSARD B., Distribution et remobilisation du plutonium dans les sédiments du prodelta du Rhône (Méditerranée nord-occidentale), Thesis, Uiversité de la Méditerranée, Marseille, 24 June 2004.

[LAN 06] LANSARD B., GRENZ C., CHARMASSON S. et al., "Potential plutonium remobilisation linked to marine sediment resuspension: First estimates based on flume experiments", *Journal of Sea Research*, vol. 55, pp. 74–85, 2006.

[LAP 71] LAPPENBUSCHT W.L., WATSON D.G., TEMPLETON W.L., "*In situ* measurement of radiation dose in the Columbia River", *Health Phys.*, vol. 21, pp. 247–251, 1971.

[LAP 06] LAPTYEV G.V., VOITSEKHOVITCH O.V., KOSTEZH A.B. et al., "Mass accumulation rates and fallout radionuclides ^{210}Pb, ^{137}Cs and ^{241}Am inventories determined in radiometrically dated abyssal sediments of the Black Sea", *IAEA-CSP-26*, IAEA, Vienna, pp. 281–282, 2006.

[LAR 04] LARSSON C.-M., "The FASSET framework for assessment of environmental impact of ionising radiation in European ecosystems. An overview", *Radiological Protection*, vol. 24, pp. A1–A12, 2004.

[LAR 08] LARSSON C.-M., "An overview of the ERICA integrated approach to the assessment and management of environmental risks from ionising contaminants", *J. Environ. Radioact.*, vol. 99, no. 9, pp. 1364–1370, 2008.

[LAR 15] LARSSON C.-M., HIGLEY K.A., REAL A., "Overview of ICRP Committee 5", *Proceedings of the Second International Symposium on the System of Radiological Protection, Annals of the ICRP*, vol. 44, 1S, pp. 47–57, 2015.

[LAZ 03] LAZORENKO G.E., POLIKARPOV G.G., OSVATH I., "Doses to Black Sea fishes and mussels from the naturally occurring radionuclide ^{210}Po", *International Conference on the Protection of the Environment from the Effects of Ionizing Radiation*, Stockholm, Sweden, 6–10 October 2003, IAEA-CN-109, IAEA, Vienna, pp. 242–244, 2003.

[LEB 95] LE BAR S., "Environmental impact of the La Hague reprocessing plant", *Environmental Impact of Radioactive Releases*, IAEA, Vienna, pp. 211–220, 1995.

[LEG 71] LE GALL P., ANCELLIN J., Facteurs de concentration du césium 137, ou cérium 144 et ou ruthénium 106 chez diverses espèces de mollusques et crustacés marins comestibles, Note CEA-N-1488, 1971.

[LEG 09] LE GUEN, B., SICLET, F., "Impact du carbone-14 autour des centrales nucléaires EDF", *Radioprotection*, vol. 44, no. 44, pp. 495–504, 2009.

[LEM 03] LEMERCIER V., MILLOT X., ANSOBORLO E. et al., "Study of uranium transfer across the blood-brain barrier", *Radiat. Prot. Dosimetry*, vol. 105, pp. 243–245, 2003.

[LI 91] LI Y.H., "Distribution patterns of the elements in the ocean – A synthesis", *Geochim. Cosmochim. Ac.*, vol. 55, no. 11, pp. 3223–3240, 1991.

[LIE 08] LIESER K.H., *Nuclear and Radiochemistry: Fundamentals and Applications*, 2nd edition, John Wiley & Sons, Hoboken, 2008.

[LIN 05] LINSLEY G., SJÖBLOM K.L., CABIANCA T., "Overview of point sources of anthropogenic radionuclides in the oceans", in LIVINGSTON H.D. (ed.), *Marine Radioactivity*, Elsevier, Amsterdam, 2005.

[LIU 16] LIU X., WANG W.X., "Time changes in biomarker responses in two species of oyster transplanted into a metal contaminated estuary", *Sci. Total Environ.*, vol. 544, pp. 281–290, 2016.

[LIV 00] LIVINGSTON H.D., POVINEC P.P., ITO T. et al., "The behaviour of plutonium in the Pacific Ocean", in KUDO A. (ed.), *Radioactivity in the Environment: Plutonium in the Environment*, Elsevier, Amsterdam, 2000.

[LIV 02] LIVINGSTON H.D., POVINEC P.P., "A millennium perspective on the contribution of global fallout radionuclides to ocean science", *Health Phys.*, vol. 82, no. 5, pp. 656–668, 2002.

[LOB 19] LOBEL L.K., LOBEL P.S., "Current status of the US military atolls in the Pacific: Johnston and Wake", in SHEPPARD, C. (ed.), *World Seas: An Environmental Evaluation. Volume II: The Indian Ocean to the Pacific*, Elsevier, Amsterdam, 2019.

[LOT 95] LOTTS J.W., STEWART A.J., "Minnow can acclimate to total residual chlorine", *Environ. Toxicol. Chem.*, vol. 14, pp. 1365–1374, 1995.

[LUC 04] LUCEY J.A., GOUZY A., BOUST D. et al., "Geochemical fractionation of plutonium in anoxic Irish Sea sediments using an optimised sequential extraction protocol", *Appl. Radiat. Isotopes*, vol. 60, pp. 379–385, 2004.

[LUO 08] LUOMA S., RAINBOW P.S., *Metal Contamination in Aquatic Environment*, Cambridge University Press, Cambridge, 2008.

[LYN 95] LYNN N., TIMMS S., WARDEN J. et al., "Radionuclide release from submarine reactors dumped in the Kara Sea", *Environmental Impact of Radioactive Releases*, IAEA, Vienna, pp. 825–827, 1995.

[MAC 98] MACKENZIE A.B., COOK G.T., MCDONALD P. et al., "The influence of mixing timescales and re-dissolution processes on the distribution of radionuclides in Northeast Irish Sea sediments", *J. Environ. Radioact.*, vol. 39, pp. 35–53, 1998.

[MAD 12] MADIGAN D.J., BAUMANN Z., FISHER N.S., "Pacific bluefin tuna transport Fukushima- derived radionuclides from Japan to California", *Proceedings of the National Academy of Sciences*, vol. 109, no. 24, pp. 9483–9486, 2012.

[MAD 13] MADIGAN D.J., BAUMANN Z., SNODGRASS O.E. et al., "Radiocesium in Pacific bluefin tuna *Thunnus orientalis* in 2012 validates new tracer technique", *Environ. Sci. Technol.*, vol. 47, no. 5, pp. 2287–2294, 2013.

[MAK 20] MAKMUR M., SUSENO H., PRIHATININGSIH W.R. et al., "Ecological risk assessment of plutonium in primary inlet of Indonesian through flow (ITF)", *IOP Conf. Series: Earth and Environmental Science*, vol. 584, no. 012019, 2020.

[MAR 74] MARCHAND M., "Considérations sur les formes physico-chimiques du cobalt, manganèse, zinc, chrome et fer dans une eau de mer enrichie ou non en matière organique", *J. Cons. Int. Explor. Mer.*, vol. 35, pp. 130–142, 1974.

[MAR 75] MARTIN J.L., Recherches sur le métabolisme des métaux chez les Crustacés Décapodes marins ; leurs rapports avec la mue et la reproduction, PhD thesis, Université d'Aix-Marseille II, CEA-R-4689 report, 1975.

[MAR 99] MARO D., HEBERT D., GANDON R. et al., "Dosage par spectrométrie gamma de l'iode 129 dans les échantillons biologiques marins et terrestres. Application à des algues prélevées le long des côtes de la Manche", *Radioprotection*, vol. 34, no. 1, pp. 13–24, 1999.

[MAR 02] MARIGOMEZ I., SOTO M., CAJARAVILLE M. et al., "Cellular and subcellular distribution of metals in molluscs", *Microsc. Res. Tech.*, vol. 56, pp. 358–392, 2002.

[MAS 95a] MASON A.Z., JENKINS K.D., "Metal detoxification in aquatic organisms", in TESSIER A., TURNER D.R. (eds), *Metal Speciation and Bioavailability in Aquatic Systems*, John Wiley & Sons, Chichester, 1995.

[MAS 95b] MASSON M., VAN WEERS A., GROOTHUIS R.E. et al., "Time series for seawater and seaweed of ^{99}Tc and ^{125}Sb originating from releases at La Hague", *J. Mar. Syst.*, vol. 6, pp. 397–413, 1995.

[MAS 02] MASQUÉ P., SANCHEZ-CABEZA J.A., BRUACH J.M. et al., "Balance and residence times of ^{210}Pb and ^{210}Po in surface waters of the northwestern Mediterranean Sea", *Cont. Shelf Res.*, vol. 22, pp. 2127–2146, 2002.

[MAT 04] MATISHOV D.G., MATISHOV G.G., *Radioecology in Northern European Seas*, Springer-Verlag, Berlin, 2004.

[MAT 06] MATTILA J., KANKAANPAA H., ILUS E., "Use of ^{137}Cs, ^{239,240}Pu and ^{210}Pb in sediment accumulation studies in the Baltic Sea", *IAEA-CSP-26*, IAEA, Vienna, pp. 405–408, 2006.

[MAT 19] MATISHOV G.G., KASATKINA N.E., USYAGINA L.S., "Technogenic radioactivity of waters in the Central Arctic Basin and adjacent water areas", *Dokl. Earth Sci.*, vol. 485, pp. 288–292, 2019.

[MCC 01] MCCUBBIN D., LEONARD K.S., BAILEY T.A. et al., "Incorporation of organic tritium (3H) by marine organisms and sediment in the Severn estuary/Bristol Channel (UK)", *Mar. Pollut. Bull.*, vol. 42, pp. 852–863, 2001.

[MCC 02] McCubbin D., Leonard K.S., Brown J. et al., "Further studies of the distribution of technetium-99 and caesium-137 in UK and European coastal waters", *Cont. Shelf Res.*, vol. 22, no. 10, pp. 1417–1445, 2002.

[MCD 93] McDonald P., Baxter M.S., Fowler S.W., "Distribution of radionuclides in mussels, winkles and prawns. Part 1. Study of organisms under environmental conditions using conventional radio-analytical techniques", *J. Environ. Radioact.*, vol. 18, pp. 181–202, 1993.

[MCD 01] McDonald P., Vives i Battle J., Bouscher A. et al., "The availability of plutonium and americium in Irish Sea sediments for re-dissolution", *Sci. Total Environ.*, vol. 267, pp. 109–123, 2001.

[MCG 72] McGregor J.F., Newcombe H.B., "Dose-response relationships for yields of major eye malformations following low dose of radiation to trout sperm", *Radiat. Res.*, vol. 49, pp. 155–169, 1972.

[MCG 96] McGee E.J., Johanson K.J., Keatinge M.J. et al., "An evaluation of ratio systems in radioecological studies. *Health Phys.*, vol. 70, pp. 215–221, 1996.

[MCK 87] McKee B.A., DeMaster D.J., Nitrouer C.A., "Uranium geochemistry on the Amazon shelf: Evidence for uranium release from bottom sediments", *Geochim. Cosmochim. Acta*, vol. 51, pp. 2779–2786, 1987.

[MCM 00] McMahon C.A., Leon Vintro L., Mitchell P.I. et al., "Oxidation-state distribution of plutonium in surface and subsurface waters at Thule, northwest Greenland", *Appl. Radiat. Isot.*, vol. 52, pp. 697–703, 2000.

[MCM 05] McMahon C.A., Fegan M., Wong J. et al., "Transfer of conservative and non-conservative radionuclides from the Sellafield nuclear fuel reprocessing plant to the coastal waters of Ireland", *Radioprotection*, vol. 40, Suppl. 1, pp. S629–S634, 2005.

[MEN 09] Menager M.T., Garnier-Laplace J., Goyffon M. (eds), *Toxicologie nucléaire environnementale et humaine*, Lavoisier, Tec&Doc, Paris, 2009.

[MET 10] Metivier H., *Plutonium. Mythes et réalités*, EDP Sciences, Les Ulis, 2010.

[MET 16] Metian M., Poul I.S., Hédouin L. et al., "Differential bioaccumulation of ^{134}Cs in tropical marine organisms and the relative importance of exposure pathways", *J. Environ. Radioact.*, vol. 152, pp. 127–135, 2016.

[MIQ 93] Miquel J.C., Heyraud M., Cherry R.D., "^{210}Pb as a dietary indicator in the Antarctic", *Mar. Biol.*, vol. 115, pp. 161–171, 1993.

[MIR 17] Miroshnikov A.Y., Laverov N.P., Chernov R.A. et al., "Radioecological investigations on the Northern Novaya Zemlya Archipelago", *Okeanologiya*, vol. 57, no. 1, pp. 227–237, 2017.

[MIT 77] Mitchell N.T., "Radioactivity in surface waters of the British Isles, 1975", *Tech. Rep. Fish Radiobiol. Lab., FRL*, vol. 12, 1977.

[MIT 95] MITCHELL P.I., VIVES I., BATTLE J. et al., "Recent observations on the physico-chemical speciation of plutonium in the Irish Sea and the Western Mediterranean", *Appl. Radiat. Isotopes*, vol. 46, no. 11, pp. 1175–1190, 1995.

[MIT 99] MITCHELL P.I., CONDREN O.M., VINTRO L.L. et al., "Trends in plutonium, americium and radiocaesium accumulation and long-term bioavailability in the western Irish Sea mud basin", *J. Environ. Radioact.*, vol. 44, nos 2–3, pp. 223–251, 1999.

[MIY 55] MIYAKE Y., SUGIURA Y., KAMEDA K., "On the distribution of radioactivity in the sea around Bikini Atoll in June, 1954", *Pap. Meteorol. Geophys.*, vol. 5, nos 3–4, pp. 253–262, 1955.

[MOO 00] MOORE W.S., "Determining coastal mixing rates using radium isotopes", *Cont. Shelf Res.*, vol. 20, pp. 1993–2007, 2000.

[MOO 10] MOORE W.S., "A reevaluation of submarine groundwater discharge along the southeastern coast of North-America", *Glob. Biogeochem. Cycles*, vol. 24, 1029/2009GB003747, 2010.

[MOR 82] MORONEY J.R., COOPER M.B., Environmental radiation at the Monte Bello Islands from nuclear weapons tests conducted in 1952 and 1956, Report, Australian Radiation Laboratory, ARL/TR049, December 1982.

[MOT 01] MOTHERSILL C., LYNG F., MULFORD A. et al., "Effect of low doses of ionizing radiation on cells cultured from the hematopoietic tissue of the Dublin Bay prawn, *Nephrops norvegicus*", *Radiat. Res.*, vol. 156, pp. 241–250, 2001.

[MOU 90] MOUCHEL J.M., "Fractionation of selected metals between the dissolved and particulate phases in the Rhone River estuary", in MARTIN J.M., BARTH H. (eds), *Water Pollution Research Report 20*, Commission of the European Communities, Brussels, 1990.

[MUI 13] MUIKKU M., SIRKKA L., BARNETT C. et al., "Blueprint of the European radioecology ALLIANCE functioning", Program COMET (Contract Number: Fission-2012–3.4.1–604974), deliverable (D-NO. 2–7), 2013.

[MUI 18] MUIKKU M., BERESFORD N.A., GARNIER-LAPLACE J. et al., "Sustainability and integration of radioecology – Position paper", *J. Radiol. Prot.*, vol. 38, p. 152, 2018.

[MUL 99] MULSOW S., COQUERY M., DOVLETE C. et al., "Radionuclide concentrations in underground waters of Mururoa and Fangataufa Atolls", *Sci. Total Environ.*, vols 237–238, pp. 287–300, 1999.

[NAK 63] NAKATANI R.E., FOSTER R.F., "Effect of chronic feeding of ^{90}Sr-Y on rainbow trout", in SCHULTZ V., KLEMENT A.W. (eds), *Radioecology*, Reinhold Publ. Corp., New York, 1963.

[NAK 66] NAKATANI R.E., "Biological response of rainbow trout (*Salmo gairdneri*) ingesting zinc 65", *Disposal of Radioactive Wastes into Seas, Oceans and Surface Waters*, IAEA, Vienna, 809–822, 1966.

[NAK 75] NAKAHARA M., KOYANAGI T., SAIKI M., "Concentration of radioactive cobalt by seaweeds in the food chain", *Impacts of Nuclear Releases into the Aquatic Environment*, IAEA, Vienna, pp. 301–312, 1975.

[NAK 77] NAKAMURA R., SUZUKI Y., UEDA T., "Distribution of radionuclides among green alga, marine sediments and sea water", *J. Radiat. Res.*, vol. 18, pp. 322–330, 1977.

[NAS 70] NASSOGNE A., "Le rôle des Invertébrés marins dans le cumul et le transport des radionucléides", *Rev. Int. Océanogr. Méd.*, vol. 20, pp. 79–99, 1970.

[NAS 71] NAS, Radioactivity in the marine environment, National Academy of Science, National Research Council, Washington, 1971.

[NCP 91] NCPR, Effects of ionizing radiation on aquatic organisms, National Council on Radiation Protection and Measurements, NCRP report no. 109, Bethesda, 1991.

[NEL 84] NELSON D., CAREY A.E., BOWEN V.T., "Plutonium oxidation state distribution in the Pacific Ocean during 1980–1981", *Earth Planet Sci. Lett.*, vol. 68, pp. 422–430, 1984.

[NG 05] NG T.Y.T., AMIARD-TRIQUET C., RAINBOW P.S. et al., "Physico-chemical form of trace metals accumulated by phytoplankton and their assimilation by filter-feeding invertebrates", *Mar. Ecol. Progr. Ser.*, vol. 299, pp. 179–191, 2005.

[NIE 99] NIELSEN S.P., BENGTSON P., BOJANOWSKY R. et al., "The radiological exposure of man from radioactivity in the Baltic Sea", *Sci. Total Environ.*, vols 237–238, pp. 133–141, 1999.

[NIE 18a] NIES H., "Pollution with radioactive substances", in SALOMON M., MARKUS T., (eds), *Handbook on Marine Environment Protection: Science, Impacts and Sustainable Management*, Springer, Cham, 2018.

[NIE 18b] NIES H., "Origin and management of radioactive substances in the marine environment", SALOMON M., MARKUS T., (eds), *Handbook on Marine Environment Protection: Science, Impacts and Sustainable Management*, Springer, Cham, 2018.

[NIK 97] NIKITIN A.I., SALBU B., STRAND P. et al., Joint Russian-Norwegian collaboration on radioactive contamination from dumped nuclear waste in the Kara Sea – 3 years with expeditions to the dumping sites: Goals and investigations, NO9700024, pp. 125–128, 1997.

[NOA 03] NOAA, New priorities for the 21st century: NOAA's strategic vision, Report, U.S. Department of Commerce, National Oceanic and Atmospheric Administration, Silver Spring, 2003.

[NOL 91] NOLAN C., DAHLGAARD H., "Accumulation of metal radiotracers by *Mytilus edulis*", *Mar. Ecol. Prog. Ser.*, vol. 70, pp. 165–174, 1991.

[NOS 80] NOSHKIN V.E., EAGLE R.J., WONG K.M. et al., Transuranic concentrations in reef and pelagic fish from the Marshall Islands, UCRL-84213, September 1980, IAEA-SM-248/146, Impacts of Radionuclide Releases into the Marine Environment, Vienna, 6–10 October 1980.

[NOS 81] NOSHKIN V.E., EAGLE R.J., WONG K.M. et al. Concentrations of radionuclides in reef and lagoon pelagic fish from the Marshall Islands, Technical report UCID-19028, California University, 1981.

[NOS 97] NOSHKIN V.E., ROBISON W.L., WONG K.M. et al., "Past and present levels of some radionuclides in fish from Bikini and Enewetak Atolls", *Health Phys.*, vol. 73, no. 1, pp. 49–65, 1997.

[NOS 98] NOSHKIN V.E., ROBISON W.L., EAGLE R.J. et al., Radionuclides in sediments and seawater at Rongelap Atoll, Report, UCRL-LR-130250, Lawrence Livermore National Laboratory, March 1998.

[NOZ 98] NOZAKI Y., YAMADA M., NAKANISHI T. et al., "The distribution of radionuclides and some trace metals in the water columns of the Japan and Bonin trenches", *Oceanol. Acta*, vol. 21, no. 3, pp. 469–484, 1998.

[OEC 92] OECD, Report of the OECD Workshop on the extrapolation of laboratory aquatic toxicity data on the real environment, OECD Environment Monography, no. 59, 1992.

[OLS 82] OLSEN C.R., CUTSHALL N.H., LARSEN I.L., "Pollutant-particle associations and dynamics in coastal marine environments: A review", *Mar. Chem.*, vol. 11, pp. 501–533, 1982.

[OLS 03] OLSEN Y.S., VIVES i BATTLE J., "A model for the bioaccumulation of ^{99}Tc in lobsters (*Homarus gammarus*) from west Cumbrian coast", *J. Environ. Radioact.*, vol. 67, pp. 219–233, 2003.

[OMS 19] OMS P.E., Transferts multi-échelles des apports continentaux dans le golfe de Gascogne, Thesis, Université de Bretagne occidentale – Brest, 2019.

[OSP 92] OSPAR, Convention for the protection of the marine environment of the North-East Atlantic, 1992.

[OSP 09] OSPAR, Towards the radioactive substances strategy objectives, Third Periodic Evaluation, 2009.

[OSP 10] OSPAR, Chapter 6: Radioactive substances, Quality Status Report, 2010.

[OSP 17] OSPAR, Summary of progress towards the objective of the radioactive substances strategy, 2017.

[OSP 18] OSPAR, Rapport annuel 2017–2018, 2018.

[OSP 19] OSPAR, French implementation report of PARCOM Recommendation 91/4 on radioactive discharges, 2019.

[OSP 20] OSPAR, Annual report and assessment of discharges of radionuclides from the non-nuclear sectors in 2018, 2020.

[OST 63] OSTERBERG C., Radioactivity in oceanic organisms, Thesis, Oregon State University, June 1963.

[OST 64] OSTERBERG C., PEARCY W., CURL JR. H., "Radioactivity and its relationship to the oceanic food chains", *J. Mar. Res.*, vol. 22, pp. 2–12, 1964.

[OST 65] OSTERBERG C., CUTSHALL N., CRONIN J., "Chromium-51 as a radioactive tracer of Columbia River water at sea", *Science*, vol. 150, pp. 1585–1587, 1965.

[OUE 13] OUEST FRANCE, "Radioactivité en rade : faible impact des sous-marins", available at: https://brest.maville.com/actu/actudet_-radioactivite-en-rade-faible-impact-des-sous-marins_52692-2313410_actu.Htm, 25 March 2013.

[OUG 04] OUGHTON D.H., SKIPPERUD L., FIFIELD L.K. et al., "Accelerator mass spectrometry measurement of ^{240}Pu/^{239}Pu isotope ratios in Novaya Zemlya and Kara Sea sediments", *Appl. Radiat. Isot.*, vol. 61, pp. 249–253, 2004.

[OUG 16] OUGHTON D.H., "Ethical foundations of environmental radiological protection", *Proceedings of the Third International Symposium on the System of Radiological Protection, Annals of the ICRP*, vol. 45, no. 1S, pp. 345–357, 2016.

[OUV 72] OUVRARD R., "La Radioactivité marine", *Rayonnements ionisants*, vol. 1, pp. 15–29, 1972.

[OWE 11] OWENS S.A., BUESSELER K.O., SIMS K.W.W., "Re-evaluation the ^{238}U-salinity relationship in seawater. Implications for the ^{238}U-^{234}Th disequilibrium method", *Mar. Chem.*, vol. 127, pp. 31–39, 2011.

[PAQ 91] PAQUET F., Etude expérimentale des biocinétiques de l'américium-241 chez le homard *Homarus gammarus*. Analyse des mécanismes d'accumulation et de détoxication au niveau subcellulaire, Thesis, Université de Nantes, 1991.

[PAT 75] PATEL B., "Field and laboratory comparability", *Design of Radiotracer Experiments in Marine Biological Systems*, IAEA, Vienna, pp. 211–239, 1975.

[PAT 86] PATTI F., MASSON M., VERGNAUD G. et al., "Activités du Technétium 99 mesurées dans les eaux résiduaires, l'eau de mer et deux bioindicateurs (littoral de la Manche, 1983)", in DESMET G., MYTENNAERE C. (eds), *Technetium in the Environment*, Elsevier, London, 1986.

[PAT 90] PATTI F., JEANMAIRE L., MASSON M. et al., "Temporal variations of iodine-127, potassium-210 and technetium-99 concentrations in *Fucus serratus* in the English Channel", *Journal of Radioanalytical and Nuclear Chemistry*, vol. 142, no. 2, pp. 467–480, 1990.

[PEA 99] PEARCE J., "High tritium levels in Severn fish", *Mar. Pollut. Bull.*, vol. 38, pp. 3–4, 1999.

[PEN 58] PENDLETON R.C., HANSON W.C., "Absorption of cesium 137 by components of an aquatic community", *Peaceful Uses of Atomic Energy*, United Nations, New York, vol. 18, pp. 419–422, 1958.

[PEN 73] PENTREATH R.J., WOODHEAD D.S., JEFFERIES D.F., "Radioecology of the plaice (*Pleuronectes platessa* L.) in the Northeast Irish Sea", in NELSON D.J. (ed.), *Radionuclides in Ecosystems*, USAEC, 1973.

[PEN 97] PENNDERS R.L.J., JANSSEN M.P.M., KOSTER H.W. et al., "Uptake of polonium-210 by mussels from effluents emitted by different phosphorus plants", in DESMET G., BLUST R.J., COMANS R.N.N. et al. (eds), *Freshwater and Estuarine Radioecology*, Elsevier, Amsterdam, 1997.

[PEN 09] PENTREATH R.J., "Radioecology, radiobiology, and radiological protection: Frameworks and fractures", *J. Environ. Radioact.*, vol. 100, pp. 1019–1026, 2009.

[PES 76] PESRET F., "Fixation de radioéléments sur des carbonates en milieu marin", *Rapp. Comm. Int. Mer. Méd.*, vol. 23, pp. 131–132, 1976.

[PES 81] PESCH G.G., YOUNG J.S., Baseline study of effects of ionizing radiation on the chromosomes of the marine worm, *Neanthes arenaceodentata*, Technical report PB-82-131517, Surveillance and Emergency Preparedness Division, Office of Radiation Programs, ORP/SEPD-81–1. Washington, D.C., 1981.

[PET 06] PETTERSSON H.B.L., SALIH I., HERRMANN J., "137Cs accumulation in coastal sediments in Sweden", *IAEA-CSP-26*, Vienna, pp. 409–415, 2006.

[PIR 70] PIRO A., "Physico-chemical states of some trace elements in seawater which are of interest from radiocontamination standpoint", *Rev. Intern. Océanogr. Méd.*, vol. 20, pp. 133–149, 1970.

[POL 66] POLIKARPOV G.G., *Radioecology of Aquatic Organisms*, Reinhold, New York, 1966.

[POV 95] POVINEC P.P., OSVATH I., BAXTER M., "Les océanographes et les mers arctiques : mise au point du dossier radiologique", *AIEA Bull.*, vol. 2, pp. 31–35, 1995.

[POV 96] POVINEC P.P., FOWLER S.W., BAXTER M., "Chernobyl & the marine environment: The radiological impact in context", *IAEA Bull.*, vol. 38, no. 1, pp. 18–22, 1996.

[POV 99] POVINEC P.P., WOODHEAD D., BLOWERS P. et al., "Marine radioactivity assessment of Mururoa and Fangataufa atolls", *Sci. Total Environ.*, vols 237–238, pp. 249–267, 1999.

[POV 03] POVINEC P.P., BAILLY DU BOIS P., KERSHAW P.J. et al., "Temporal and spatial trends in the distribution of ^{137}Cs in surface waters of Northern European Seas – A record of 40 years of investigations", *Deep Sea Res. Pt II: Topic. Stud. Oceanogr.*, vol. 50, nos 17–21, pp. 2785–2801, 2003.

[POV 05] POVINEC P.P., AAKROG A., BUESSELER K.O. et al., "^{90}Sr, ^{137}Cs and ^{239,240}Pu concentration surface water time series in the Pacific and Indian Oceans – WOMARS results", *J. Environ. Radioact.*, vol. 81, pp. 63–87, 2005.

[POV 06] POVINEC P.P., AARKROG A., BUESSELER K.O. et al., "Average ^{3}H, ^{90}Sr, ^{137}Cs and ^{239,240}Pu concentrations in surface waters of the Atlantic Ocean – WOMARS collaboration", *IAEA-CSP-26*, Vienna, pp. 38–42, 2006.

[POV 13] POVINEC P.P., HIROSE K., AOYAMA M., *Fukushima Accident: Radioactivity Impact on the Environment*, Elsevier, Amsterdam, 2013.

[PRE 71] PRESTON A., WOOD P.C., "Monitoring the marine environment", *Proceedings of the Royal Society of London. Series B, Biological Sciences*, vol. 177, no. 1048, A Discussion on Biological Effects of Pollution in the Sea (Apr. 13, 1971), pp. 451–462, 1971.

[PRI 20] PRIHATININGSIH W.R., SUSENO H., MAKMUR M. et al., "A review on status of marine radioecology in Indonesia after Fukushima accident", *IOP Conf. Series: Earth and Environmental Science*, vol. 584, p. 012041, 2020.

[PRI 58] PRITCHARD D.W., "Factors affecting the dispersal of fission products in estuarine and inshore environment", *Peaceful Uses of Atomic Energy*, United Nations, New York, vol. 18, pp. 410–413, 1958.

[PRI 59] PRITCHARD D.W., GOLDBERG E.D., SCHAEFER M.B. et al., *Radioactive Waste Disposal from Nuclear-Powered Ships*, Nat. Acad. Sci./Nat. Res. Council, Publ. No. 658, 1959.

[PRU 72] PRUTER A.T., ALVERSON D.L., *The Columbia River Estuary and Adjacent Ocean Waters, Bioenvironmental Studies*, University of Washington Press, Seattle, 1972.

[PUT 17] PUTRA D.I.P., SUSENO H., "Marine radioecology in Bali and Lombok waters: Present status of ^{137}Cs and natural radionuclides in Lombok Strait", *Omni-Akuatika*, vol. 13, no. 2, pp. 180–189, 2017.

[QIU 17] QIU X., UNDAP S.L., HONDA M. et al., "Pollution of radiocesium and radiosilver in wharf roach (*Ligia* sp.) by the Fukushima Dai-ichi Nuclear Power Plant accident", *J. Radioanal. Nucl. Chem.*, vol. 311, pp. 121–126, 2017.

[QUE 65] QUENEUDEC J.P., "Le rejet à la mer de déchets radioactifs", *Annuaire français de droit international*, vol. 11, pp. 750–782, 1965.

[RAD 99] RADAKOVITCH O., CHARMASSON S., ARNAUD M. et al., "^{210}Pb and caesium accumulation in the rhone delta sediments", *Estuar. Coast. Shelf Sci.*, vol. 48, no. 1, pp. 77–92, 1999.

[RAF 98] RAFFESTINT D., ARMINGAUD F., SCHNEIDERT T. et al., "Statistical analysis of accident data associated with sea transport", *Int. J. Radioact. Mater. Transp.*, vol. 9, no. 2, pp. 103–109, 1998.

[RAI 99] RAINBOW P.S., "Bioaccumulation of trace metals: Biological significance", *Océanis*, vol. 25, pp. 547–561, 1999.

[RAI 02] RAINBOW P.S., "Trace metal concentrations in aquatic invertebrates: Why and so what?", *Environ. Pollut.*, vol. 120, pp. 497–507, 2002.

[RAI 06] RAINBOW P.S., POIRIER L., SMITH B.D. et al., "Trophic transfer of trace metals: Subcellular compartmentalization in a polychaete and assimilation by a decapod crustacean", *Mar. Ecol. Progr. Ser.*, vol. 308, pp. 91–100, 2006.

[RAI 07] RAINBOW P.S., AMIARD J.C., AMIARD-TRIQUET C. et al., "Trophic transfer of trace metals: Subcellular compartmentalization in bivalve prey, assimilation by a gastropod predator and in vitro digestion simulations", *Mar. Ecol. Progr. Ser.*, vol. 348, pp. 125–138, 2007.

[RAI 10] RAINBOW P.S., SMITH B.D., "Trophic transfer of trace metals: Subcellular compartmentalization in bivalve prey and comparative assimilation efficiencies of two invertebrate predators", *J. Exp. Mar. Biol. Ecol.*, vol. 390, pp. 143–148, 2010.

[RAI 11] RAINBOW P.S., LUOMA S.N., WANG W.X., "Trophically available metal: A variable feast", *Environ. Pollut.*, vol. 159, pp. 2347–2349, 2011.

[RAV 67] RAVERA O., "The effect of X-rays on the demographic characteristics of *Physa acuta* (Gastropoda: Basommatophora)", *Malacologia*, vol. 5, pp. 95–109, 1967.

[RAY 20] RAY D., LEARY P., LIVENS F. et al., "Controls on anthropogenic radionuclide distribution in the Sellafield-impacted Eastern Irish Sea", *Sci. Total Environ.*, vol. 743, p. 140765, 2020.

[REI 91] REINFELDER J.R., FISHER N.S., "The assimilation of elements ingested by marine copepods", *Science*, vol. 251, pp. 794–796, 1991.

[REN 73] RENFRO W.C., "Transfer of 65Zn from sediments by marine polychaete worms", *Mar. Biol.*, vol. 21, pp. 305–316, 1973.

[REN 11] RENSHAW J.C., HANDLEY-SIDHU S., BROOKSHAW D.R., "Pathways of radioactive substances in the environment", in HESTER R.E., HARRISON R.M. (eds), *Nuclear Power and the Environment*, Issues in Environmental Science and Technology no. 32, RSC Publishing, Cambridge, 2011.

[RHO 67] RHOADS D.C., "Biogenic reworking of intertidal and subtidal sediments in Barnstable Harbor and Buzzards Bay, Massachusetts", *J. Geol.*, vol. 75, pp. 461–476, 1967.

[RIC 56] RICE T.R., "The accumulation and exchange of strontium by marine planktonic algae", *Limnol. Oceanogr.*, vol. 1, pp. 123–128, 1956.

[RIC 08] RICHARDS Z.T., BEGER M., PINCA S. et al., "Bikini Atoll coral biodiversity resilience five decades after nuclear testing", *Mar. Pollut. Bull.*, vol. 56, pp. 503–515, 2008.

[ROB 71] ROBERTSON D.E., "Influence of the physic-chemical forms of radionuclides and stable trace elements in seawater in relation to uptake by the marine biosphere", *2nd Colloque OCDE-ENEA*, Paris, pp. 21–84, 1971.

[ROB 81] ROBISON W.L., NOSHKIN V.E., PHILLIPS W.A. et al., The Northern Marshall Islands radiological survey: Radionuclide concentrations in fish and clams and estimated doses via the marine pathway, UCkL-52853 Pt. 3, 1981.

[ROB 98] ROBISON W.L., NOSHKIN V.E., "Radionuclide characterization and associated dose from long lived radionuclides in close-in fallout delivered to the marine environment at Bikini and Enewetak Atolls", *IAEA-SM-354/76 XA9951206*, 1998.

[ROM 01] ROMANA L.A., GONZALEZ J.L., "Estuarine contaminants: Fate and environmental risks", in BRÉCHIGNAC F., HOWARD B.J. (eds), *Radioactive Pollutants: Impact on the Environment*, EDP Sciences, Les Ulis, 2001.

[RUD 08] RUDRUD R.W., KROEKER J.W., LESLIE H.Y., "La guerre des tortues : culture, guerre et tortues de mer aux Îles Marshall", *Ressources marines et traditions, Bulletin de la CPS*, no. 21, pp. 1–29, January 2008.

[RUG 60] RUGH R., "General biology: Gametes, the developing embryo, and cellular differentiations", in ERRERA M., FORSSBERG A. (eds), *Mechanisms in Radiobiology, vol. II, Multicellular Organisms*, Academic Press, New York, 1960.

[RYA 97] RYAN T.P., DOWDALL A.M., MCGARRY A.T., "Polonium-210 uptake by *Mytilus edulis* (L.) in Irish estuarine and inshore waters", in DESMET G., BLUST R.J., COMANS R.N.N. et al. (eds), *Freshwater and Estuarine Radioecology*, Elsevier, Amsterdam, 1997.

[RYA 99] RYAN T.P., DOWDALL A.M., MCGARRY A.T. et al., "^{210}Po in *Mytilus edulis* in the Irish marine environment", *J. Environ. Radioact.*, vol. 43, no. 3, pp. 325–342, 1999.

[SAL 93] SALOMON J.C., BRETON M., "An atlas of long-term currents in the Channel", *Oceanol. Acta*, vol. 16, pp. 439–448, 1993.

[SAL 97a] SALBU B., NIKITIN A.I., STRAND P. et al., Joint Russian-Norwegian collaboration on radioactive contamination from dumped nuclear waste in the Kara Sea Results from the 1994 expedition to Abrosimov and Stepovogo Fjords, NO9700025, pp. 129–134, 1997.

[SAL 97b] SALBU B., NIKITIN A.I., STRAND P. et al., "Radioactive contamination from dumped nuclear waste in the Kara Sea – Results from the joint Russian-Norwegian expeditions in 1992–1994", *Sci. Total Environ.*, vol. 202, pp. 185–198, 1997.

[SAL 01] SALBU B., LIND O.C., BORRETZEN P. et al., "Advanced speciation techniques for radionuclides associated with colloids and particles", in BRÉCHIGNAC F., HOWARD B.J., (eds), *Radioactive Pollutants: Impact on the Environment*, EDP Sciences, Les Ulis, 2001.

[SAL 11] SALBU B., LIND O.C., "Radioactive particles released into the environment from nuclear events", in KALMYKOV S.N., DENECKE M.A. (eds), *Actinide Nanoparticle Research*, Springer-Verlag, Berlin, 2011.

[SAL 18] SALBU B., KASHPAROV V., LIND O.C. et al., "Challenges associated with the behaviour of radioactive particles in the environment", *J. Environ. Radioact.*, vol. 186, nos 101–115, 2018.

[SAN 04] SANTUCCI P., VOIGT G., "Arguments for revising radioecological transfer factors: How to improve and extend current syntheses", *Second ECORAD International Radioecology Conference*, 6–10 September, 2004, Aix-en-Provence, France, IRSN Collection Series, vol. 39, Suppl. 1, 2004.

[SAN 08] SANTOS I.R., BURNETT W.C., GODOY J.M., "Radionuclides as tracers of coastal processes in Brazil: Review, synthesis, and perspectives", *Brazilian Journal of Oceanography*, vol. 56, no. 2, pp. 115–131, 2008.

[SAN 19] SANCHO C., GARCÍA-TENORIO R., "Radiological evaluation of the transuranic remaining contamination in Palomares (Spain): A historical review", *J. Environ. Radioact.*, vol. 203, pp. 55–70, 2019.

[SAR 75] SARAIVA M.C., L'utilisation d'une algue nano planctonique comme organisme-test en Molysmologie marine. Quelques réponses de *Dunaliella biocculata* Butcher 1959, à l'irradiation gamma et à la contamination par le chrome et le cadmium, Thesis, Université de Paris-6, no. A0 11503 CNRS, 1975.

[SAT 19a] SATOH Y., KAKIUCHI H., UEDA S. et al., "Concentrations of iodine-129 in coastal surface sediments around spent nuclear fuel reprocessing plant at Rokkasho, Japan, during and after its test operation", *J. Radioanal. Nucl. Chem.*, vol. 322, pp. 2019–2024, 2019.

[SAT 19b] SATOH Y., KAKIUCHI H., UEDA S. et al., "Concentrations of iodine-129 in livestock, agricultural, and fishery products around spent nuclear fuel reprocessing plant in Rokkasho, Japan, during and after its test operation", *Environ. Monit. Assess.*, vol. 191, p. 61, 2019.

[SAZ 02] SAZYKINA T.G., KRYSHEV I.I., "Methodology for radioecological assessment of radionuclides permissible levels in the seas – Protection of human and marine biota", *Radioprotection*, vol. 37, no. 1, pp. 899–902, 2002.

[SCH 63] SCHULTZ V., KLEMENT A.W., "Radioecology", *Proceedings of the First National Symposium on Radioecology*, Colorado State University, Fort Collins, Colorado, September 10–15, 1961, Reinhold Pub. Corp., New York, 1963.

[SCH 74] SCHELL W.R., SAUZAY G., PAYNE B.R., "World distribution of environmental tritium", *Physical Behaviour of Radioactive Contaminants in the Atmosphere*, IAEA, Vienna, pp. 375–400, 1974.

[SCH 01] SCHNEIDER, T., ARMINGAUD, F., TABARRE, M., "Statistical analysis of accident data associated with sea transport (data from 1994–1997)", *Rep. CEPN-IPSN, NTE/99/02, IAEA-TECDOC-1231*, Annex 1, pp. 67–80, 2001.

[SCH 06] SCHLOSSER P., NEWTON B., SPIELER A., "Distribution of tritium (^{3}H) in the Southern Ocean: Implications for formation rates, circulation patterns and mean residences of major water masses", *IAEA-CSP-26*, Vienna, p. 37, 2006.

[SEY 11] SEYVE C., "Les rejets radioactifs des installations nucléaires", *GR21 de la SFEN*, available at: http://www.sfen.org/Les-rejets-radioactifs-des,734, 2011.

[SHI 98] SHIMADA A., SHIMA A., "Combination of genomic DNA fingerprinting into the medaka specific-locus test system for studying environmental germ-line mutagenesis", *Mutat. Res.*, vol. 399, pp. 149–165, 1998.

[SHI 04a] SHI D., WANG W.X., "Modification of trace metal accumulation in the green mussel *Perna viridis* by exposure to Ag, Cu, and Zn", *Environ. Pollut.*, vol. 132, pp. 265–277, 2004.

[SHI 04b] SHIMADA A., SHIMA A., "Transgenerational genomic instability as revealed by a somatic mutation assay using the medaka fish", *Mutat. Res.*, vol. 552, pp. 119–124, 2004.

[SHO 84] SHOLKOVITZ E.R., MANN D.R., "The pore water chemistry of $^{239,240}Pu$ and $^{13}7Cs$ in sediments of Buzzards Bay, Massachusetts", *Geochim. Cosmochim. Acta*, vol. 48, no. 5, pp. 107–1114, 1984.

[SHU 19] SHURYAK I., "Review of microbial resistance to chronic ionizing radiation exposure under environmental conditions", *J. Environ. Radioact.*, vol. 196, pp. 50–63, 2019.

[SHU 20] SHURYAK I., "Review of resistance to chronic ionizing radiation exposure under environmental conditions in multicellular organisms", *J. Environ. Radioact.*, vol. 212, p. 106128, February 2020.

[SIV 95] SIVINTSEV Y.V., DANILYAN V., VYSOTSKIJ V., "Environmental impact of liquid radioactive releases into the Sea of Japan in October 1993", *Environmental Impact of Radioactive Releases*, IAEA, Vienna, pp. 820–821, 1995.

[SJO 95] SJOEBLOM K.L., LINSLEY G.S., The International Arctic Seas Assessment Project (IASAP): Interim progress report, Environmental Impact of Radioactive Releases, IAEA, Vienna, pp. 155–164, 1995.

[SJO 98] SJOEBLOM K.L., LINSLEY G.S., "International Arctic Seas Assessment Project (IASAP)", *Radiat. Prot. Dosimetry*, vol. 75, nos 1–4, pp. 263–268, 1998.

[SJO 99] SJOBLOM K.L., SALO U.A., BEWERS J.M. et al., "International Arctic Seas Assessment Project", *Sci. Total Environ.*, vol. 237–238, pp. 153–166, 1999.

[SKW 97] SKWARZEC B., "Polonium, uranium and plutonium in the southern Baltic Sea", *Ambio*, vol. 26, no. 2, pp. 113–117, 1997.

[SKW 01] SKWARZEC B., STRUMINSKA D.I., BORYLO A., "Bioaccumulation and distribution of plutonium in fish from Gdansk Bay", *J. Environ. Radioact.*, vol. 55, pp. 167–178, 2001.

[SMA 73a] SMALL L.F., FOWLER S.W., "Turnover and vertical transport of zinc by the Euphausiid *Meganyctiphanes norvegica* in the Ligurian Sea", *Mar. Biol.*, vol. 18, pp. 284–290, 1973.

[SMA 73b] SMALL L.F., FOWLER S.W., KECKES S., "Flux of zinc through a Macroplanktonic crustacean", *Radioactive Contamination of the Marine Environments*, IAEA, Vienna, pp. 437–452, 1973.

[SMA 97] SMALL L.F., "Interactions of radionuclides with marine biota", *Strategies and Methodologies for Applied Marine Radioactivity Studies*, IAEA Training Courses Series no. 7, Vienna, 1997.

[SMI 98] SMITH D.L., KNOWLES J.F., WINPENNY K., "The accumulation, retention and distribution of ^{95m}Tc in crab (*Cancer pagurus*) and lobster (*Homarus gammarus*). A comparative study", *J. Environ. Radioact.*, vol. 40, pp. 113–135, 1998.

[SMI 00] SMITH J.N., ELLIS K.M., POLYAK L. et al., "^{239,240}Pu transport into the Arctic Ocean from underwater nuclear tests in Chernaya Bay, Novaya Zemlya", *Cont. Shelf Res.*, vol. 20, pp. 255–279, 2000.

[SMI 01] SMITH V., FEGAN M., POLLARD D., RYAN T.P., "Technetium 99 in the Irish marine environment", *J. Environ. Radioact.*, vol. 56, pp. 269–284, 2001.

[SMI 05] SMITH J.T., VOITSEKOVITCH O.V., KONOPLEV A.V. et al., "Radioactivitity in aquatic systems", SMITH J.T., BERESFORD N.A. (eds), *Chernobyl. Catastrophe and Consequences*, Springer-Verlag, Berlin, 2005.

[SMI 11a] SMITH J.N., MCLAUGHIN F.A., SMETHIE JR. W.M. et al., "Iodine-129, ^{137}Cs and CFC-11 tracer transit time distributions in the Arctic Ocean", *J. Geophys. Res.*, vol. 116, C04024, pp. 1–19, 2011.

[SMI 11b] SMITH K., LEÓN VINTRÓ L., MITCHELL P.I. et al., "Plutonium measurements at NEA's North East Atlantic dumpsites", *Radioprotection*, vol. 46, no. 6, pp. S271–S275, 2011.

[SMI 12] SMITH, K., JACKSON, D., WOOD, M.D., Demonstrating compliance with protection objectives for non-human biota within post-closure safety cases for radioactive waste repositories, BIOPROTA, version 3.0 (final), 2012.

[SMI 17] SMITH J.N., ROSSI V., BUESSELER K.O. et al., "Recent transport history of Fukushima radioactivity in the Northeast Pacific Ocean", *Environ. Sci. Technol.*, vol. 51, no. 18, pp. 10494–10502, 2017.

[SOU 71] SOURNIA A., "Mesure de la productivité primaire des océans par la méthode du ^{14}C", *Energie nucléaire et ses applications biologiques à Madagascar. Terre Malgache*, vol. 12, pp. 251–267, 1971.

[SOY 11] SOYFER V., GORYACHEV V., SALYUK A. et al., "The new results of experimental studies of fission products release from spent nuclear fuel of the icebreaker 'Lenin' reactor dumped in the Kara Sea", *Radioprotection*, vol. 46, no. 6, pp. S363–S369, 2011.

[STA 17] STARK K., GOMEZ-ROS J., VIVES i BATLLE J. et al., "Dose assessment in environmental radiological protection: State of the art and perspectives", *J. Environ. Radioact.*, vols 175–176, pp. 105–114, 2017.

[STE 00] STEPNOWSKI P., SKWARZEC B., "A comparison of ^{210}Po accumulation in molluscs from the southern Baltic, the coast of Spitsbergen and Sasek Wielki Lake in Poland", *J. Environ. Radioact.*, vol. 49, no. 2, pp. 201–208, 2000.

[STE 05] STEWART G.M., FOWLER S.W., TEYSSIÉ J.L. et al., "Contrasting transfer of polonium-210 and lead-210 across three trophic levels in marine plankton", *Mar. Ecol. Prog. Ser.*, vol. 290, pp. 27–33, 2005.

[STE 07] STEPANETS O., BORISOV A., LIGAEV A. et al., "Radioecological investigations in shallow bays of the Novaya Zemlya Archipelago in 2002–2005", *J. Environ. Radioact.*, vol. 96, pp. 130–137, 2007.

[STE 09] STEPANETS O., BORISOV A., LIGAEV A. et al., "Geoecological investigations of artificial radionuclides in the aquatic system of the Ob River and Kara Sea", *Radioprotection*, vol. 44, no. 5, pp. 311–314, 2009.

[STO 02] STONE R., "Radioecology's coming of age – or its last gasp?" *Science*, vol. 297, pp. 1800–1801, 2002.

[STR 73] STRAND J.A., FUJIHARA M.P., TEMPLETON W.L. et al., "Suppression of *Chondroccocus columnaris* immune response in rainbow trout sub-lethality exposed to tritiated water during embryogenesis", *Radioactive Contamination of the Marine Environments*, IAEA, Vienna, pp. 543–550, 1973.

[STR 95] STRAND P., NIKITIN A., RUDJORD A.L. et al., "Dumping of radioactive waste in the Barents Sea and the Kara Sea", *Environmental Impact of Radioactive Releases*, IAEA, Vienna, pp. 143–153, 1995.

[STR 14] STRAND, P., AONO T., BROWN J.E. et al., "Assessment of Fukushima-derived radiation doses and effects on wildlife in Japan", *Environ. Sci. Technol. Lett.*, vol. 1, pp. 198–203, 2014.

[STR 15] STRADY E., HARMELIN VIVIEN M., CHIFFOLEAU J.F. et al., "210Po and 210Pb trophic transfer within the phytoplankton–zooplankton–anchovy/sardine food web: A case study from the Gulf of Lion (NW Mediterranean Sea)", *J. Environ. Radioact.*, vol. 143, pp. 141–151, 2015.

[STR 17] STRAND P., SUNDELL-BERGMAN S., BROWN J.E. et al., "On the divergences in assessment of environmental impacts from ionising radiation following the Fukushima accident", *J. Environ. Radioact.*, vols 169–170, pp. 159–173, 2017.

[SUC 86] SUCHANEK T.H., COLIN P.L., MCMURTRY G.M. et al., "Bioturbation and redistribution of sediment radionuclides in Enewetak Atoll lagoon by Callianassid Shrimp: Biological aspects", *Bull. Mar. Sci.*, vol. 38, no. 1, pp. 144–154, 1986.

[SUZ 08] SUZUKI T., KABUTO S., AMANO H. et al., "Measurement of iodine-129 in seawater samples collected from the Japan Sea area using accelerator mass spectrometry: Contribution of nuclear fuel reprocessing plants", *Quater. Geochronol.*, vol. 3, pp. 268–275, 2008.

[SWA 99] SWARZENSKI P.W., MCKEE B.A., SORENSEN K. et al., "Pb-210 and Po-210, manganese and iron cycling across the O2/H2S interface of a permanently anoxic fjord: Framvaren, Norway", *Mar. Chem.*, vols 67, nos 3–4, pp. 199–217, 1999.

[SWI 85] SWIFT D.J., "The accumulation of ^{95m}Tc from sea water by juvenile lobsters (*Homarus gammarus* L.)", *J. Environ. Radioact.*, vol. 2, pp. 229–243, 1985.

[SZE 87a] SZEFER P., "Concentration of uranium and thorium by some Baltic Seaweeds and lake plants", *Mar. Poll. Bull.*, vol. 18, no. 8, pp. 439–442, 1987.

[SZE 87b] SZEFER P., WENNE R., "Concentration of uranium and thorium in molluscs inhabiting Gdansk Bay, Baltic Sea", *Sci. Total Environ.*, vol. 65, pp. 191–202, 1987.

[SZY 12] SZYMCZAK R., "Tropical radiochemical oceanography", in TWINING J.R. (ed.), *Tropical Radioecology*, Elsevier, Amsterdam, 2012.

[TAK 19] TAKATA H., JOHANSEN M.P., KUSAKABE M. et al., "A 30–year record reveals re-equilibration rates of 137Cs in marine biota after the Fukushima Dai-ichi nuclear power plant accident: Concentration ratios in pre- and post-event conditions", *Sci. Total Environ.*, vol. 675, pp. 694–704, 2019.

[TAP 13] TAPIO S., "Ionizing radiation effects on cells, organelles and tissues on proteome level", in LESZCZYNSKI D. (ed.), *Radiation Proteomics. The Effects of Ionizing and Non-ionizing Radiation on Cells and Tissues*, Springer, Dordrecht, 2013.

[THE 97a] THEODORAKIS C.W., BLAYLOCK B.G., SHAGART L.R., "Genetic ecotoxicology I: DNA integrity and reproduction in mosquitofish exposed *in situ* to radionuclides", *Ecotoxicology*, vol. 6, pp. 205–218, 1997.

[THE 97b] THEODORAKIS C.W., SHUGART L.R., "Genetic ecotoxicology II: Population genetic structure in mosquito fish exposed *in situ* to radionuclides", *Ecotoxicology*, vol. 6, pp. 335–354, 1997.

[THE 98] THEODORAKIS C.W., SHUGART L.R., "Genetic ecotoxicology III: The relationship between DNA strand breaks and genotype in mosquito fish exposed to radiation", *Ecotoxicology*, vol. 7, pp. 227–236, 1998.

[THE 99] THEODORAKIS C.W., ELBL T., SHUGART L.R., "Genetic ecotoxicology IV: Survival and DNA strand breakage is dependent on genotype in radionuclide-exposed mosquitofish", *Aquat. Toxicol.*, vol. 45, pp. 279–291, 1999.

[THI 99] THIESSEN K.M., THORNE M.C., MAUL P.R. et al., "Modelling radionuclide distribution and transport in the environment", *Environ. Pollut.*, vol. 100, pp. 151–177, 1999.

[THO 88] THOMPSON P.M., "Environmental monitoring for radionuclides in marine ecosystems; are species other than man protected adequately?", *J. Environ. Radioact.*, vol. 7, no. 3, pp. 275–283, 1988.

[THØ 06] THØRRING H., BROWN J.E., IOSJPE M. et al., "The EPIC exposure assessment methodology – A case study for Arctic marine systems", *IAEA-CSP-26*, Vienna, pp. 233–236, 2006.

[THØ 09] THØRRING H., BROWN J.E., HOSSEINI A., "Characterisation of background dose-rates for marine environments", *Radioprotection*, vol. 44, no. 5, pp. 595–600, 2009.

[TOC 95] TOCQUET N., Biocinétiques d'un élément transuranien, le ^{238}Pu, et d'une terre rare, le ^{152}Eu, chez le homard *Homarus gammarus* (organes et niveau cellulaire), Modalités des transferts (accumulation et détoxication), CEA-R- 5698 report, 1995.

[TOP 10] TOPÇUOĞLU S., ERGÜL H.A., BELIVERMIŞ M. et al., "Monitoring of radionuclide concentrations in marine algae, mussel and sediment samples from the Turkish marine environment during the period of 2001–2009", *J. Black Sea/Mediterranean Environ.*, vol. 16, no. 3, pp. 285–293, 2010.

[UCH 17] UCHIDA S., TAGAMI K., "Comparison of coastal area sediment-seawater distribution coefficients (K_d) of stable and radioactive Sr and Cs", *Appl. Geochem.*, vol. 85, Part B, pp. 148–153, 2017.

[UDD 15] UDDIN S., ABA A., FOWLER S.W. et al., "Radioactivity in the Kuwait marine environment – Baseline measurements and review", *Mar. Pollut. Bull.*, vol. 100, no. 2, pp. 651–661, 2015.

[UDD 17] UDDIN S., FOWLER S.W., BEHBEHANI M. et al., "^{210}Po bioaccumulation and trophic transfer in marine food chains in the northern Arabian Gulf", *J. Environ. Radioact.*, vol. 174, pp. 23–29, 2017.

[UDD 18] UDDIN S., BEHBEHANI M., "Concentrations of selected radionuclides and their spatial distribution in marine sediments from the northwestern Gulf, Kuwait", *Mar. Pollut. Bull.*, vol. 127, pp. 73–81, 2018.

[UNE 07] UNEP-CBD, In-depth review of the application of the ecosystem approach, UNEP/CBD/SBSTTA/12/2, 2007.

[UNS 93] UNSCEAR, Report to the General Assembly with Scientific Annexes, United Nations Scientific Committee on the Effects of Atomic Radiation, United Nations Publication, New York, 1993.

[UNS 94] UNSCEAR, Sources and effects of ionizing radiation, United Nations Scientific Committee on the Effects of Atomic Radiation, United Nations, New York, 1994.

[UNS 96] UNSCEAR, Source and effects of ionizing radiation, Report to the General Assembly with Scientific Annex, United Nations, New York, 1996.

[UNS 08] UNSCEAR, "Sources and effects of ionizing radiation", vol. II, Annex E, *Effects of Ionizing Radiation on Non-human Biota*, United Nations, New York, pp. 221–313, 2008.

[UNS 10] UNSCEAR, Rapport du Comité scientifique des Nations Unies pour l'étude des effets des rayonnements ionisants, Report, UNSCEAR, Cinquante-septième session, 2010.

[UNS 14] UNSCEAR, Source and effects of ionizing radiation, UNSCEAR 2013 Report, vol. I: Report to the General Assembly, Scientific Annex A: Levels and effects of radiation exposure due to the nuclear accident after the 2011 great east-Japan earthquake and tsunami, United Nations, New York, April 2014.

[UNS 18] UNSCEAR, Evaluation of data on thyroid cancer in regions affected by the Chernobyl accident, A white paper to guide the Scientific Committee's future programme of work, UNSCEAR, 2018.

[USD 02] USDOE, A graded approach for evaluating radiation doses to aquatic and terrestrial biota, DOE-STD-1153, 2002.

[VAK 94] VAKULOVSKY S.M., NIKITIN A.L., CHUMICHEV V.B. et al., "Cesium-137 and strontium-90 contamination of water bodies in the areas affected by releases from the Chernobyl nuclear power plant accident: An overview", *J. Environ. Radioact.*, vol. 23, pp. 103–122, 1994.

[VAL 00] VALKOVIC V., *Radioactivity in the Environment*, Elsevier, Amsterdam, 2000.

[VAN 69] VAN WYCK POPE S., Antimony- 124 in the Lower Columbia River, Thesis, Oregon State University, 1969.

[VAN 75a] VAN WEERS A.W., "Uptake of cobalt 60 from sea water and from labelled food by the common shrimp *Crangon crangon* (L.)", *Impacts of Nuclear Releases into the Aquatic Environments*, IAEA, Vienna, pp. 349–361, 1975.

[VAN 75b] VANDERPLOEG H.A., PARZYCK D.C., WILCOX W.H. et al., Bioaccumulation factors for radionuclides in freshwater biota, ORNL-5002, ORNL, Oak Ridge, 1975.

[VAN 17] VANDENHOVE H., MUIKKU M., VIVES i BATLLE J. et al., COMET PROJECT FINAL REPORT – Advancement in science, integration and sustainability of European Radioecology, Program COMET (Contract Number:604974), DELIVERABLE (D-no. 1.5), 2017.

[VAN 18] VANDENHOVE H., BRADSHAW C., BERESFORD N.A. et al., "ALLIANCE perspectives on integration of humans and the environment into the system of radiological protection", *Proceedings of the Fourth International Symposium on the System of Radiological Protection, Annals of the ICRP*, vol. 47, nos 3–4, pp. 285–297, 2018.

[VAR 06] VARINLIOGLU A., KUCUKCEZZAR R., KOSE A., "Determination of natural and artificial radioactivity levels in sediment in the Marmara Sea", *IAEA-CSP-26*, Vienna, 2006.

[VEL 11] VELISCEK CAROLAN, J., HUGHES, C.E., HOFFMANN, E.L., "Dose assessment for marine biota and humans from discharge of ^{131}I to the marine environment and uptake by algae in Sydney, Australia", *J. Environ. Radioact.*, vol. 102, pp. 953–963, 2011.

[VER 18] VERNON, E.L., SMITH J.T., JHA A.N., "Relative comparison of tissue specific bioaccumulation and radiation dose estimation in marine and freshwater bivalve molluscs following exposure to phosphorus-32", *J. Environ. Radioact.*, vol. 192, pp. 312–320. 2018.

[VIN 01] VINTRÓ L.L., MITCHELL P.I., NIELSEN S.P., "Marine environment", in VAN DER STRICHT E., KIRCHMANN R. (eds), *Radioecology. Radioactivity and Ecosystems*, UIR, Liège, 2001.

[VIV 07] VIVES i BATLLE J., WILSON R.C., MCDONALD P., "Allometric methodology for the calculation of biokinetic parameters for marine biota", *Sci. Total Environ.*, vol. 388, pp. 256–269, 2007.

[VIV 09] VIVES i BATLLE J., WILSON R.C., WATTS S.J. et al., "Derivation of allometric relationships for radionuclides in marine phyla", *Radioprotection*, vol. 44, no. 5, pp. 47–52, 2009.

[VIV 12] VIVES i BATLLE J., COPPLESTONE D., JONES S.R. "Allometric methodology for the assessment of radon exposures to terrestrial wildlife", *Sci. Total Environ.*, vols 427–428, pp. 50–59, 2012.

[VIV 13] VIVES i BATLLE J., "Dynamic modelling of the radiological impact of the Fukushima accident on marine biota", *International Symposium on the Occasion of the 50th Anniversary of the Belgian Association for Radiological Protection BVS-ABR*, 10 April 2013.

[VIV 14] VIVES i BATLLE J., AONO T., BROWN J.E. et al., "The impact of the Fukushima nuclear accident on marine biota: Retrospective assessment of the first year and perspectives", *Sci. Total Environ.*, vol. 487, pp. 143–153, 2014.

[VIV 18] VIVES i BATLLE J., AOYAMA M., BRADSHAW C. et al., "Marine radioecology after the Fukushima Dai-ichi nuclear accident: Are we better positioned to understand the impact of radionuclides in marine ecosystems?", *Sci. Total Environ.*, vol. 618, pp. 80–92, 2018.

[WAL 93] WALLBRINKS P.J., MURRAY A.S., "Use of fallout radionuclides as indicators of erosion processes", *Hydrol. Process*, vol. 7, pp. 297–304, 1993.

[WAL 03a] WALLACE W.G., LEE B.G., LUOMA S.N., "Subcellular compartmentalization of Cd and Zn in two bivalves I. Significance of metal-sensitive fractions (MSF) and biologically detoxified metal (BDM)", *Mar. Ecol. Progr. Ser.*, vol. 249, pp. 183–197, 2003.

[WAL 03b] WALLACE W.G., LUOMA S.N., "Subcellular compartmentalization of Cd and Zn in two bivalves II. The significance of trophically available metal (TAM)", *Mar. Ecol. Progr. Ser.*, vol. 257, pp. 125–137, 2003.

[WAN 99] WANG W.X., FISHER N.S., "Assimilation efficiencies of chemical contaminants in aquatic invertebrates. A synthetis", *Environ. Toxicol. Chem.*, vol. 18, pp. 2034–2045, 1999.

[WAN 02] WANG W.X., "Interactions of trace metals and different marine food chains", *Mar. Ecol. Prog. Ser.*, vol. 243, nos 295–309, 2002.

[WAT 59] WATSON D.G., GEORGE L.A., HACKETT P.L., "Effects of chronic feeding of phosphorus 32 on rainbow trout", *Ann. Rep. 1958*, USAEC, vol. 2, pp. 1125–1130, 1959.

[WAT 73] WATSON D.G., TEMPLETON W.L., "Thermoluminescent dosimetry of aquatic organisms", in NELSON D.J. (ed.), *Radionuclides in Ecosystems*, USAEC, 1973.

[WAT 15] WATERS C.N., SYVITSKI J.P.M., GAŁUSZKA A. et al., "Can nuclear weapons fallout mark the beginning of the Anthropocene Epoch?", *Bulletin of the Atomic Scientists*, vol. 71, no. 3, pp. 46–57, 2015.

[WEN 17] WENDEL C.C.S., LIND O.C., FIFIELD L.K. et al., "No Fukushima Dai-ichi derived plutonium signal in marine sediments collected 1.5 – 57 km from the reactors", *Appl. Radiat. Isot.*, vol. 129, pp. 180–184, 2017.

[WHI 82] WHICKER F.W., SCHULTZ V., *Radioecology Nuclear Energy and the Environment*, CRC Press, Boca Raton, 1982.

[WHI 88] WHITEHEAD N.E., BALLESTRA S., HOLM E., HUYNH-NGOC L., "Chernobyl radionuclides in shellfish", *J. Environ. Radioact.*, vol. 7, pp. 107–121, 1988.

[WHI 03] WHICKER F.W., HINTON T.G., "Effects of ionising radiation on plants and animals: What we now know and still need to learn", in IAEA (ed.), *Protection of the Environment from Ionising Radiation. The Development and Application of a System of Radiation Protection for the Environment*, IAEA-CSP-17, Vienna, 2003.

[WIL 75] WILSON D.W., NG Y.C., ROBISON W.L., Evaluation of plutonium at Enewetak Atoll, Report UCRL-76436, Lawrence Livermore Laboratory, 1975.

[WIL 01] WILLIAMS J.L., RUSS R.M., MCCUBBIN D., KNOWLES J.F., "An overview of tritium behaviour in the Severn Estuary (UK)", *J. Radiol. Prot.*, vol. 21, pp. 337–344, 2001.

[WOL 97] WOLF S., BATES J.K., BUCK E.C. et al., "Physical and chemical characterization of actinides in soil from Johnston Atoll", *Environ. Sci. Technol.*, vol. 31, pp. 467–471, 1997.

[WOO 71] WOODHEAD D.S., "The biological effects of radioactive waste", *Proc. Roy. Soc. London*, vol. 177B, pp. 423–437, 1971.

[WOO 73] WOOHEAD D.S., "Levels of radioactivity in the marine environment and the dose commitnent to marine organisms", *Radioactive Contamination of the Marine Environment*, Seattle, 10–14 July 1972, IAEA, Vienna, pp. 499–523, 1973.

[WOO 74] WOODHEAD D.S., "The estimation of radiation dose rates to fish in contaminated environments, and the assessment of the possible consequences", *Population Dose Evaluation and Standards for Man and his Environment*, IAEA, Vienna, pp. 555–575, 1974.

[WOO 84] WOODHEAD D.S., "Contamination due to radioactive materials", *Mar. Ecol.*, vol. 5, pp. 1111–1287, 1984.

[WOO 97a] WOODHEAD D.S., "The behaviour of radionuclides discharged from Sellafield in UK coastal waters", in IAEA (ed.), *Strategies and Methodologies for Applied Marine Radioactivity Studies*, IAEA Training Courses Series no. 7, Vienna, 1997.

[WOO 97b] WOODHEAD D.S., "Environmental dosimetry and radiation effects", in IAEA (ed.), *Strategies and Methodologies for Applied Marine Radioactivity Studies*, IAEA Training Courses Series no. 7, Vienna, 1997.

[WOO 98] WOODHEAD D., "Report by an International Advisory Committee: The radiological situation at the Atolls of Mururoa and Fangataufa, Vienna, 30 June–3 July 1998", *J. Radiol. Prot.*, vol. 18, pp. 213–214, 1998.

[WOO 99] WOODHEAD D., "The radiological situation at the Atolls of Mururoa and Fangataufa", *J. Radiol. Prot.*, vol. 19, no. 2, pp. 194–195, 1999.

[WOO 11] WOOD M.D., COPPLESTONE D., "Temporal changes in gamma dose rates in the Esk Estuary, UK", *Radioprotection*, vol. 46, no. 6, pp. S301–S307, 2011.

[WRE 74] WRENN M.E., Environmental levels of plutonium and the transplutonium elements, Pu and other transuranium elements: Sources, environmental distribution and biomedical effects, Report WASH-1359, pp. 89–114, 1974.

[WRE 76] WRENN M.E., JINKS S.M., "The dosimetric implications of release to the aquatic environment from the nuclear power industry", in MILLER M.W., STANNARD J.N. (eds), *Environmental Toxicity of Aquatic Radionuclides: Models and Mechanisms*, Ann Arbor Sci. Publ. Inc., Ann Arbor, 1976.

[XIN 17] XING S., HOU X., ALDAHAN A. et al., "Water circulation and marine environment in the Antarctic traced by speciation of ^{129}I and ^{127}I", *Sci. Rep.*, vol. 7, p. 7726, 2017.

[YAM 94] YAMAMOTO M., ABE T., KUWABARA J. et al., "Polonium-210 and Lead-210 in marine organisms: Intake levels for Japanese", *J. Radioanal. Nucl. Chem.*, vol. 178, no. 1, pp. 81–90, 1994.

[YAM 00] YAMAMOTO K., KITAMURA T., SHIBATA K. et al., "Study of the accident environment during sea transport of nuclear material: Probabilistic safety analysis of plutonium transport from Europe to Japan", *Int. J. Radioact. Mater. Transp.*, vol. 11, no. 3, pp. 217–237, 2000.

[YAM 01] YAMAMOTO K., KITAMURA T., SHIBATA K. et al., "Study of accident environment during sea transport of nuclear material, probabilistic safety analysis of plutonium transport from Europe to Japan", *IAEA-TECDOC-1231*, Annex 4, Vienna, pp. 107–117, 2001.

[YAN 15] YANG B., HA Y., JIN J., "Assessment of radiological risk for marine biota and human consumers of seafood in the coast of Qingdao, China", *Chemosphere*, vol. 135, pp. 363–369, 2015.

[YES 06] YESIN N.V., KOS'YAN R.D., "Field study of sedimentation and radionuclide pollution of the Black Sea", *IAEA-CSP-26*, Vienna, pp. 277–278, 2006.

[YIO 02] YIOU F., RAISBECK G.M., CHRISTENSEN G.C. et al., "$^{127}I/^{129}I$, $^{129}I./^{137}Cs$ and $^{129}I/^{99}Tc$ in the Norwegian coastal current from 1980 to 1998", *J. Environ. Radioact.*, vol. 60, pp. 61–71, 2002.

[ZER 09] ZERBIB J.C., CORYN P., GOUMONDY J.P., "Analyse critique des rejets des usines de La Hague. Année 2006", *Appréciation par le GRNC de l'estimation des doses présentée dans le rapport annuel de surveillance de l'environnement d'AREVA-NC La Hague, quatrième avis du GRNC, Année 2006*, pp. 9–67, October 2009.

[ZER 16] ZERBIB J.C., GUILLEMETTE A., Les iodes radioactifs, ACRO technical file, June 2016.

[ZHA 06] ZHANG L., WANG W.X., "Significance of subcellular metal distribution in prey in influencing the trophic transfer of metals in a marine fish", *Limnol. Oceanogr.*, vol. 51, pp. 2008–2017, 2006.

[ZIE 18] ZIELINSKI K., "Sources, transport and sinks of radionuclides in marine environments", in ZIELINSKI T., SAGAN I., SUROSZ W. (eds), *Interdisciplinary Approaches for Sustainable Development Goals*, Springer, Cham, 2018.

[ZLO 70] ZLOBIN V.S., MOKANO O.V., "Mechanisms of the accumulation of ^{239}Pu and ^{210}Po by the brown alga *Ascophyllum nodosum* and marine phytoplankton", *Radiobiologya*, vol. 10, pp. 584–589, *AEC-Tr-7205*, pp. 160–169, 1970.

[ZLO 71a] ZLOBIN V.S., "Active phase of assimilation of ^{239}Pu by the marine Algae *Ascophyllum nodosum*", *AEC-Tr-7418*, pp. 207–217, 1971.

[ZLO 71b] ZLOBIN V.S., PERLYUK M.F., "Photosynthesis and the mechanism of the action of cyanide on cell respiration and ^{239}Pu accumulation by algae", *AEC-Tr-7418*, pp. 195–206, 1971.

Index

G, H

I, J, L

M, N

O, P

R, S

T, U, V

Other titles from

iSTE

in

Ecological Science

2022

AMIARD Jean-Claude
Management of Radioactive Waste
(Radioactive Risk Set – Volume 5)

GODET Laurent, DUFOUR Simon, ROLLET Anne-Julia
The Baseline Concept in Biodiversity Conservation: Being Nostalgic or Not in the Anthropocene Era

GRISON Claude, CASES Lucie, LE MOIGNE Mailys, HOSSAERT-MCKEY Martine
Photovoltaism, Agriculture and Ecology From Agrivoltaism to Ecovoltaism

LÉVÊQUE Christian
Biodiversity Erosion: Issues and Questions

2021

FLEURENCE Joël
Microalgae: From Future Food to Cellular Factory

AMIARD Jean-Claude
Disarmament and Decommissioning in the Nuclear Domain
(Radioactive Risk Set – Volume 4)

JOLY Fernand, BOURRIÉ Guilhem
Mankind and Deserts 1: Deserts, Aridity, Exploration and Conquests
Mankind and Deserts 2: Water and Salts
Mankind and Deserts 3: Wind in Deserts and Civilizations

MUTTIN Frédéric, THOMAS Hélène
Marine Environmental Quality

PIOCH Sylvain, SOUCHE Jean-Claude
Eco-design of Marine Infrastructures: Towards Ecologically-informed Coastal and Ocean Development

2020

BRUSLÉ Jacques, QUIGNARD Jean-Pierre
Fish Behavior 1: Eco-ethology
Fish Behavior 2: Ethophysiology

LE FLOCH Stéphane
Remote Detection and Maritime Pollution: Chemical Spill Studies

MIGON Christophe, NIVAL Paul, SCIANDRA Antoine
The Mediterranean Sea in the Era of Global Change 1: 30 Years of Multidisciplinary Study of the Ligurian Sea
The Mediterranean Sea in the Era of Global Change 2: 30 Years of Multidisciplinary Study of the Ligurian Sea

ROSSIGNOL Jean-Yves
Climatic Impact of Activities: Methodological Guide for Analysis and Action

2019

AMIARD Jean-Claude
Industrial and Medical Nuclear Accidents: Environmental, Ecological, Health and Socio-economic Consequences
(Radioactive Risk SET – Volume 2)
Nuclear Accidents: Prevention and Management of an Accidental Crisis
(Radioactive Risk SET – Volume 3)

BOULEAU Gabrielle
Politicization of Ecological Issues: From Environmental Forms to Environmental Motives

DAVID Valérie
Statistics in Environmental Sciences

GIRAULT Yves
UNESCO Global Geoparks: Tension Between Territorial Development and Heritage Enhancement

KARA Mohamed Hichem, QUIGNARD Jean-Pierre
Fishes in Lagoons and Estuaries in the Mediterranean 2: Sedentary Fish
Fishes in Lagoons and Estuaries in the Mediterranean 3A: Migratory Fish
Fishes in Lagoons and Estuaries in the Mediterranean 3B: Migratory Fish

OUVRARD Benjamin, STENGER Anne
Incentives and Environmental Policies: From Theory to Empirical Novelties

2018

AMIARD Jean-Claude
Military Nuclear Accidents: Environmental, Ecological, Health and Socio-economic Consequences
(Radioactive Risk SET – Volume 1)

FLIPO Fabrice
The Coming Authoritarian Ecology

GUILLOUX Bleuenn
Marine Genetic Resources, R&D and the Law 1: Complex Objects of Use

KARA Mohamed Hichem, QUIGNARD Jean-Pierre
Fishes in Lagoons and Estuaries in the Mediterranean 1: Diversity, Bioecology and Exploitation

2016

BAGNÈRES Anne-Geneviève, HOSSAERT-MCKEY Martine
Chemical Ecology

2014

DE LARMINAT Philippe

Climate Change: Identification and Projections

Printed and bound by CPI Group (UK) Ltd, Croydon, CR0 4YY
13/09/2026
14961041-0004